Lecture Notes in Computer Science 13988

Founding Editors

Gerhard Goos
Juris Hartmanis

Editorial Board Members

The series Lecture Notes in Computer Science (LNCS), including its subseries Lecture Notes in Artificial Intelligence (LNAI) and Lecture Notes in Bioinformatics (LNBI), has established itself as a medium for the publication of new developments in computer science and information technology research, teaching, and education.

LNCS enjoys close cooperation with the computer science R & D community, the series counts many renowned academics among its volume editors and paper authors, and collaborates with prestigious societies. Its mission is to serve this international community by providing an invaluable service, mainly focused on the publication of conference and workshop proceedings and postproceedings. LNCS commenced publication in 1973.

Colin Johnson · Nereida Rodríguez-Fernández ·
Sérgio M. Rebelo
Editors

Artificial Intelligence in Music, Sound, Art and Design

12th International Conference, EvoMUSART 2023
Held as Part of EvoStar 2023
Brno, Czech Republic, April 12–14, 2023
Proceedings

Springer

Editors
Colin Johnson ⓘ
University of Nottingham
Nottingham, UK

Nereida Rodríguez-Fernández ⓘ
University of A Coruña
A Coruña, Spain

Sérgio M. Rebelo ⓘ
University of Coimbra
Coimbra, Portugal

ISSN 0302-9743   ISSN 1611-3349 (electronic)
Lecture Notes in Computer Science
ISBN 978-3-031-29955-1   ISBN 978-3-031-29956-8 (eBook)
https://doi.org/10.1007/978-3-031-29956-8

This Springer imprint is published by the registered company Springer Nature Switzerland AG
The registered company address is: Gewerbestrasse 11, 6330 Cham, Switzerland

Preface

EvoMUSART 2023—the 12th International Conference on Artificial Intelligence in Music, Sound, Art and Design—took place on April 12–14, 2023, in Brno, Czech Republic, as part of Evo*, the leading European Event on Bio-Inspired Computation.

Following the success of previous events and the importance of the field of Artificial Intelligence, specifically, evolutionary and biologically inspired (artificial neural network, swarm, alife) music, sound, art and design, EvoMUSART has become an Evo* conference with independent proceedings since 2012.

Although the use of Artificial Intelligence for artistic purposes can be traced back to the 1970s, the use of Artificial Intelligence for the development of artistic systems is a recent, exciting and significant area of research. There is a growing interest in the application of these techniques in fields such as: visual art and music generation, analysis and interpretation; sound synthesis; architecture; video; poetry; design; and other creative tasks.

The main goal of EvoMUSART 2023 was to bring together researchers who are using Artificial Intelligence techniques for artistic tasks, providing the opportunity to promote, present and discuss ongoing work in the area. As always, the atmosphere was fun, friendly and constructive.

EvoMUSART has grown steadily since its first edition in 2003 in Essex, UK, when it was one of the Applications of Evolutionary Computing workshops. Since 2012, it has been a full conference as part of the Evo* co-located events.

EvoMUSART 2023 received 55 submissions. The peer-review process was rigorous and double-blind. The international Programme Committee, listed below, was composed of 63 members from 19 countries. EvoMUSART continues to provide useful feedback to authors: among the papers sent for full review, there were on average 3 reviews per paper. The number of accepted papers was 20 long talks (36% acceptance rate) and 7 posters accompanied by short talks, meaning an overall acceptance rate of 49%.

As always, the EvoMUSART proceedings cover a wide range of topics and application areas, including generative approaches to music, visual art and design. This volume of proceedings collects the accepted papers.

As in previous years, the standard of submissions was high and good quality papers had to be rejected. We thank all authors for submitting their work, including those whose work was not accepted for presentation on this occasion.

The work of reviewing is done voluntarily and generally with little official recognition from the institutions where reviewers are employed. Nevertheless, professional reviewing is essential to a healthy conference. Therefore we particularly thank the members of the Programme Committee for their hard work and professionalism in providing constructive and fair reviews.

EvoMUSART 2023 was part of the Evo* 2023 event, which included three additional conferences: EuroGP 2023, EvoCOP 2023 and EvoAPPS 2023. Many people helped to make this event a success.

We thank SPECIES, the Society for the Promotion of Evolutionary Computation in Europe and its Surroundings, for handling the coordination and financial administration.

We thank the local organising team led by Jiri Jaros and Lukas Sekanina (Brno University of Technology, Czech Republic) and also Brno University of Technology, Czech Republic, for their patronage of the event.

We thank João Correia (University of Coimbra, Portugal) for Evo* publicity, José Francisco Chicano García (University of Málaga, Spain) for the Evo* website, Nuno Lourenço (University of Coimbra, Portugal) for the submission system coordination and all involved in the organisation of Evo* 2023.

We thank the Evo* invited keynote speakers, Marek Vácha and Evelyne Lutton for their inspiring and enlightening keynote talks.

We thank the steering committee of EvoMUSART for the advice, support and supervision. In particular, we thank Juan Romero (University of A Coruña, Spain) and Penousal Machado (University of Coimbra, Portugal).

Finally, we would like to express our most heartfelt thanks to Anna Esparcia-Alcázar (SPECIES, Europe), for her dedicated work and coordination of the event. Without her work, and the work of Jennifer Willies in earlier years, Evo* would not enjoy its current level of success as the leading European event on Bio-Inspired Computation.

April 2023   Colin Johnson
                                        Nereida Rodríguez-Fernández
                                                   Sérgio M. Rebelo

Organization

Conference Chairs

Colin Johnson   University of Nottingham, UK
Nereida Rodríguez-Fernández   University of A Coruña, Spain

Publication Chair

Sérgio M. Rebelo   University of Coimbra, Portugal

Program Committee

Mauro Annunziato   ENEA, Italy
Peter Bentley   University College London, UK
Gilberto Bernardes   University of Porto, Portugal
Ulysses Bernardet   Aston University, UK
Daniel Bisig   Zurich University of the Arts, Switzerland
Tim Blackwell   Goldsmiths, University of London, UK
Jean-Pierre Briot   Sorbonne Université - CNRS, France & Pontifical
                          Catholic University of Rio de Janeiro, Brazil
Andrew Brown   Griffith University, Australia
Marcelo Caetano   McGill University, Canada
Amílcar Cardoso   University of Coimbra, Portugal
Luz Castro Pena   University of A Coruña, Spain
Vic Ciesielski   RMIT University, Australia
Simon Colton   Queen Mary University of London, UK
Michael Cook   Falmouth University, UK
João Correia   University of Coimbra, Portugal
Pedro M. Cruz   Northeastern University, USA
Camilo Cruz Gambardella   Monash University, Australia
João Miguel Cunha   University of Coimbra, Portugal
Hans Dehlinger   University of Kassel, Germany
Georgios Diapoulis   Chalmers University of Technology, Sweden
Edward Easton   Aston University, UK
Arne Eigenfeldt   Simon Fraser University, Sweden
Frederic Fol Leymarie   Goldsmiths, University of London, UK
José Fornari   University of Campinas, Brazil
Philip Galanter   Texas A&M University, USA

Contents

Long Talks

Long Talks

LooperGP: A Loopable Sequence Model for Live Coding Performance Using GuitarPro Tablature

Sara Adkins$^{(\boxtimes)}$, Pedro Sarmento, and Mathieu Barthet

Centre for Digital Music, Queen Mary University of London, London, UK
sara.adkins65@gmail.com, {p.p.sarmento,m.barthet}@qmul.ac.uk

Abstract. Despite their impressive offline results, deep learning models for symbolic music generation are not widely used in live performances due to a deficit of musically meaningful control parameters and a lack of structured musical form in their outputs. To address these issues we introduce LooperGP, a method for steering a Transformer-XL model towards generating loopable musical phrases of a specified number of bars and time signature, enabling a tool for live coding performances. We show that by training LooperGP on a dataset of 93,681 musical loops extracted from the DadaGP dataset [22], we are able to steer its generative output towards generating 3x as many loopable phrases as our baseline. In a subjective listening test conducted by 31 participants, LooperGP loops achieved positive median ratings in originality, musical coherence and loop smoothness, demonstrating its potential as a performance tool.

Keywords: Controllable Music Generation · Sequence Models · Live Coding · Transformers · AI Music · Loops · Guitar Tabs

1 Introduction

Sequence models such as the Pop Music Transformer [11] and Transformer-XL [8] are able to generate symbolic polyphonic music that maintains a coherent musical structure over several minutes of audio. However, these types of models present several issues that prevent them from being viable in the context of a live performance. They offer no control over important musical parameters such as key signature, time signature, instrumentation, genre and duration. While the outputs generated may be musically sound, they are difficult to incorporate into a performance if these structural music parameters are uncontrollable. Furthermore, the slow speed of inference makes it very difficult to develop a usable interface for running sequence models in traditional performance settings.

Live coding, an artform in which a performer writes code that synthesizes music in real time, has the potential to be a performance environment in which sequence models are viable. The live coding language can be thought of as a score notation tool, where edits are evaluated in real time [14]. In a typical performance, the performer will alter the score by generating a symbolic pattern through algorithmic composition techniques [6] or hard-coding. The pattern is

© The Author(s), under exclusive license to Springer Nature Switzerland AG 2023
C. Johnson et al. (Eds.): EvoMUSART 2023, LNCS 13988, pp. 3–19, 2023.
https://doi.org/10.1007/978-3-031-29956-8_1

then typically added to the audio mix and repeated as a loop [16]. Rather than traditional pattern generation techniques, a sequence model could be used to generate pattern options for the performer to choose from.

In this paper, we address the issue of controllable generative music by presenting a training and inference algorithm for steering a Transformer model towards generating loopable phrases of a specified length, key and time signature. We then evaluate the success of our loop model, coined as LooperGP, by comparing the number of viable loops generated to a baseline model. Finally, we present results from a listening test that compares participants' ratings of creativity, likability, coherence and smoothness between human- and machine-made loops in order to subjectively evaluate LooperGP and identify areas for improvement. Our loop model and inference controls are intended to enable the use of deep generative models in a live coding context by addressing the constraints of ensuring model outputs fit within the context of an ongoing musical performance. LooperGP evaluations in live performance will be explored in future work.

2   Related Work

2.1   Sequence Models for Music Generation

A variety of deep learning architectures such as Recurrent Neural Networks, Variational Autoencoders and Generative Adversarial Networks have been used to generate symbolic music [5,12], and Transformers have been particularly successful at modeling temporal dependencies [8]. As demonstrated by the Music Transformer in [11], the self-attention mechanism of the Transformer architecture allows the model to refer back to previously generated motifs and phrases, making it a promising architecture for symbolic music generation. The Pop Music Transformer paper further shows that the way in which musical data is stored can have a significant effect on output quality. For instance, [11]'s results demonstrated that encoding musical knowledge such as metrical position into the training data resulted in better beat and meter salience.

2.2   Controllable Music Generation

A large barrier stifling the adoption of symbolic music generation systems in mainstream music creation is a lack of control. Often the only available control is the temperature parameter [2], used to select a token from the model's probability distribution. One exception to this is the DeepBach system [9], which allows the user to impose constraints on generation by fixing specific notes in a Bach-style chorale. Given these note constraints, the bi-directional long short-term memory (LSTM) model fills in the rest of the chorale using Gibbs sampling. This level of customization would be useful in a live coding context for generating phrases with a good loopable cadence, as start and endpoint notes could be hard-coded. However, the DeepBach method is constrained to a fixed receptive field and has only been tested on Bach chorales, potentially limiting its use in performance of modern genres.

Vanilla Transformers often expand upon a motif they are primed with, but this behavior is not guaranteed. The Theme Transformer in [23] introduces a new model architecture that conditions the Transformer encoder on a specific motif/theme. The training data is manipulated such that "theme-composition pairs" are extracted from each training example by clustering similar measures together. By training the model on these pairs, it learns how motifs are expanded upon throughout a piece and is rewarded for repeating and varying themes. This ability to expand upon existing material is useful in live performance, as it ensures generated output is related to existing music material.

2.3 Live Coding

[20] defines live coding as "the act of programming a computer under concert conditions," where we are specifically interested in programs that produce musical output. While traditional music performance involves specifying a stream of music on a note by note or chord by chord basis, live coding involves performing at a higher level of abstraction through score-level control over the creation of music [20]. A large focus of live coding practice is on control over generative processes; a performer controls the parameters of a generative algorithm that creates the musical output such that writing software becomes an act of performing [6].

Prior work which focused on incorporating deep learning (DL) into live coding performances includes Cibo [24], a text-based sequence-to-sequence model that generates Tidal Cycles [16] code, and RaveForce [13], a pipeline for translating synthesizer parameters derived from a DL model to a SuperCollider server using Open Sound Control (OSC) [15].

2.4 DadaGP Dataset

A generative model used in a live coding context will ideally be capable of outputting music from a variety of modern genres to match the styles typically found at these types of performances. The DadaGP dataset [22] contains over 26,000 songs covering 700 genres and a dedicated encoder represent songs as tokens compatible with sequence models. The dataset emphasizes rock and metal, but also contains jazz, pop, classical and electronic dance music (EDM). Furthermore, expressive performance information for string instruments such as tremolo, palm muting and hammer-ons/pull-offs are included in the token set, giving it an advantage over MIDI. The song diversity and token expressivity of the DadaGP dataset make it an ideal training corpus candidate to base music generation on for live coding performances.

[22] also presents a Transformer-XL model with 8 attention heads and 12 self-attention heads trained on the DadaGP dataset. We use this model as a baseline and starting point for our LooperGP model presented in the next section.

3 Methodology

The Theme Transformer in [23] demonstrates promising results in reorganizing training data to steer the model towards certain types of outputs. The Pop

Music Transformer in [11] showed that including music information retrieval (MIR) features such as beat position, bar position and chord labels in the tokens improved higher level music features for the generated content like beat salience. Combining these ideas, we extract "loopable phrases" from the training data to create a new training corpus consisting only of segments that loop naturally. Each loop will be stored as an ordered list of DadaGP tokens [22], where the tokens describe the score and performance details for each activated instrument in the loop. The goal is to further steer the model towards generating loopable phrases during inference, so they can be incorporated into a live coding performance.

3.1   Defining a Loop

Previous work on loop extraction has classified a loop as a short segment of audio that transitions seamlessly when repeated [7,21]. Seamless "loopability" is ultimately a subjective consideration from listeners, but we can focus on the structural attributes of the composition to identify repeatable sections. For our purposes, we consider a segment of a song to be loopable if it is bookended by a repeated phrase of a minimum length[1]. An example loop from AC/DC's "You Shook Me All Night Long" is shown in Fig. 1.

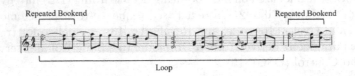

Fig. 1. An example loop from AC/DC's "You Shook Me All Night Long." Only the guitar part is shown, though the loop is multi-track in the DadaGP dataset.

3.2   Extracting Loops

To identify all the loops in a given song, we must first identify all the repeated phrases, or "bookends" as shown in Fig. 1. To extract the repetitions from a given song, we use the correlative matrix approach presented in [10]. The DadaGP dataset [22] is provided in a text format designed for the multi-track tablature editing software Guitar Pro. A Guitar Pro formatted song is first converted into a list M of active note sets as shown in Eq. 1. $N(t,d)$ represents the set of all active notes at tick time t with duration d, such that $t_0 = 0$ and $t_n = t_{n-1}+d_{n-1}$.

$$M = [N(t_0, d_0), N(t_1, d_0), ..., N(t_n, d_n)] \qquad (1)$$

Let C_s and D_s be $L \times L$ matrices for song s with melody line M_s, where L is the length of M_s. C_s is populated as a correlation matrix for notes as shown

---

[1] We focus here on loops where the exact same content is repeated, but it is worth noting that a more general definition could encompass loops where certain types of musical variations can occur across repetitions (e.g. modulation).

in Eq. 2, where $C_s(i, j)$ represents the number of notes in the repeated segment that ends at both $M_s[i]$ and $M_s[j]$. Similarly, $D_s(i, j)$ is a correlation matrix that stores the duration (in ticks) of a repeated segment that ends at both $M_s[i]$ and $M_s[j]$.

$$C_s[i, j] = C_s[i - 1, j - 1] + 1 \text{ if } M_s[i] = M_s[j] \text{ else } 0 \qquad (2)$$

The matrices C_s and D_s allow us to identify loop candidates and filter repeated phrases by both duration and note length. The recursive nature of the correlation matrix ensures that for any given cell $C_s[i, j] > 0$; it follows that $C_s[i - 1, j - 1] = C_s[i, j] - 1$. Therefore, we can identify the starting points of all repetitions of a minimum length L_{min} by searching for cells in the correlation matrix where $C_s[i, j] = L_{min}$. The two starting indices of the repeated phrase $[s_i : s_j]$ are calculated in Eq. 3, and form a potential loop $p_{i,j}$ as defined in Eq. 4.

$$[s_i, s_j] = (i - L_{min}, j - L_{min}) \text{ iff } C_s[i, j] = L_{min} \qquad (3)$$

$$p_{i,j} = M[s_i : s_j] \qquad (4)$$

Given a list of all potential loops $p_{i,j}$ and their corresponding endpoints $[s_i, s_j]$, we can filter out loops that are trivial, too lengthy, or have repeated bookends that are too short. The following length parameters are configurable:

- Minimum Repetition Notes (L_{min}): minimum number of notes in bookend
- Minimum Repetition Beats (RD_{min}): minimum number of beats in bookend
- Minimum Loop Bars (LB_{min}): minimum number of bars between bookends
- Maximum Loop Bars (LB_{max}): maximum number of bars between bookends

We filter for LB_{min} and LB_{max} by comparing the timestamp difference between $M[s_i]$ and $M[s_j]$, and converting this tick duration to number of bars using the time signature metadata for the song. Next, we filter for RD_{min} by locating the end of each loop bookend. Given a loop with endpoints $[s_i, s_j]$, we can find the end of the repeated sub-phrase by traversing along the diagonal of C_s until we hit a decrease in cell value as shown in Eq. 5. This decrease indicates that the matching between s_i and s_j has ended, and we have found the end of the bookend. We can then use Eq. 6 to lookup the duration of the bookend in ticks and filter out loops with duration shorter than desired.

$$n_{end} = n|(C_s[i + n, j + n] < C_s[i + n - 1, j + n - 1]) \qquad (5)$$

$$D_{ticks} = D_s[i + n_{end} - 1, j + n_{end} - 1] \qquad (6)$$

A final filtering strategy to improve extracted loops' musical interest is to filter by note density, in order to filter out loops made up primarily of rests or long held notes. We defined note density (ρ) for a loop $p_{i,j}$ as the average number of note onsets in a measure, scaled by the number of instrumental tracks. This definition is formalized in Eq. 7, where B is the number of bars in the loop and T is the number of instrument tracks. A ρ value of 4.0 for instance would mean each track in $p_{i,j}$ has on average four note onsets per measure.

$$\rho_{p_{i,j}} = len(p_{i,j}) \times \frac{T}{B} \tag{7}$$

In addition to loops extracted using the correlative matrix algorithm, we can also identify built-in loops by simply searching for repeat signs in the training corpus and filtering by the duration in beats. The full loop dataset is then made up of both extracted and built-in loops. Table 1 shows how different combinations of loop extraction parameters affect the number of loops identified. We chose the bolded parameter settings (total of 93,681 loops), as it was the median of all the parameter combinations explored.

Table 1. Loop Extraction Parameter Statistics.

Bars	Density	L min	RD min	Loops	Avg. Loops Per Song
4-8	4	4	4	79,066	3.02
4-8	3	4	4	80,557	3.008
4-8	4	2	4	82,244	3.14
4-8	3	2	4	83,787	3.20
4-16	4	4	4	88,842	3.34
4-16	3	4	4	90,512	3.46
4-8	4	4	2	92,325	3.52
4-16	4	2	4	92,422	3.53
4-8	**3**	**4**	**2**	**93,681**	**3.59**
4-16	3	2	4	94,138	3.60
4-8	4	2	2	99,331	3.79
4-8	3	2	2	100,930	3.89
4-16	4	4	2	104,910	4.01
4-16	3	4	2	106,607	4.08
4-16	4	2	2	113,547	4.34
4-16	3	2	2	115,347	4.41

Table 2. Loop Dataset Statistics.

Data Format	Files	Max Tokens Per Song	Avg. Tokens Per Song	Size
Hard	66,637	12,253	1,376	9.13GB
Barred	23,024	43,954	2,270	5.19GB
DadaGP	26,158	51,997	4,456	9.56GB

3.3 Training

We experimented with two different loop training dataset formats. The first format ("Barred Repeats") surrounds each loop in a given song with `repeat_open` and `repeat_close` tokens, then concatenates these barred loops into a single file.

The second format ("Hard Repeats") stores each loop in a separate file, where the loop is manually repeated 4 times. We hypothesized that by training on the Barred Repeats dataset format, the model would learn where to place repeat open/close tokens during generation to best form musically coherent loops. The Hard Repeats model may better learn smooth transitions due to the multiple repetitions of each loop. However, as shown in Table 2, the Hard Repeats storage format is much less space efficient than Barred Repeats, and therefore takes longer to run through an epoch during training. We used the Pop Music Transformer-XL architecture presented in [22] to train LooperGP. For each data format, we trained one model from scratch and a second starting from epoch 200 of the DadaGP model. The pretrained DadaGP model has already learned how to create structurally coherent music, but early testing revealed that it tended to switch to different thematic sections at random. By training this model further on a loop dataset, we hoped to teach the model to bias its generation towards coherent loops that we can then extract using the filtering procedure presented in Sect. 3.2. The results of the different training configurations are covered in the results section. For all four models, we used a learning rate of 0.0001, with 15% dropout and an Adam optimizer as used by [22].

3.4 Controllable Inference

In a live coding performance, it is important that the performer has control over the musical parameters of generation. In order to fit a generated loop into an ongoing performance, the loop must match the key and time signature of what is currently playing. It is also essential to control the length of the loop in bars, to ensure the phrase lines up (or intentionally doesn't line up) with other patterns that are currently active in the set.

[22] observed that priming inference with one or more notes from the desired instruments forming the root chord of the key is generally successful at generating music in that key and instrumentation, at least over a short duration. Figure 2 shows how priming with an eighth note of drums (unpitched), guitar (A3) and bass (A2) results in the model continuing with the established instrumentation in the key of A minor for several measures.

Fig. 2. Example output generated from the DadaGP model, with primer shown in the black box. Chord symbols were analyzed by hand.

We further explored how to control time signature and duration during inference. The DadaGP Transformer model operates by generating a set number of tokens [22], but these tokens do not have any relationship to duration. To address this unpredictability, we modified the inference method to keep track of cumulative duration by summing up the tick values of the `wait:xxx` tokens and artificially inserting `new_measure` tokens after a specified amount of ticks to enforce a measure duration. If the tick duration of a wait token crosses a bar boundary, we shorten it to fit within the measure. We also keep a cumulative count of the `new_measure` tokens, and artificially insert an "end" token to replace the last `new_measure` token when the desired number of bars is reached. A shortcoming of this approach is that it does not allow us to differentiate between time signatures with the same measure duration, e.g. between 3/4 and 6/8.

After LooperGP has generated a desired number of bars, the next step is to extract loops using our filtering procedure. By setting the LB_{min} and LB_{max} parameters to the same value, we can specify the exact length of loops to output. As LooperGP was trained specifically on loop data, the generated outputs should contain significantly more loops compared to the baseline model. We compare the loop generation rate between different versions of LooperGP and the baseline DadaGP model in the results section.

4   Results

4.1   Training Performance

The four model configurations described in Sect. 3.3 were each trained for 20 epochs (due to resource constraints) with an 85/15 training/validation split on the loop dataset. Cross-entropy training and validation loss curves are shown in Fig. 3. As expected, the pretrained model configurations have a much flatter loss curve than the scratch models, as they have already converged on the original DadaGP dataset. Validation loss decreases over time in all but the Pretrained Barred Repeats model.

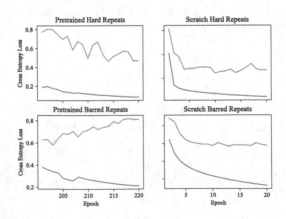

Fig. 3. Training loss (blue) and Validation loss (orange) across 20 epochs for the 4 LooperGP model configurations (Color figure online)

4.2   Loop Performance

To evaluate the quality of output generated by each of the model configurations, we generated 25 excerpts from each model and calculated the average number of loops extracted per excerpt and the average loop note density as defined in Eq. 7. Each model was configured to generate 16 bars of 4/4 music, primed with a random tempo and instrument configuration extracted from the training data. We then ran the loop extraction algorithm from Sect. 3.2 with the following parameters: $L_{min} = 4$, $RD_{min} = 2$, $LB_{min} = 4$ and $LB_{max} = 4$, chosen as the median of Table 1.

The results from the inference analysis are shown in Table 3. It is immediately clear that the Hard Repeats model trained from scratch was not able to generate meaningful loops; very few were extracted and the average note density was extremely low. Manually viewing the generated results from this model revealed most of the outputs were comprised of long held notes and excessive rests, indicating it was not able to sufficiently learn musical phrases. The Pretrained Hard Loops model did not have this issue with excessive rests, however it failed to perform better than the baseline DadaGP model, generating only 1 more valid loop across the total 25 excerpts.

Table 3. Loop Extraction Statistics Between Models.

Model	Loops Found	Avg. Loops	Avg. Note Density
DadaGP Baseline	20	0.80	6.03
Pretrained Hard	21	0.84	6.84
Scratch Hard	3	0.12	0.25
Pretrained Barred	88	3.52	6.96
Scratch Barred	77	3.08	9.80

The Barred Repeats models both performed well above the baseline in terms of number of loops generated. In both cases three or more 4-bar loops were extracted from each 16 bars of generation, with high average note density. While these results are promising, they do not give us any indication as to the musical quality of the extracted loops. To prepare stimuli for a subjective listening test, we continued to train the Pretrained Barred Repeats model, stopping at epoch 40 with a Cross-Entropy loss of 0.148.

4.3   Subjective Evaluation

To evaluate the subjective quality of loops generated by LooperGP, we conducted an online listening study which had participants evaluate machine-generated loops and human-composed loops extracted from the training data. Participants listened to 50 excerpts, each containing a 4-bar loop that repeated four times. 25 of the excerpts were human-composed, and the other 25 were machine-generated[2]. Excerpts were presented to participants in a random order, without revealing the generation method. Participants were asked the following list of questions about each excerpt, and also filled out the Goldsmiths Music Sophistication Index (MSI) survey [19]:

1. Have you heard the music in this excerpt before? (Y/N)
2. Do you think the music in this loop has been produced by a human or by a machine? (Human/Machine)
3. Please rate to what extent you agree or disagree with the following statements: (7-point Likert scale)
   Q1: The music in this loop is original and musically creative.
   Q2: I like the music in this loop.
   Q3: Within the loop, the music is coherent rhythmically, melodically and harmonically.
   Q4: The transition between loop repetitions is smooth rhythmically, melodically and harmonically.
4. Briefly explain the reasoning behind your ratings. You may list keywords rather than using full sentences.

Listening Test Excerpts. The 25 human-made loops were randomly sampled from the set of all 4-bar loops extracted from the DadaGP training data. The other 25 excerpts were machine-generated by the LooperGP model. To generate an excerpt, a primer tempo, instrumentation and 1st note were randomly sampled from the 4-bar loop DadaGP dataset, then inference was run for 16 bars of 4/4. Next, all 4-bar loops were extracted from the generated output and a random one was chosen as the final excerpt. This procedure was repeated 25 times. All excerpts were rendered as MP3s (variable 260 kbps) using the default virtual instruments in GuitarPro 7. Participants were instructed to focus on the quality of the music composition and not on the quality of the virtual instruments or the music production mix when answering questions.

Participants. We recruited 31 participants in total, six of whom responded to a department-wide participant invitation email, and 25 from the data collection website Prolific. The inclusion criteria were access to headphones, normal hearing, no amusia (tone deafness), and having music as a hobby. Participants were roughly evenly split between genders, with 17 male and 14 female participants.

Ordinal Data Transformation. The Likert scale data collected from the four statements in Question 3 is ordinal data. Opinions are mixed as to whether or not Likert data can be treated as continuous data in statistical tests [4] and [25]. The mean value is not necessarily a useful metric, as participants may

---

[2] Link to listening test excerpts: https://drive.google.com/drive/folders/1I0MCPYjj 8nXqKkmDN-d-C2ETOHJpCZyn?usp=share_link.

not assume a constant distance between Likert items [26]. To work around this issue, we implemented the Snell scaling procedure proposed in [26], which uses Maximum Likelihood Estimation to place the ordinal Likert categories as points on a continuous scale so that mean-based parametric methods may be applied. Applying their algorithm to our listening test data, we mapped the seven Likert values (Strongly Disagree, Disagree, Slightly Disagree, Neutral, Slightly Agree, Agree, Strongly Agree) to the values shown in Fig. 4.

Rating Results and Mixed Effects Models. We used a linear mixed-effects model (LMM) [17] to test the effect of the generation type on the results from the survey questions. A LMM was chosen as it is suitable for repeated measures and individual differences can be modeled by using different random intercepts for each participant. Participants' MSI scores [19] were included as an additional independent variable to evaluate the relationship between music background and ratings. Finally, we used participant ID (PID) as a random effect to account for variation in rating tendencies between individual participants. The mixed effects model was implemented using the random intercept model from the StatsModel 0.13.5 Python library [1]. In all four questions, generation type had a significant effect on the rating results, with $p < .001$. The MSI score did not have a significant effect on any of the dependent variables, with all $p \geq 0.22$. Table 4 shows the breakdown of regression results by question.

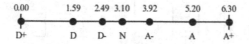

Fig. 4. 7-point Likert scale values converted to a continuous scale (Snell method [26]).

Table 4. LMM Results for Fixed Effect (Coefficients, Standard Errors, and p-value) and Random Effect (PID Variance and Standard error)

	Coef.				Std. Err.				$P > |z|$			
	Q1	Q2	Q3	Q4	Q1	Q2	Q3	Q4	Q1	Q2	Q3	Q4
Intercept	3.388	3.463	4.019	3.966	0.584	0.581	0.495	0.643	0.001	0.001	0.001	0.001
Gen. Type	-0.072	-0.897	-0.621	-0.565	0.072	0.082	0.070	0.069	0.001	0.001	0.001	0.001
MSI	0.009	0.005	0.006	0.008	0.007	0.007	0.006	0.008	0.220	0.446	0.315	0.310
PID Var.	0.420	0.403	0.292	0.522	0.086	0.075	0.063	0.110	-	-	-	-

To visualize the rating difference between generation type, we plotted the Likert distributions shown in Fig. 5 using the original Likert data. With the exception of Q2:Machine ("I like the music in this loop"), the median of all response categories are positive. For Likert questions 2–4, the human-composed excerpts are rated approximately one category higher than the machine-generated excerpts. The median ratings for Q1 ("The music in this loop is original and musically creative") were equivalent across generation type. Mean, variance and standard deviation for the Snell-scaled [26] ratings are shown in Table 5.

Musical Turing Test Results. We evaluated participants ability to discern between human- and machine-made loops using a logistic regression model. As with the Likert rating questions, both generation type and MSI score were used as independent variables in the model. To ensure participants' answers would not be biased towards human-made loops due to familiarity, we discarded 83 (out of 1550) datapoints where the participants indicated they recognized the music in the loop. We used the logistic regression implementation from the StatsModel Python library [1] to calculate the model results. Generation type was found to have a significant effect on perceived generation source (Human or Machine), with $p < 0.01$. MSI score did not have a significant effect on perceived generation source, with $p \geq 0.28$. To further discern how well participants were able to differentiate between human- and machine-made loops, we plotted the confusion matrix in Table 6. Though generation type was found to have a significant effect on perceived generation source, the hit rate is only 10% above chance. Participants correctly predicted generation type about 60% of the time, and the distribution is similar for both human and machine sources.

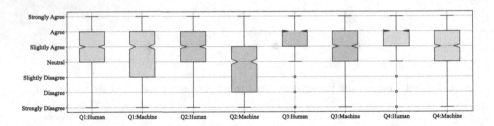

Fig. 5. Comparing 7-point Likert rating distributions between human and machine generated loops for each evaluation question.

Table 5. Likert Rating Statistics. H: Human; M: Machine.

	Median		Mean		Var		Std Dev	
	H	**M**	**H**	**M**	**H**	**M**	**H**	**M**
Q1	Slightly Agree	Slightly Agree	4.09	3.38	2.24	2.62	1.50	1.62
Q2	Slightly Agree	Neutral	3.89	3.00	2.93	3.04	1.71	1.74
Q3	Agree	Slightly Agree	4.50	3.88	1.83	2.55	1.35	1.60
Q4	Agree	Slightly Agree	4.60	4.03	1.94	2.73	1.39	1.65

Table 6. Confusion Matrix of Ground Truth and Predicted Generation Type.

	Human	Machine
Human	0.6	0.4
Machine	0.42	0.58

Free Response Results. Participants were given a text box to explain their ratings for each excerpt in the listening test. We visualized the top 40 most frequently occurring words in answers for each generation type using a word cloud generated by the WordCloud Python library [18]. The following words were used as stop words and removed from the cloud as they occurred frequently but were not informative: loop, melody, sounds and feels. As shown in Fig. 6, "simple", "boring" and "repetitive" are used more often to describe machine excerpts than human ones. On the other hand, "interesting", "smooth" and "coherent" are roughly equal size in both types of excepts. "Good" is a top word in both categories, but is more often used to describe human excerpts.

Fig. 6. Frequency-based word clouds for machine-generated (left) and human-generated (right) excerpts from listening test free response.

5   Discussion

Our approach to restructuring the DadaGP dataset was overall successful in encouraging loopable outputs. However, there was a surprising difference in output quality between the Hard Repeats and Barred Repeats dataset formats. The Pretrained Hard Repeats model produced the same amount of loops on average as the baseline model, indicating the model was unable to learn what features constitute a loopable phrase. The Hard Repeats model performed even worse when trained from scratch, falling into a local minimum in which the best it could do was generate rests and the occasional long note. Seemingly, the model learned to focus on the easier low density tracks while ignoring the pertinent features in the high density tracks of a song.

The Barred Repeats model was much more successful at generating loops during inference, with both the pretrained and scratch models generating, on average, 3x as many loops as the baseline. Despite high performance on this loop metric, the validation loss of the pretrained Barred Repeats model strictly increased over time, indicating that it was over fitting the training data. Evidently the model learned to place repeat bars, but it did not learn to place them in meaningful places. Unlike the pretrained version, the scratch Barred Repeats model did converge and generated loops with a higher average note density, indicating that it was successful at learning musically meaningful loops.

Our subjective listening test allowed us to evaluate the quality of generated loops beyond the objective note density metric. Our statistical evaluation showed a significant effect of generation type on all four Likert rating questions, indicating that the generated outputs are distinguishable from human-composed ones. However, it is promising that median ratings were positive across all questions and suggests our method has potential as a performance tool.

Breaking down the Likert ratings by question, it is peculiar that Q1, which asked for ratings of originality and musicality, had the most similar ratings across generation types. Musicality and originality are particularly human characteristics that AI models traditionally have difficulty emulating, so this result is surprising. It is possible that unexpected musical outputs were interpreted as "original" by participants, inflating ratings for machine-made excerpts. Q2, which asks participants if they like an excerpt, had the lowest median rating for both machine and human excerpts. Though we asked participants to focus on the musical score rather than the quality of the digital instruments, for preference questions such as this one it may have been difficult for participants not to be influenced by the synthetic timbres and lack of human expression in the renditions of both the human and machine excerpts, biasing the scores negatively. Q3 and Q4, which asked about musical coherence and loop-point smoothness respectively, had very similar rating distributions as shown in Fig. 5, hinting that perhaps participants had trouble differentiating between the meaning of musical coherence and loop-point smoothness. This may be in part caused by the presence of sub-loops within some of the excerpts. For instance, with our extraction algorithm, it is possible a 4-bar loop could actually be made up of a repeated 2-bar loop. This may have made it difficult for participants to focus on the exact loop point when answering Q4 since the amount of repetitions varied.

The musical Turing test data showed a significant effect of generation type on perceived generation srouce, but results were still close to chance for both types. While the usefulness of Turing tests in evaluating the musicality of algorithmic composition systems is disputed [3], in our case it is a useful metric of how well LooperGP is able to mimic the loopability of the training data. In other words, it offers insight into whether LooperGP has learned to place repeat bars in convincing places. Given our results, LooperGP is able to somewhat convincingly emulate human composers, but there is room for improvement.

The sample size for the listening test remains fairly small (31 participants who each assessed 25 items per condition, so a total of about 775 observations per condition). However, for fixed effects, power in LMMs does not necessarily increase as the total sample of observations increases [17].

6   Conclusion and Future Work

As the motivation behind LooperGP is for it be used in live coding, we shall discuss its potential in this context. Reference [6] argues that for a generative process to be useful in a live coding environment, it must be succinct to call, applicable to a variety of musical circumstances, modifiable and require limited temporal scope. Given that LooperGP focuses on loopability of output and can be primed with a specific instrumentation, key/time signature and bar length, it can be configured to fit a wide variety of musical circumstances. Our algorithm requires no temporal scope, as the model does not have any temporal dependencies. A downside of LooperGP, and many other DL methods for music generation, is a lack of modification control. Once a loop has been generated, there is no way to algorithmically adjust it. However, it could be feasible to generate variations on a generated loop by passing it back through the system as a primer, as observed in [23]. Finally, an inference of LooperGP can be easily generated via a succinct Python function call, as the only required parameters are key, duration and bar length.

Overall, LooperGP was able to increase the frequency of high-density loopable outputs compared to the baseline. Subjective ratings showed a positive response to the loops overall, but also identified some key areas for improvement in reducing simplicity and repetitiveness. LooperGP exhibits many of the characteristics required for a generative process to be useful in live coding, with a specific focus on generating controllable output able to fit most musical circumstances. The duration control and loopable generation of LooperGP make it a viable option for use in a live performance context, although this should be the object of further assessment with live coders.

To address the listening test comments that machine-made loops were "repetitive" and "boring", future work could focus on increasing the average complexity of extracted loops by adjusting the loop extraction algorithm to identify and filter out sub-loops. On the training side, more of the hyperparameter space could be explored, such as increasing dropout or decreasing layer size, to prevent the early overfitting demonstrated by some of the training setups.

Finally, an exciting area for future work is integrating the LooperGP system into a live coding framework such as Tidal Cycles [16] or SuperCollider [15]. The OSC pipeline for communicating between Python and SuperCollider presented in [13] could be used to prototype this concept without requiring the model to be exported to a new environment. It is also important that LooperGP be wrapped into a usable interface with a succinct API. Ideally, the performer should be able to cherry-pick a loop from a set of potential options generated by the model and automatically transpose it to their desired key signature. Once LooperGP has been wrapped into a performer-friendly interface, its potential as a live coding feature can be truly evaluated by incorporating it into a concert and collecting feedback from audience members and performers.

Acknowledgements. This work has been partly supported by the EPSRC UKRI Centre for Doctoral Training in Artificial Intelligence and Music (Grant no. EP/S022694/1).

References

1. Statsmodels (2022). https://github.com/statsmodels/statsmodels. Accessed 14 Aug 2022
2. Ackley, D.H., Hinton, G.E., Sejnowski, T.J.: A learning algorithm for Boltzmann machines. Cognit. Sci. 147–169 (1985)
3. Ariza, C.: The interrogator as critic: the turing test and the evaluation of generative music. Comput. Music. J. **33**, 48–70 (2009)
4. Bishop, P.A., Herron, R.L.: Use and misuse of the likert item responses and other ordinal measures. Int. J. Exer. Sci. **8**, 297–302 (2015)
5. Briot, J.P., Hadjeres, G., Pachet, F.D.: Deep Learning Techniques for Music Generation, vol. 1. Springer, Cham (2020). doi: https://doi.org/10.1007/978-3-319-70163-9
6. Brown, A.R., Sorensen, A.: Interacting with generative music through live coding. Contemp. Music. Rev. **28**, 17–29 (2009)
7. Chandna, P., Ramires, A., Serra, X., Gómez, E.: Loopnet: Musical loop synthesis conditioned on intuitive musical parameters. In: ICASSP 2021–2021 IEEE International Conference on Acoustics, Speech and Signal Processing (ICASSP), pp. 3395–3399. IEEE (2021)
8. Dai, Z., Yang, Z., Yang, Y., Carbonell, J., Le, Q.V., Salakhutdinov, R.: Transformer-xl: Attentive language models beyond a fixed-length context. arXiv preprint arXiv:1901.02860 (2019)
9. Hadjeres, G., Pachet, F., Nielsen, F.: Deepbach: a steerable model for bach chorales generation. In: International Conference on Machine Learning, pp. 1362–1371 (2017)
10. Hsu, J.L., Liu, C.C., Chen, A.L.: Discovering nontrivial repeating patterns in music data. IEEE Trans. Multimedia **3**, 311–325 (2001)
11. Huang, Y.S., Yang, Y.H.: Pop music transformer: beat-based modeling and generation of expressive pop piano compositions. In: Proceedings of the 28th ACM International Conference on Multimedia, pp. 1180–1188 (2020)
12. Ji, S., Luo, J., Yang, X.: A comprehensive survey on deep music generation: Multi-level representations, algorithms, evaluations, and future directions. arXiv preprint arXiv:2011.06801 (2020)
13. Lan, Q., Tørresen, J., Jensenius, A.R.: Raveforce: a deep reinforcement learning environment for music. In: Proceedings of the SMC conferences, pp. 217–222. Society for Sound and Music Computing (2019)
14. Magnusson, T.: Sonic Writing: Technologies of Material, Symbolic, and Signal Inscriptions. Bloomsbury Publishing USA (2019)
15. McCartney, J.: Supercollider: A new real-time sound synthesis language. In: Proceedings of the International Computer Music Conference, pp. 257–258 (1996)
16. McLean, A., Wiggins, G.: Tidal-pattern language for the live coding of music. In: Proceedings of the 7th Sound and Music Computing Conference, pp. 331–334 (2010)
17. Meteyard, L., Davies, R.A.: Best practice guidance for linear mixed-effects models in psychological science. J. Mem. Lang. **112**, 104092 (2020)

18. Mueller, A.: Word cloud (2022). https://github.com/amueller/word_cloud. Accessed: 14 Aug 2022
19. Müllensiefen, D., Gingras, B., Musil, J., Stewart, L.: The musicality of non-musicians: an index for assessing musical sophistication in the general population. PLoS ONE **9**(2), e89642 (2014)
20. Nilson, C.: Live coding practice. In: Proceedings of the 7th International Conference on New Interfaces for Musical Expression, pp. 112–117 (2007)
21. Ramires, A., et al.: The freesound loop dataset and annotation tool. arXiv preprint arXiv:2008.11507 (2020)
22. Sarmento, P., Kumar, A., Carr, C., Zukowski, Z., Barthet, M., Yang, Y.H.: DadaGP: a dataset of tokenized GuitarPro songs for sequence models. In: Proceedings of the 22nd International Social for Music Information Retrieval Conference (2021)
23. Shih, Y.J., Wu, S.L., Zalkow, F., Muller, M., Yang, Y.H.: Theme transformer: symbolic music generation with theme-conditioned transformer. IEEE Trans. Multimedia 1–1 (2022)
24. Stewart, J., Lawson, S.: CIBO: an autonomous tidalcyles performer. In: Proceedings of the Fourth International Conference on Live Coding, p. 353 (2019)
25. Sullivan, G.M., Artino, A.R.: Analyzing and interpreting data from likert-type scales. J. Graduate Med. Educ. **5**, 541–542 (2013)
26. Wu, C.H.: An empirical study on the transformation of likert-scale data to numerical scores. Appl. Math. Sci. **1**, 2851–2862 (2007)

Chordal Embeddings Based on Topology of the Tonal Space

Anton Ayzenberg[1]([✉])[iD], Maxim Beketov[1][iD], Aleksandra Burashnikova[2],
German Magai[1][iD], Anton Polevoi[3], Ivan Shanin[4], and Konstantin Sorokin[1][iD]

[1] Higher School of Economics, Moscow, Russia
{aayzenberg,mbeketov,gmagaj,ksorokin}@hse.ru
[2] Skolkovo Institute of Science and Technology, Moscow, Russia
Aleksandra.Burashnikova@skoltech.ru
[3] Lomonosov Moscow State University, Moscow, Russia
s02210194@gse.cs.msu.ru
[4] Federal Research Center "Computer Science and Control",
Russian Academy of Sciences, Moscow, Russia

Abstract. In the classical western musical tradition, the mutual simultaneous appearance of two tones in a melody is determined by harmony, i.e. the ratio of their frequencies. To perform NLP-based methods for MIDI file analysis, one needs to construct vector embeddings of chords, taking mutual harmonicity into account. Previous works utilising this idea were based on the notion of Euler's Tonnetz. Being a beautiful topological model describing consonance relations in music, the classical Tonnetz has a certain disadvantage in that it forgets particular octaves. In this paper, we introduce the mathematical generalisation of Tonnetz taking octaves into account. Based on this model, we introduce several types of metrics on chords and use them to construct chordal embeddings. These embeddings are tested on two types of tasks: the chord estimation task, based on the Harmony Transformer model, and the music generation task, provided on the basis of TonicNet.

Keywords: Tonnetz · Chordal embedding · Lattice models · Topology in music

1 Introduction

Treatment of music as an instance of a natural language, and application of NLP methods in Music Information Retrieval (MIR) is the classical idea covered in many papers [1–4]. Working in the MIDI domain, one deals with pitches (piano notes) and their collections,—the chords. In a sense, pitches or chords play the

The work of A. Ayzenberg, M. Beketov, G. Magai, and K. Sorokin was supported by the HSE University Basic Research Program. A. Burashnikova was supported by the Analytical center under the RF Government (subsidy agreement 000000D730321P5Q0002, Grant No. 70-2021-00145 02.11.2021).

C. Johnson et al. (Eds.): EvoMUSART 2023, LNCS 13988, pp. 20–33, 2023.
https://doi.org/10.1007/978-3-031-29956-8_2

role of words. If one wants to apply ML algorithms based on continuous optimisation, one needs to construct vector embedding of chords in the first place. In NLP, the famous embeddings such as bag-of-words [5] were constructed from the observation that pairs of words appearing often in the same context should be represented by close points in the Euclidean space. Word2vec is another type of embedding used in MIR [6]. The resulting metric structure on words is highly complicated: it is unlikely, that it will ever be described mathematically. The case of music is simpler. The western classical tradition, stemming back to Pythagoras, theoretically describes which pairs of sounds should appear simultaneously in a melody. The pair of sounds is considered consonant if the ratio of their frequencies is a fraction with a small numerator and denominator. This regularity, that ordinary natural languages lack, gives rise to many studies in music theory involving topological notions, including orbifolds [7,8], persistent homology, potency spaces [9], group theory [10], formal concepts [11], and many more.

One of the basic models of musical consonance was proposed by Euler [12] (see [13]) by the name Tonnetz and is now widely used both in music theory [8,14] and in MIR [15–19]. Some generalizations of Tonnetz recently gained the attention of mathematicians [20]. Euler's Tonnetz is a 2-dimensional simplicial complex, whose vertices are the 12 notes (pitch classes) of one octave, and triangles are major and minor chords—the most common consonant combinations. Surprisingly at first glance, this simplicial complex is homeomorphic to a 2-dimensional torus. This observation supports the idea that the musical harmony inside a single octave has the metric structure of a torus. This metric structure was extended to a pseudo-metric on chords in [2] and used further to build chordal vector embeddings. The approach has a certain disadvantage in that it collapses all octaves into a single octave, thus forgetting important information stored in the original MIDI.

In this paper, we introduce a mathematical model that we called the tonal space. It is similar to Tonnetz in the sense that it tracks consonance, but it also takes the whole musical scale into account, instead of dealing with a single octave. Our model is based on the space of multiples, introduced by Oleynikov [21], and can be considered as its natural further development. Topologically, the tonal space is homeomorphic to the thickened torus, the cartesian product of a torus with an interval. The construction of this space and the algorithm to pass from musical scale to this space and back is considered the main theoretical contribution of this article.

We propose two directions where the constructed mathematical model of the tonal space can prove useful. First direction is related to computer science: the metric structure on the tonal space can be used to construct vector embeddings of chords. In this direction we extend and complement the work [2]. The constructed vector embeddings are tested on the task of next chord prediction using the LSTM network. The approach has shown certain improvement in [22], accuracy compared to the previous results.

The second direction is of potential interest to musicians. Transformations (diffeomorphisms) of the tonal space provide a universal way to transform a given melodic sequence to another sequence, preserving harmonicity relations. This idea may be used in combinatorial music generation similar to the idea realized in HexaChord (see [17]) for the classical Tonnetz. As before, instead of working in a single octave, we treat all octaves at once, which may provide bigger variability for music generation.

2   Tonal Space: Tonnetz Revisited

2.1   Euler's Tonnetz

We adopt the following convention. Notes (otherwise called pitches) represent consecutive sounds, and the shift by one halftone corresponds to increasing frequency $2^{1/12}$ times. Taking logarithms of frequency, the whole musical scale becomes represented by the set $\mathbb{Z}$ of integers. The hearing range for humans is between 20 and 20000 Hz, therefore only a finite range of $\mathbb{Z}$ appears in applications. We assume that the perceived range is given by the set $[128] = \{0, 1, \ldots, 127\}$; this convention is compatible with the standard MIDI format.

The octave shift (doubling the frequency) corresponds to adding 12 halftones in $\mathbb{Z}$. In the music theory, the pitches are considered modulo octave shifts. The elements of the cycle set $C_{12} = \mathbb{Z}/12\mathbb{Z}$ are called pitch classes. In C_{12}, there exist certain intervals: perfect fifths $\{x, x + 7\}$ (equiv., $\{x, x + 5\}$), major thirds $\{x, x+4\}$, minor thirds $\{x, x+3\}$, major triads $\{x, x+4, x+7\}$, and minor triads $\{x, x+3, x+7\}$. In general, a chord is a subset of either $\mathbb{Z}$ or C_{12}.

A simplicial complex is a higher-dimensional generalization of a graph, see [23]. Consider the 2-dimensional abstract simplicial complex Tonnetz on the vertex set C_{12} with triangles determined by all major and minor triads. It is known [13,20] that Tonnetz is a triangulation of a 2-torus T^2. The edges of Tonnetz are perfect fifths, major and minor thirds, therefore the distance on Tonnetz represents the harmonic closeness of pitch classes. There is a freedom, however, in the choice of edge lengths: lengths of perfect fifths, major and minor thirds can be considered as hyperparameters. The metric on Tonnetz can be further extended to a (pseudo)metric on meaningful musical chords. This idea was implemented in [2] to construct chordal embeddings.

2.2   Tonal Graph

Euler's Tonnetz has a certain deficiency: it only allows to treat pitch classes, not pitches themselves. The problem is to analyze which topological space models harmonic relations between pitches. The first approach that we propose is to introduce a translation-invariant metric on $[128]$, compatible with musical theory, and after that use the machinery of applied topology to understand the shape of this finite metric space. In the following arguments we abuse the terminology by translating from equal temperament scale to just intonation: we implicitly replace frequencies' ratios by their suitable rational approximations where it makes sense.

The first version of metric on [128] is based on the notion of Benedetti height and Tenney distance. If the frequencies of two pitches give the ratio p/q (with small coprime p, q), then we set the distance equal to $\log_2(pq)$. For example, one octave has length $\log_2 2 = 1$, while the perfect fifth $3/2$ is $\log_2 6$. The metric defined this way is represented by the graph on [128], where two pitches are connected with an edge if their frequencies differ 2 or 3 or 5 times, i.e. has the form $\{x, x + 12\}$, $\{x, x + 19\}$ or $\{x, x + 28\}$ respectively. The lengths of all different types of edges are set to $d_2 = \log_2 2 = 1$, $d_3 = \log_2 3$, and $d_5 = \log_2 5$. The lengths d_2, d_3, d_5 are the hyperparameters which can be adjusted in the other variants of the metric. For example, instead of Tenney distance one can set $d_2 = d_3 = d_5 = 1$ (equidistant case). Another option is to set the lengths of all intervals of the form $\{x, x + \delta\}$ where δ belongs to $\{\pm12, \pm19, \pm28, \pm12 \pm 19, \pm12 \pm 28, \pm19 \pm 28, \pm12 \pm 19 \pm 28\}$ equal to 1 (Chebyshev case). In this case the perfect interval $\{x, x + 7\}$, for example, has the same small length as the octave.

Given a finite metric space, there is a standard technique to estimate its topological shape, based on the notion of persistent homology. We refer to [24, 25] for the introduction to this technique. Persistent homology of the set of pitches with Chebyshev metric resembles the homology of a 2-torus. This observation supports the idea that the space of all pitches has something in common with Euler's Tonnetz.

2.3 Tonal Space

To explain the homological observation of the previous subsection, we look at the tonal graph from a different perspective. The construction of this subsection was motivated by the book of Oleynikov [21] on music theory. Let us introduce the Euclidean space in which the frequency changes $\times 2$, $\times 3$, and $\times 5$ represent parallel shifts by vectors of small length. This idea is proposed in [21] by the name "space of multiplicities". Let $\mathbb{Z}^3 \subset \mathbb{R}^3$ be the lattice generated by the basis $\{e_2, e_3, e_5\}$. In this model the shift by e_2 (resp. e_3, e_5) will represent multiplication of frequency by 2 (resp. 3, 5). To make things precise, we consider the colouring $f \colon \mathbb{Z}^3 \to \mathbb{Z}$ of lattice points by pitches: we set the rules

$$
\begin{aligned}
f(0) &= 0, \\
f(v + e_2) &= f(v) + 12, \\
f(v + e_3) &= f(v) + 19, \\
f(f + e_5) &= f(v) + 28.
\end{aligned}
\tag{1}
$$

for any point $v \in \mathbb{Z}^3$ of the lattice. It is easily seen that $f \colon \mathbb{Z}^3 \to \mathbb{Z}$ is a homomorphism of abelian groups, given by

$$
f(xe_2 + ye_3 + ze_5) = 12x + 19y + 28z.
\tag{2}
$$

The point with integral coordinates (x, y, z) is mapped to the pitch $12x + 19y + 28z$. To define the topology on the whole set of pitches, we proceed with this lattice model. The lattice $\mathbb{Z}^3$ itself is not the correct model for the tonal space since some points of the lattice are coloured with the same pitch (since we identified equal temperament and just intonation). Let $v_1 = (x_1, y_1, z_1)$ and $v_2 = (x_2, y_2, z_2)$ be two integral points such that $f(v_1) = f(v_2)$. In this case, the points v_1 and v_2 should be identified in the model. Since f is a homomorphism, the condition $f(v_1) = f(v_2)$ is equivalent to $v_1 - v_2 \in \mathrm{Ker}\, f$. Denoting the kernel $\mathrm{Ker}\, f$ by L, we see that the correct model for the tonal space should the quotient $\widetilde{\mathrm{Tonn}}^{\mathbb{Z}}_L = \mathbb{Z}^3/L$.

Notice that L is a sublattice of rank 2, given as the set of integral solutions to the homogeneous linear Diophantine equation

$$12x + 19y + 28z = 0. \tag{3}$$

Solving this equation by standard methods we found that the lattice L is generated by the vectors

$$v_1 = (3, 4, -4) \text{ and } v_2 = (-4, 4, -1). \tag{4}$$

There is a freedom in the choice of fundamental solutions, however, we chose these particular vectors since they have a small Euclidean norm and they are almost orthogonal.

Algebraically, $\mathbb{Z}^3/L = \mathrm{Im}\, f = \mathbb{Z}$ by the fundamental theorem on the homomorphisms. This is not surprising, since $\mathbb{Z}$ is indeed the set of pitches, by the definition. However, an important thing is that $\mathbb{Z}^3$ has a metric induced from the ambient Euclidean space $\mathbb{R}^3$, and this metric gives "the quotient metric" on $\mathbb{Z}^3/L$ according to the following construction. Let $[v]$, $[w]$ be the classes of vectors $v, w \in \mathbb{R}^3$, and let d be any metric on $\mathbb{R}^3$ (the natural choice is Euclidean L_2-norm, but we also checked other variants described in Subsect. 3.1). The quotient metric is defined by $d([v], [w]) = \min_{v', w'} d(v', w')$ where v' runs over $v + L$ and w' runs over $w + L$. Since d on $\mathbb{R}^3$ is translation-invariant, the expression can be simplified $d([v], [w]) = \min_{a \in L} d(v - w, a)$, where only one variable a is optimised over the lattice L.

The problem of computing the distances on $\mathbb{Z}^3/L$ is equivalent to the problem of finding the closest lattice point of L to a given point in $\mathbb{R}^3$. This general problem is known in the literature, see e.g. [26]. However, the theory is well developed for Euclidean distances (in this case Voronoi diagrams are widely used: Voronoi cells for L_2-metric are convex, so the problem can be reduced to convex optimisation). There seem to be no good algorithms to find the closest lattice point in L_∞-metric, however, a brute force solution applies: one can check all lattice points within a reasonable finite range. The arguments given above support the following statement.

Proposition 1. *The correct (continuous) topological model for the tonal space is the quotient* $\widetilde{\mathrm{Tonn}}_L = \mathbb{R}^3/L$, *where L is the lattice of integral solutions to Diophantine equation (3). As a topological space, $\widetilde{\mathrm{Tonn}}_L$ is homeomorphic to $T^2 \times \mathbb{R}^1$.*

Proof. Observe that $\mathbb{R}^3$ is the natural continuous replacement for the discrete model $\mathbb{Z}^3$. Since the points differing by $v \in L$ should be identified in the model, the tonal space is the quotient $\widetilde{\mathrm{Tonn}}_L = \mathbb{R}^3/L$. Since $\mathrm{rk}\, L = 2$, its real span $L_{\mathbb{R}} = L \otimes \mathbb{R}$ is a 2-dimensional plane in $\mathbb{R}^3$, so we have $\mathbb{R}^3/L \cong L_{\mathbb{R}}^{\perp} \oplus (L_{\mathbb{R}}/L) \cong \mathbb{R} \times T^2$ (the quotient of 2-plane by 2-lattice is a 2-torus, while the transversal direction to L is not affected by the quotient). $\qquad \square$

Remark 1. Roughly speaking, the 2-torus T^2 appearing as the factor of $\widetilde{\mathrm{Tonn}}_L$ is an analogue of Euler's Tonnetz, while the complementary direction $\mathbb{R}$ corresponds to "octave shifts". This may help imagination, however, mathematically this is incorrect. One can notice that Eq. 3 determining L has nothing to do with the numbers $3, 4, 5$ appearing in the construction of the classical Tonnetz. The classical Tonnetz T^2 arises from $\widetilde{\mathrm{Tonn}}_L$ if one takes its additional quotient by the first coordinate axis $\langle (1, 0, 0) \rangle$ (recall that this axis corresponds to octave-shifts), but this construction cannot be reversed.

Now we restrict to the topological model for the hearing range. Consider the space between two parallel planes (the layer):

$$\Pi = \{(x, y, z) \in \mathbb{R}^3 \mid 0 \leqslant 12x + 19y + 28z \leqslant 127\}. \tag{5}$$

All integral points in Π are coloured with pitches, which are perceived by humans. The lattice L acts on Π by linear shifts. Taking quotient by the lattice L, we get the correct topological model of the tonal space of perceived sounds.

Definition 1. *The space* $\mathrm{Tonn}_L = \Pi/L$ *is called* the tonal space.

Again Tonn_L has a metric induced from $\mathbb{R}^3$. If $\mathbb{R}^3$ is endowed with the standard Euclidean metric, then Tonn_L becomes a 3-dimensional Riemannian manifold with a boundary.

Proposition 2. *Assume that* $\mathbb{R}^3$ *is given a standard Euclidean metric. Then the following holds true:*

1. *As a topological space,* Tonn_L *is homeomorphic to the thickened torus* $T^2 \times D^1$, *where* $D^1 \cong [0, 1]$ *is the closed interval.*
2. *As a Riemannian manifold,* Tonn_L *is isometric to* $\mathbb{T}^2 \times \mathbb{I}$, *where* $\mathbb{T}^2$ *is a flat torus generated by the parallelogram of vectors of lengths* ≈ 6.4 *and* ≈ 5.74 *with angle* $\approx 77.43°$, *and* $\mathbb{I}$ *is the interval of length* ≈ 3.54.

Proof. Similar to the proof of Proposition 1, we have $\Pi = L_{\mathbb{R}} \times D^1$, and $\mathrm{Tonn}_L = \Pi/L = (L_{\mathbb{R}}/L) \times D^1 \cong T^2 \times D^1$. The metric structure is determined as follows. The length of D^1 in the decomposition above is the width of the layer Π that is the distance between two parallel planes, it equals $\frac{127}{\sqrt{12^2 + 19^2 + 28^2}} \approx 3.54$. The metric parameters of the torus $L_{\mathbb{R}}/L$ are determined by a fundamental parallelogram of the lattice L, for example, the one spanned by the vectors v_1, v_2, given in (4). Their lengths and the angle are easily computed.

3 Embeddings of Pitches and Chords

Here we pursue an idea similar to the one of [2]. We construct several embeddings of chords, which are further used in the Sect. 4. There are two main differences between our approach and the approach of [2].

1. In [2], the metric was defined on pitch-classes modulo octave, while we analyse the whole chromatic scale as explained in the introduction.
2. In [2], the metric was extended from particular pitch classes to chords in a way, that produces a distance function that is not a metric in a strict mathematical sense. In our paper, we propose several ideas resulting in actual metrics.

3.1 Metric on Pitches

As described in Subsect. 2.3, the metric on pitches is defined by representing each pitch by a point in $\mathbb{Z}^3$ (up to adding the lattice L) and inducing the metric from $\mathbb{R}^3$. We tested 4 options for the underlying metric in $\mathbb{R}^3$ listed below.

1. Euclidean distance $d_2(u, v) = \sqrt{\sum(u_i - v_i)^2}$.
2. Skew Euclidean metric $d_{s2}(u, v)$ defined by the quadratic form

$$G = \begin{pmatrix} 1 & 1/2 & 3/2 \\ 1/2 & 1 & 1 \\ 3/2 & 1 & 3 \end{pmatrix} \tag{6}$$

that is $d_{s2}(u, v) = \sqrt{(u - v)G(u - v)^\top}$. The choice of this particular matrix is explained below.
3. Pure Chebyshev metric on $\mathbb{R}^3$: $d_\infty(u, v) = \max(|u_1 - v_1|, |u_2 - v_2|, |u_3 - v_3|)$.
4. Modified Chebyshev metric: $d_{l\infty}(u, v) = \max(|u_1 - v_1|\log_2 2, |u_2 - v_2|\log_2 3, |u_3 - v_3|\log_2 5)$. This metric is more compatible with the Tenney distance and Benedetti height described in Subsect. 2.2.

There is particular reasoning behind the choice of Chebyshev variants of a metric similar to the one described in Subsect. 2.2. In musical theory, the interval $\frac{3}{2}$, the perfect fifth, plays a prominent role. It is reasonable to assert that pitches differing by this interval should be at the same distance as pitches differing 3 times in frequency. In our $\mathbb{Z}^3$-lattice model, the perfect fifth corresponds to shifting along the vector $(-1, 1, 0)$. Both Chebyshev metrics satisfy $d(u, u + (-1, 1, 0)) = d(u, u + (0, 1, 0))$, while the ordinary euclidean metric does not satisfy this property.

The same phenomenon motivates the choice of the matrix G in the definition of d_{s2}. It can be checked that in this particular metric, all frequencies' ratios $2, 3, \frac{3}{2}, \frac{5}{2}, \frac{5}{4}, \frac{6}{5}$ known in musical theory (they correspond to the shifts by $12, 19, 7, 16, 4, 3$ halftones respectively) have lengths 1. They form a regular tetrahedron in $\mathbb{R}^3$ exactly in the case when G is given by (6).

Next, each of the listed metrics determines a quotient metric on a torus $\mathbb{R}^3/L$ as described in the construction of the quotient metric. This metric defines a harmonic distance between pitches.

3.2   Metric on Chords

A chord is a collection of pitches. In the end, we want to have a vector embedding of chords not just particular pitches. To generate such embeddings we propose two general strategies.

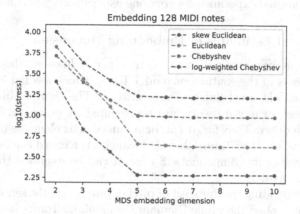

Fig. 1. Logarithms of stresses in MDS for various types of metrics on pitches.

(1) Generate an embedding of pitches in $\mathbb{R}^d$ using multidimensional scaling (MDS) and one of the harmonic metrics described above. For each ambient dimension d, we compute the stress, that is the difference between the original metric and the (euclidean) metric of the embedded data. Analysing how the logarithm of stress depends on d, we choose the appropriate ambient dimension as the value d_{min} after which the stress stabilises. The graphs are shown in Fig. 1.
After we get an embedding $\{x_0, \ldots, x_{127}\} \subset \mathbb{R}^d$ of pitches, we make use of the affine structure of $\mathbb{R}^d$. A chord $A \subset \{0, 1, \ldots, 127\} = [128]$ can be represented by the barycenter of the pitches of which it is composed: $x_A = \left(\sum_{i \in A} x_i\right) / |A|$. The embedding of each chord can be calculated on the run, so it is not required to store the embedding of the whole collection of chords.

(2) Alternatively, we first extend the metric from pitches to chords, then embed chords using MDS. The whole power-set $2^{[128]}$ is too huge, so at first we list only the most common chords, expected in a composition. Generally, we follow the algorithm of chord generation proposed in [2], but extended it to the whole hearing range, and included some additional chords.

To extend the metric from particular pitches to subsets of pitches, we use the distance formula on subsets proposed in [27]. After the distances between chords are computed, we apply MDS. To choose the appropriate embedding dimension, it was very costly to check all dimensions one by one. To estimate the dimension, we formed a sequence of subcollections of chords and found a trend describing the dependence of the optimal embedding dimension on the size of the input set. The experiment has shown that the optimal embedding dimension in our case increases roughly as the logarithm of the size of the dataset. Extrapolating the trend, we found the value 10, which was used as the embedding dimension of the set of chords.

The computational experiments were done using strategy (2) described above.

3.3 Theoretical Estimates of Embedding Dimension

Notice that all pitches form a discrete metric space. However, they can be considered as elements of the continuous model $Tonn_L$ which is a flat 3-dimensional Riemannian manifold according to Proposition 2. The chords which are meaningful in the sense of music theory, are represented by points on $Tonn_L$ which are close to each other. Thus far, it can be assumed that elements of every chord lie in a small local chart of $Tonn_L$. In this local chart, a chord can be represented by the barycenter of its components. So far, it can be assumed that all chords belong to $Tonn_L$ as well.

Instead of embedding the finite set of pitches into $\mathbb{R}^d$ (in the sense of computer science), one can embed the whole Riemannian manifold $Tonn_L$ isometrically (in the sense of differential geometry), and then pick up the particular values of this embedding for all pitches and chords lying on $Tonn_L$.

The problem of isometrical embeddings is the classical problem of differential geometry. There always exists an embedding of a compact n-dimensional manifold M into $\mathbb{R}^{2n}$ according to the celebrated Whitney theorem, see [28]. Speaking of isometric embeddings, Tompkins [29] proved that a flat n-dimensional torus cannot be isometrically embedded into the space of dimension less than $2n$. The famous result of Nash [30] states that any smooth compact n-dimensional manifold can be isometrically embedded into Euclidean space of dimension $n(3n + 11)/2$. The general upper bound of Nash is too big to be applied in our situation. Pinkall [31] constructed isometric embeddings of a flat 2-torus $\mathbb{T}^2$ into $\mathbb{R}^4$. These results imply that a flat thickened torus $Tonn_L \cong \mathbb{T}^2 \times \mathbb{I}^2$ can be isometrically embedded into $\mathbb{R}^5$, and not less.

Notice that according to Fig. 1, the choice of any of the listed 4 metrics for MDS results in the effective embedding dimension equal to 5. The theory described above asserts that it is indeed the best dimension available theoretically. In the following, we concentrate on the skew Euclidean metric as the one giving the lowest stress compared to the diameter of the set.

4   Experimental Setup and Results

In this section, we introduce the empirical evaluations of the theoretically found embeddings by building them into two existing music understanding models. The first one is Harmony Transformer [32] for estimation of the chords over the feature set proposed by MGMcGill Billboard project [33] and the second one is TonicNet [22] for the sequential modeling of polyphonic music with GRU neural networks, estimated on the JSB Chorales benchmark [34].

4.1 Chord Estimation

The chord estimation task could be considered as the sequence labelling task that assigns chord labels for each event of the sequence. Because of the sequential and structural nature of the music, the state-of-the-art approaches proposed for Natural Language Processing (NLP) problems are also applied for various music estimation and generation tasks. The most of them are based on the Recurrent Neural Networks [22,35], and popular nowadays Transformer architecture [32,36].

We skipped the technical aspects regarding the architecture of Harmony Transformer and implementation details as you could find them in the original paper [32] as well as all the necessary information about the MGMcGill Billboard benchmark in [33]. For the estimation of the proposed embeddings on the chord labeling task, we try to approximate the output embeddings from the Harmony Transformer encoder to the theoretical embeddings provided for each major and minor triad. To do this, we add "the head" with a dense layer into the Harmony Transformer, which provides an approximation of embeddings as the regularisation on the training step. Then, the minimised loss function is modified to the next one:

$$\mathcal{L} = \mathcal{L}_{HT} + \alpha \cdot \mathcal{L}_{APP} \qquad (7)$$

In Eq. 7, by $\mathcal{L}_{HT}$ we suggest Harmony Transformer loss from the original paper and by $\mathcal{L}_{APP}$, we consider the loss for the approximation part and represented by the mean squared error between the theoretical ground-truth embeddings and embeddings from Harmony Transformer encoder after the approximation step. The parameter α is responsible for the part of the approximation loss that we take into account over the total loss.

The effectiveness of the proposed embeddings is estimated on the comparison of the evolution of validation accuracy during training over the original Harmony Transformer model and the modified Harmony Transformer with the head of approximation. The output results are presented in the Fig. 2.

From these results it could be seen the tendency between the ratio of α and smoothness of the evolution of validation accuracy; it means that more attention to the chords from the tonal space (greater α) comes to more accurate predictions and stabilise the training process.

4.2 Automatic Music Generation

Speaking about the music generation, it is difficult to provide a relevant formalisation of the problem statement. We could try to create some new music or we could do it based on the set of rules or even based on the existed composition for making a remix. Moreover, it is not clear how to estimate the generated melodies. There are few popular models, such as [37,38], that demonstrated the impressive results in music generation, but unfortunately all of them required huge resources, which comes to the difficulties with the reproduction.

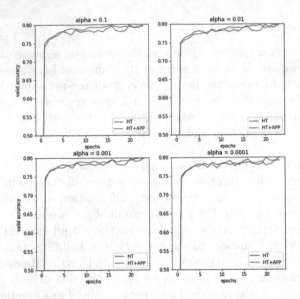

Fig. 2. Evolution of the validation accuracy for different alphas. Alpha here is α from Eq. 7. HT - Harmony Transformer, HT+APP - Harmony Transformer with approximation.

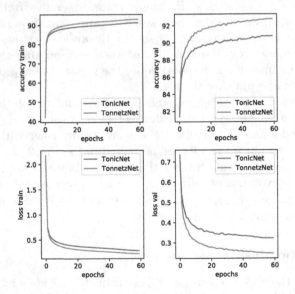

Fig. 3. Comparison of the TonnetzNet and TonicNet performances.

As we are currently interested in the influence of the chord embeddings on the performance of the model, for our experiments we chose the TonicNet, which explores sequential modelling of polyphonic music based on Johann Sebastian Bach chorales (JSB chorales). For the evaluation of the effect, we add the proposed information about the chords represented by the vectors into the latent space of TonicNet (the modified version further is called TonnetzNet). The output results are presented in Fig. 3. The effectiveness of the proposed embeddings is estimated on the comparison of the evolution of accuracy and loss function during training and validation over the original TonicNet model and the modified TonnetzNet. As we can see, the training error decreases in both cases, However, the speed of convergence as well as the rate of accuracy growth is much faster with TonnetzNet.

4.3   Music Transformation

One can use the mathematical model of the tonal space for combinatorial music transformation. The general idea is the following. We consider any continuous transformation (a homeomorphism) $f: \text{Tonn}_L \rightarrow \text{Tonn}_L$ of a thickened torus. The continuity of the map guarantees that consonant pairs of pitches get mapped to the pairs which are consonant as well. Therefore, applying this transformation to the melody, we hope to get another harmonical melody. Since there exist a variety of nontrivial homeomorphisms, we get a variety of combinatorial transformations. Some of them can be used by composers in their theoretical work.

Recall that the set of pitches $\widetilde{\text{Tonn}}_L^{\mathbb{Z}}$ naturally embeds into the continuous model Tonn_L as a discrete set. The practical problem is that, given a homeomorphism f, there is no guarantee, that a pitch maps to a pitch. We propose two solutions to this problem. Code realizations and the practical examples of applications can be found in [39].

First solution produces a combinatorial transformation: each pitch $i \in \widetilde{\text{Tonn}}_L^{\mathbb{Z}}$ gets mapped to the pitch closest to $f(i) \in \text{Tonn}_L$ in the tonal metric.

Second solution adopts a version of smoothening. Each image $f(i)$ gets distributed between the closest pitches with certain weights. To define weights continuously, we triangulate the space Tonn_L (subdivide into the union of tetrahedra with vertices in $\widetilde{\text{Tonn}}_L^{\mathbb{Z}}$). Then, given a point $f(i) \in \text{Tonn}_L$, we find a tetrahedron containing this point, and take its vertices with the weights equal to barycentric coordinates of $f(i)$ in this tetrahedron. The weights of $f(i)$ computed for all pitches i are stored in a square matrix. This matrix is applied to the column vector of intensities of pitches which consecutively appears in a piano roll.

5   Conclusion

In this paper, we introduce the tonal space which takes different octaves into account. The metric structure of this space is used to construct vector embeddings of chords. The effectiveness of this approach is demonstrated on several computational models. The continuous transformations of the tonal space can be used to automatically compose new chordal sequences from a given sequence preserving consonance relations.

References

1. Zhu, H., Niu, Y., Fu, D., Wang, H.: MusicBERT: a self-supervised learning of music representation. In: Proceedings of the 29th ACM International Conference on Multimedia, MM 2021, New York, NY, USA, pp. 3955–3963. Association for Computing Machinery (2021)
2. Aminian, M., Kehoe, E., Ma, X., Peterson, A., Kirby, M.: Exploring musical structure using tonnetz lattice geometry and LSTMs. In: Krzhizhanovskaya, V.V., Závodszky, G., Lees, M.H., Dongarra, J.J., Sloot, P.M.A., Brissos, S., Teixeira, J. (eds.) ICCS 2020. LNCS, vol. 12138, pp. 414–424. Springer, Cham (2020). https://doi.org/10.1007/978-3-030-50417-5_31
3. Chou, Y.-H., Chen, I.-C., Chang, C.-J., Ching, J., Yang, Y.-H.: MidiBERT-piano: large-scale pre-training for symbolic music understanding. ArXiv, vol. abs/2107.05223 (2021)
4. Prang, M., Esling, P.: Signal-domain representation of symbolic music for learning embedding spaces. ArXiv, vol. abs/2109.03454 (2021)
5. Zhang, Y., Jin, R., Zhou, Z.-H.: Understanding bag-of-words model: a statistical framework. Int. J. Mach. Learn. Cybernet. **1**, 43–52 (2010)
6. Chuan, C.-H., Agres, K., Herremans, D.: From context to concept: exploring semantic relationships in music with word2vec. Neural Comput. Appl. **32**(4), 1023–1036 (2018). https://doi.org/10.1007/s00521-018-3923-1
7. Tymoczko, D.: The geometry of musical chords. Science **313**(5783), 72–74 (2006)
8. Tymoczko, D.: The generalized Tonnetz. J. Music Theory **56**, 1–52 (2012)
9. Budney, R., Sethares, W.A.: Topology of musical data. J. Math. Music **8**, 73–92 (2013)
10. Papadopoulos, A.: Mathematics and group theory in music. In: Handbook of Group Actions, vol. II Ji, L., Papadopoulos, A., Yau, S.-T., vol. 32 of Advanced Lectures in Mathematics, pp. 525–572, International Press and Higher Education Press (2015)
11. Freund, A., Andreatta, M., Giavitto, J.-L.: Lattice-based and topological representations of binary relations with an application to music. In: Annals of Mathematics and Artificial Intelligence, vol. 73, pp. 311–334 (2014)
12. Euler, L.: Tentamen novae theoriae musicae ex certissismis harmoniae principiis dilucide expositae. Saint Petersburg: Saint Petersburg Academy (1739)
13. Cohn, R.: Introduction to neo-Riemannian theory: a survey and a historical perspective. J. Music Theory **42**, 167 (1998)
14. Gollin, E.: Some aspects of three-dimensional "onnetze". J. Music Theory **42**, 195 (1998)
15. Bigo, L., Andreatta, M.: Topological structures in computer-aided music analysis. In: Computational Music Analysis, pp. 57–80. Springer, Cham (2016). https://doi.org/10.1007/978-3-319-25931-4_3
16. Karystinaios, E., Guichaoua, C., Andreatta, M., Bigo, L., Bloch, I.: Musical genre descriptor for classification based on Tonnetz trajectories. Journées d'Informatique Musicale, (Strasbourg, France), October 2020
17. Bigo, L., Ghisi, D., Spicher, A., Andreatta, M.: Representation of musical structures and processes in simplicial chord spaces. Comput. Music J. **39**, 9–24 (2015)
18. Bergomi, M.G., Baratè, A., Di Fabio, B.: Towards a topological fingerprint of music. In: Bac, A., Mari, J.-L. (eds.) CTIC 2016. LNCS, vol. 9667, pp. 88–100. Springer, Cham (2016). https://doi.org/10.1007/978-3-319-39441-1_9
19. Bergomi, M.G., Baratè, A.: Homological persistence in time series: an application to music classification. J. Math. Music **14**, 204–221 (2020)

20. Jevti'c, F.D., Živaljević, R.T.: Generalized Tonnetz and discrete Abel-Jacobi map. Topological Methods in Nonlinear Analysis, p. 1 (2021)
21. Oleynikov, R., Linkova, M.: Music theory. Feniks, Visual representation of harmony. Rostov-on-Don (2021)
22. Peracha, O.: Improving polyphonic music models with feature-rich encoding, ArXiv, vol. abs/1911.11775 (2020)
23. Hatcher, A.: Algebraic Topology. Cambridge University Press, Cambridge (2000)
24. Edelsbrunner, H., Harer, J.: Computational Topology - an Introduction. AMS (2010)
25. Zomorodian, A., Carlsson, G.E.: Computing persistent homology. Discrete Comput. Geometry **33**, 249–274 (2005)
26. Agrell, E., Eriksson, T., Vardy, A., Zeger, K.: Closest point search in lattices. IEEE Trans. Inf. Theory **48**(8), 2201–2214 (2002)
27. Fujita, O.: Metrics based on average distance between sets. Jpn. J. Ind. Appl. Math. **30**, 1–19 (2011)
28. Skopenkov, A.: Embedding and knotting of manifolds in Euclidean spaces. In: Surveys in Contemporary Mathematics, London Math. Soc. Lect. Notes, vol. 347, pp. 248–342 (2008)
29. Tompkins, C.: Isometric embedding of flat manifolds in Euclidean space. Duke Math. J. **5**, 58–61 (1939)
30. Nash, J.F.: The imbedding problem for Riemannian manifolds. Ann. Math. **63**, 20–63 (1956)
31. Pinkall, U.: Hopf tori in S^3. Inventiones Mathematicae **81**, 379–386 (1985)
32. Chen, T.-P., Su, L.: Harmony transformer: incorporating chord segmentation into harmony recognition. In: Proceedings of the 20th International Society for Music Information Retrieval Conference, Delft, The Netherlands, pp. 259–267, ISMIR, November 2019
33. Burgoyne, J.A., Wild, J., Fujinaga, I.: An expert ground truth set for audio chord recognition and music analysis. In: Proceedings of the 12th International Society for Music Information Retrieval Conference, (Miami, United States), pp. 633–638, ISMIR, October 2011
34. Boulanger-Lewandowski, N., Bengio, Y., Vincent, P.: Modeling temporal dependencies in high-dimensional sequences: Application to polyphonic music generation and transcription, arXiv preprint arXiv:1206.6392 (2012)
35. Cífka, O., Şimşekli, U., Richard, G.: Groove2Groove: one-shot music style transfer with supervision from synthetic data. IEEE/ACM Trans. Audio Speech Language Process. **28**, 2638–2650 (2020)
36. Chen, T.-P., Su, L.: Attend to chords: improving harmonic analysis of symbolic music using transformer-based models. Trans. Int. Soc. Music Inf. Retrieval **4**, 1–13 (2021)
37. Dhariwal, P., Jun, H., Payne, C., Kim, J.W., Radford, A., Sutskever, I.: Jukebox: a generative model for music. arXiv preprint arXiv:2005.00341 (2020)
38. Payne, C.: MuseNet. OpenAI blog (2019). https://openai.com/blog/musenet. Accessed 10 Nov 2022
39. Sorokin, K., Ayzenberg, A., Beketov, M.: Python source code for melody transformations based on homeomorphisms of the tonal space. GitHub (2023). https://github.com/mopsless/style-transfer. Accessed 01 Feb 2023

Music Generation with Multiple Ant Colonies Interacting on Multilayer Graphs

Lluc Bono Rosselló[✉][iD] and Hugues Bersini[iD]

Université Libre de Bruxelles (ULB), Brussels, Belgium
lluc.bono.rossello@ulb.be

Abstract. We propose a methodology for music generation that makes use of Ant Colony Optimization (ACO) algorithms on multilayer graphs. In our methodology we first define a new multilayer graph model of music that contains several voices musical works. Then, we adapt ACO algorithms to allow multiple ant colonies to generate solutions on each layer while interacting with each other. This methodology is illustrated with some example configurations that show how music emerge as a result of the interaction of different simultaneous ant colony optimization instances.

Keywords: Ant colony optimization · Music generation · Multilayer graphs · Collective creativity

1 Introduction

Music constitutes a great field to explore the potential and limitations of artificial intelligence in terms of creativity. While great composers were able to break well-established compositions rules favouring new musical ideas, researchers in artificial intelligence try to make use of different techniques pursuing the same goal.

Pushing further the limits of AI in terms of creativity, researchers have taken different approaches that have been classified by Carnovalini et al. [5] in seven groups: markov chains, formal grammars, rule or constraint based systems, neural networks or deep learning, evolutionary or genetic algorithms, chaos or self similarity and agents based systems.

In this work we put our focus on the use of agent based systems [18] and we use swarm intelligence ideas to reproduce the social interactions existing in collective music creativity. Todd et al. [19] propose a classification of these artificial life approaches depending on the relation between the agents and the generated music: i) when the music is the result of the movement of agents which are not aware of what they produce; ii) when the music is the result of each individual creation and its survival depends on it, which is the main inspiration for genetic algorithms and

Supplementary Information The online version contains supplementary material available at https://doi.org/10.1007/978-3-031-29956-8_3.

iii) when the music is produced by each individual and has an influence on other agents production. This third approach, where both internal evaluation and external influence coexist, is described by the authors as the most promising in terms of creativity and is the approach that we follow in our work.

The main contribution of this work is to establish a methodology where each agent generates music following both its own configuration and the influence of other agents in their individual environment, being the final piece the result of these multi-agent interactions. To do so, we make use of Ant Colony Optimization (ACO) algorithms [7] which consist of simulating the behavior of ants foraging food and are widely used in different optimization problems [16].

This contribution is not the first work making use of such algorithms in the music field since its versatility has already been exploited to generate music [9,12]. In these approaches, the authors generate melodies by building paths on a graph that are optimized for a specific musical goal. However, these approaches lack of the required flexibility in their graphs to generate multiple melodies that emerge from interacting with each other. This form of collective creativity can be compared to the one generated in musical group improvisations where the music emerge from the multiple interactions, as discussed by Borgo in [4].

In this paper we address this lack of flexibility by proposing a new graph representation that provides the system with the required distributed nature. Such graph representation is based on the concept of multilayer networks [14] that has been used in diverse applications [2]. Indeed, as with our work, these multilayer models respond to the need of describing linked systems with interdependencies that cannot be expressed by a single graph representation. In our case, the innovation of a multilayer graph for music is what allows us to define different search spaces while addressing the interaction between the different voices, hence enabling the music to emerge from a sort of complex system.

Subsequently, we propose a variation of an Ant Colony Optimization (ACO) algorithm that we adapt for multilayer graphs. This adaptation allows us to define heterogeneous behaviors for the different agents while enabling their interactions.

Both proposed innovations, the multilayer representation and the related ACO adaptation, constitute a methodology that is aligned with the concept of collective creativity and allows music to emerge from its social interactions. The main focus of this work is the description of these innovations and how the components of the system can be tuned to achieve different interactions resulting in different musical outputs.

The rest of the article is organized as follows: Sect. 2 states the related work in terms of collective intelligence approaches to art and music as well as the current network science approaches to music. Section 3 describes the methodology by introducing the multilayer graph model, the ant system adaptation and the algorithm for generating multiple interconnected voices. Section 4 illustrates with some examples different configurations of the algorithm to understand the potential of the methodology. Section 5 discusses the significance of these results and proposes new research directions to follow.

2   Related Work

The literature presents different approaches to generate music from a swarm intelligence perspective. For instance, Blackwell et al. [3] took inspiration from the birds flocking to generate a multi-agent musical system. The Ant Colony Optimization (ACO) introduced by Dorigo et al. [7] is another swarm intelligence inspired heuristic that has been used in several engineering applications [16] and also in artistic applications [11] where different agents make use of the pheromone to coordinate with each other and optimize a desired artistic pattern.

Regarding music generation, ACO has been used first by Guéret et al. [12] to generate melodies. To do so, ant agents produce paths into a graph that contains MIDI events (musical objects containing attributes such as pitch or duration of the notes) according to an heuristic based on the interval distance between two consecutive notes. In such approach, the ants leave constant pheromone trails in the edges used in their solution, biasing future iterations towards previous solutions, which normally drive them to constitute cycles. Then, Geis et al. [9] go further by adapting ACO to generate harmonized baroque melodies. In their work, the authors define precise musical constraints in the graph that restrict the possible solutions to be generated. In addition they incorporate objective functions (both for melody generation and for melody harmonization) to evaluate the quality of the solutions based on given musical rules and update correspondingly the pheromone trails. These two approaches, however, can not be configured to simultaneously produce interactive melodies in an interactive manner given their limitations in the single graph representation of music.

In that sense, to establish a graph representation for our approach we consider the literature in complex networks [1]. These networks have been used to model a large number of phenomena of diverse nature [6]. Within these models we can also find several works on modelling music as complex networks. Liu et al. [15] and Ferretti [8] map melodies into networks by considering consecutive elements in music scores. Therefore, each note constitutes a node in the network and the edges encode the existent transitions in the original piece. The authors, furthermore, analyze these graphs and demonstrate how these networks can be classified as real-networks by having similar topological characteristics such as low-density, high clustering coefficient or a non-linear degree distribution. Furthermore, Liu et al. also propose in [15] to use these networks to generate music but the proposed algorithms mainly reproduce the behavior of a first order markov chain. Gómez et al. in [10] and Serrà et al. in [17] propose a different mapping of music into networks. In these approaches the authors focus on the harmonic evolution of the music and decide to consider together in one node multiple simultaneous voices, hence simplifying several voices into a single graph representation.

In our work, we take inspiration from both approaches to introduce a new model containing both design choices in order to simultaneously map melodies into individual graphs as in [8,15] and, connect them harmonically as in [10,17]. Such representation will constitute a multilayer network [14] which, to the best of our knowledge, has not been proposed by the network science literature as a representation of music.

3   Methodology

3.1   Multilayer Model

In this work we introduce a new multilayer model of music that enables a more flexible and distributed generation of melodies. This model differs from the single graph representation exploited in the works of Gueret et al. [12] and Geis et al. [9]. In fact, while such approaches have proved to be valid both for optimizing melodic rules and for harmonizing previously generated melodies, such graph representations are limited when it comes to produce interactions between different melodies.

With this multilayer model we define several search spaces co-existing in a connected manner which allows us to represent independently several musical voices as well as their interactions. By doing so we provide a model with the required flexibility to define different agent behaviors while containing the structural requirements for their interactions.

To define the multilayer model we make use (for ease of understanding) of an original musical input in MIDI notation. We first map the elements of each sequence of MIDI events (containing notes, duration, instrument and other musical information) into a different layer. This mapping is equivalent to the mapping of each voice or instrument's score into a different graph representation.

In such way we build several directed graphs by mapping the transitions from consecutive elements as proposed authors in [8,15] and illustrated in the scheme in Fig. 1a. Moreover, each possible transition is described by an edge and weighted according to its number of occurrences during the input sequence. The transition probabilities are thus encoded into the model and can be used as heuristic information later on.

Subsequently, we connect the various layers containing the individual voices. To do this, whenever two notes occur concurrently (or within a window of time t), we create undirected inter-layer edges between the two nodes as it can be seen in Fig. 1b. This mapping contains information about the correlations between various voices aspects such as harmonic relations (the interval relation between two different pitches) or rhythmic relations (when voices play rhythmically coordinated).

With such methodology we can map any original notated music in MIDI format into a multilayer discrete space with its transitions constraints, probabilities and voice correlations. It should be noted that this methodology is not confined to existing musical works, but may also be used to design any other musical space, complete with transition constraints, probabilities, and voice correlations.

To conclude, the model that we present defines a space satisfying both of our requirements for multi-agent music to emerge: i) each layer constitutes an

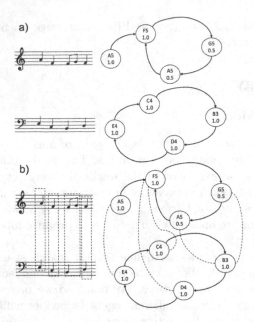

Fig. 1. Mapping scheme of two voices into a multilayer graph.

independent search space for each voice that can be exploited heterogeneously and ii) the inter-layer edges keep the different layers of the graph connected, allowing the different voices to interact.

3.2   Ants System Description

In our methodology we adapt the Ant System [7] to our defined multilayer space. Such algorithms constitute a heuristic that is used to build solutions by searching in a graph containing the problem constraints. To do so, each ant agent builds a path through the graph that will constitute a possible solution. This solution is then evaluated and, according to such evaluation, used edges will be reinforced by leaving a pheromone trail. This pheromone trail will guide future iterations and other agents towards promising solutions. Simultaneously, pheromone will evaporate along time decreasing the influence of solutions that have not been used recently.

Similarly to the existing approaches in the literature [9,12], we will make use of the Ant System to build melodies by navigating the graph using the allowed transitions. However, in our case we will use several ant colonies instances and each one of them will build melodies using their own search space defined by their layer constraints. Therefore in order to produce simultaneously several melodies, we might first consider our approach as parallel instances of ACO, each one of them on one layer of the described multilayer graph. In that sense, each search space constrains the possible transitions according to the existing directed intra-layer edges.

Regarding the objective functions, that will evaluate each ant colony's solution, heterogeneity is provided by the fact that each colony will optimize a - possibly - different objective function. This allows the methodology to be configured, for instance, as a polyphonic music generator.

To complete the adaptation of the Ant System to our multilayer model we need to provide a way for the different colonies to interact. To achieve this, we introduce a second type of pheromone, as proposed in other Ant System variations for multi-objective optimization problems [13]. This pheromone will be used by one layer colony to influence other layers colonies. To do this, the inter-layer edges will be employed to convey the information of nodes involved in the solution developed on one layer to other layers.

Therefore, the pheromone trails are used to influence two environments with two different purposes: i) the coordination of one colony towards their goal in their own layer and ii) the coordination of different colonies towards connected musical regions on other layers.

Regarding the formal implementation of this algorithm, we first define the transition probabilities (for every possible edge from i to j defined by neighbourhood N_i) biasing the random walk of each ant k through each layer in Eq. 1.

$$p_{ij}^k(t) = \frac{[\tau_{in_{ij}}]^{\alpha_{in}} \cdot [\tau_{ex_{ij}}]^{\alpha_{ex}} \cdot [\eta_{ij}]^{\beta}}{\sum_{l \in N_i^k} [\tau_{in_{il}}]^{\alpha_{in}} \cdot [\tau_{ex_{il}}]^{\alpha_{ex}} \cdot [\eta_{il}]^{\beta}}, \qquad \text{if} \quad j \in N_i^k \qquad (1)$$

These probabilities depend on the following three factors:

- Intra-pheromone Information ($\tau_{in_{ij}}$): A numerical value representing the amount of internal pheromone in such edge. This value will be updated accordingly to the use of the edge in a well-evaluated solution.
- Extra-pheromone Information ($\tau_{ex_{ij}}$): A numerical value representing the amount of external pheromone in such edge. This value will be updated according to the use of connected nodes in other layers solutions.
- Heuristic Information (η_{ij}): A numerical value representing the probability of that transition in the initialized graph. This value is generated by the existing probabilities in an original work or by a pre-configured artificial search space.

Therefore, each transition probability p_{ij} depends on the conjunction of these three parameters. In order to control the influence of each one of these parameters, we make use of α_{in}, α_{ex} and β. By doing so we can promote: the individual goal optimization (α_{in}), the dependence on other agents (α_{ex}) or the use of the heuristic musical information (β).

According to these probabilities, each ant colony will generate iteratively multiple solutions depending on the number of ants (K) defined for each layer. These solutions will be then evaluated according to an objective function $F_L(path_k)$ that can be different for each layer (L). Moreover, in our approach, we opt for an elitist variation of the Ant System and we select the best solution out of K of each iteration according to $F_L(path_k)$. This choice facilitates the optimization of individual goals when that is the desired behavior.

Then, the intra-pheromone trails are updated according to Eq. 2:

$$\tau_{in_{ij}}(t) = \begin{cases} (1 - \rho_{in}) \cdot \tau_{in_{ij}}(t-1) + \triangle\tau_{in_{ij}} & \text{if } ij \in path \\ (1 - \rho_{in}) \cdot \tau_{in_{ij}}(t-1) & \text{if } ij \notin path \end{cases} \tag{2}$$

where, after each iteration a proportion of pheromone ρ_{in} will be evaporated on all edges. Then if edge ij is part of the selected solution $(path)$, it will receive an increase of pheromone $(\triangle\tau_{in_{ij}})$ according to the result of best path evaluation $F_L(path_{best})$.

Regarding the external pheromone, it will be updated in a similar manner according to Eq. 3:

$$\tau_{ex_{ij}}(t) = \begin{cases} (1 - \rho_{ex}) \cdot \tau_{ex_{ij}}(t-1) + \sum_{l=1}^{L} \triangle\tau_{ex_{ij}} & \text{if } \exists jk : k \in path_l \\ (1 - \rho_{ex}) \cdot \tau_{ex_{ij}}(t-1) & \text{else} \end{cases} \tag{3}$$

where, after each iteration a proportion of pheromone ρ_{ex} will be evaporated on all edges. Then if edge (ij) ends in a node j connected to a node contained in another layer solution $(path_l)$, it will receive an amount of pheromone according to $F_l(path_l)$. This choice aims at guiding agents on other layers to go towards correlated nodes in their respective layers.

Finally, the objective functions $F(path)$ mentioned above are left to be designed according to the desired behavior of an agent regarding specific musical variables. Such objective functions can be considered as one parameter more to be chosen when designing a specific creation rather than a core part of the system. In such a way, these objective functions can be used, for instance, to reinforce certain rhythmic behavior, to obtain specific levels of randomness in the generated melodies or to emulate specific styles.

3.3   Algorithm Description

Once the multilayer model is described and the Ant System has been adapted to target its distributed yet interactive behaviour we can proceed to describe a run of the algorithm.

To initialize the algorithm we use a MIDI file of an original work. This file containing different voices is mapped into a multilayer network according to the procedure described in Fig. 1. Such multigraph already contains each search space with the allowed transitions (constraints). Furthermore, the edges will be already weighted with the corresponding transition probabilities that will be used as heuristic information (η_{ij}) in Eq. 1.

Then, different parameters can be configured depending on the interactions desired by the user. These decisions have influence on elements ranging from the length of the melodies, their complexity, or the dependence that one layer might have on another. Such design decisions will be discussed in Sect. 3.4 in more detail.

Then, a different melody will be produced on each layer according to the following steps:

1. An ant colony generates K solutions by producing a sequence of musical elements using a biased random walk with probabilities in Eq. 1.
2. Best solution is selected according to the layer (L) goal given by individual objective function (F_L).
3. Both pheromone trails in the layer are evaporated according to parameters ρ_{in} and ρ_{ex}.
4. The best solution on each iteration will be used to update the intra-pheromone (τ_{in}) according to the evaluation of the layer objective function.
5. Each of the other layers' best solutions is utilized to update the extra-pheromone (τ_{ex}) by using the inter-layer edges and according to their own evaluation.
6. Solution is obtained after a given number of iterations is reached or an ending condition is reached.

3.4   Design Choices

This section aims at summarizing the different configuration parameters of the algorithm as well as draw some links between such configurations and the behavior of the system. Each melody to be produced on each layer can be configured independently by the following parameters:

1. Number of ants (K): To regulate how many different melodies will be explored on each iteration in order to maximize each layer goal.
2. Number of notes (N): The maximum number of notes of the generated melodies. Short melodies will allow for more controlled results that could then be used as sub-sections of more elaborated creations.
3. Intra-pheromone coefficient (α_{in}): To regulate the importance of the internal pheromone (τ_{in}). A high value will imply that this voice will tend to follow its own goal optimization.
4. Extra-pheromone coefficient (α_{ex}): To regulate the importance of the external pheromone (τ_{ex}). A high value will imply that this voice will tend to follow the influence of other agents on its environment disregarding its own goal. This can be interpreted as some sort of follower behavior where the agent will rather play notes correlated to other agents than those interesting for its goal.
5. Heuristic coefficient (β): To regulate the importance of the heuristic information (η). A high balue will cause the agent to follow the existing transition probabilities generating melodies highly based on the graph configuration.
6. Intra-pheromone evaporation factor (ρ_{in}): To regulate the evaporation of the intra-pheromone (τ_{in}). A high value will imply that edges that have not recently been used will be quickly disregarded which could favor new regions to be explored.

7. Extra-pheromone evaporation factor (ρ_{ex}): To regulate the evaporation of the extra-pheromone (τ_{ex}). A high value will imply that edges not recently influenced will be quickly disregarded which could favor synchronization between current iterations on different layers (since they will be biased only by recent iterations).
8. Objective Function (F): To guide the behavior of an agent by establishing a goal in order to evaluate its solutions. If the objective function aims for very exploratory paths (for instance by trying to maximize the number of nodes visited) the agent will tend to create more complex melodies. Note that such objective function is not the ultimate goal of the design but one parameter that will have an influence dependent on other parameters. For instance, if parameter (α_{in}) is low, this information will not be the main driver of the agents behavior.

It should be noted that the potential of the presented methodology resides not in the configuration of one agent parameters but on the music that will emerge from different agents parameter configurations. Therefore, the aim of the system is that the user can assign roles to different agents by tuning such parameters. For instance, we could consider a two-voices creation where one agent has a high value for the intra-pheromone importance (α_{in}) and low extra-pheromone importance (α_{ex}), hence acting as a leader: while a second agent could have these parameters in a reversed configuration, hence acting as a follower.

4   Results

Based on the values of the described parameters we can model different interactions in our methodology. These interactions aim to reproduce the complex compositional process in a collective environment. In this process each agent generates music while adapting to the changes in its environment produced by other agents.

In the following section we present some examples to showcase how different agents configurations affect both their behavior and their musical output. Note that these are illustrative examples of the system, therefore their goal is not to optimize the configuration for a certain music style.

To analyze the behavior of the system and the effect of the parameter choices we decide to keep the objective functions very simple. Therefore, these functions will evaluate the generated paths according to a target number of different nodes to be visited. These objective functions, despite their simplicity and lack of musical information, are good enough to generate pleasant music which, shows the potential of this methodology. The audio files corresponding to each example can be found in the Supplementary Material.

4.1   Example 1: On the Role of Different Pheromone Types

The goal of this first example is to illustrate how different configurations of the pheromone importance α_{in} and α_{ex} modify the individual behavior of each agent. When intra-pheromone importance (α_{in}) is high, the agent manages to reach the goal defined by the objective function F; when α_{in} is 0, the agent fails in its attempt to reach its individual goal and behaves as a follower guided by extra-pheromone τ_{ex}.

Table 1. Parameter Configuration of Example 1

Parameter	Layer 0	Layer 1
Intra-pheromone factor (α_{in})	2	0
Extra-pheromone factor (α_{ex})	0	2
Heuristic information factor (β)	1	1
Target Length of Objective Function (F)	8	8

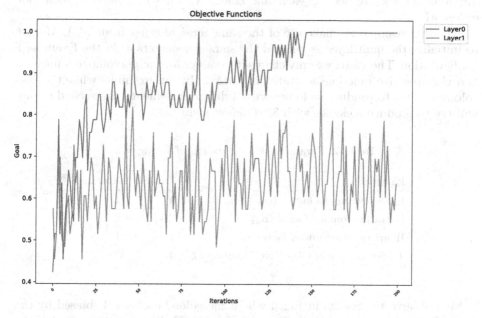

Fig. 2. Evolution of the Objective Functions for Example 1 with different intra-pheromone influence.

For this example we initialize a two voices multilayer graph by using the Piano Sonata No. 11 in A major, K. 331, 1st Movement by W.A. Mozart. In this mapping, each layer corresponds to the score of each hand of the piano. Both pheromone types are initialized to 1 in each layer, and their evaporation factors ρ_{in} and ρ_{ex} are both set to 0.1. The music will be generated by using 10 ants on each colony and layer during 200 iterations with a length of 32 musical elements. he colony parameters associated to each layer, e.g. the influence of each type pheromone, heuristic information and the target length of the objective function, can be found in Table 1.

The results in Fig. 2 show the colony on layer 0 reaching its individual goal by using intra-pheromone (τ_{in}) to guide its transitions. On the other hand, the colony on layer 1 is not basing its decisions on the optimization given by the intra-pheromone (τ_{in}) and so fails to reach its goal.

In musical terms, the high-pitched voice corresponding to the right hand of the piano produces a melody using only 8 different musical elements, while the left-hand uses more elements biased by the right hand melody.

4.2   Example 2: On the Use of Different Objective Functions

The goal of this second example is to illustrate how different optimization goals can be set for different agents and how music is the result of the trade-off between these behaviors interactions. In this case different objective functions (F) with different targets are set for each ant colony which will bias their behavior differently.

For this example we make use of the same musical input from W.A. Mozart to initialize the multilayer graph and the same parameters as in the Example 1 configuration. Therefore, we only change the values for the pheromone influences and the objective functions as stated in Table 2. In this example, while the first colony will try to produce melodies with 4 different elements, the second colony will try to produce melodies with 32 different elements.

Table 2. Parameter Configuration of Example 2

Parameter	Layer 0	Layer 1
Intra-pheromone factor (α_{in})	1	1
Extra-pheromone factor (α_{ex})	2	2
Heuristic information factor (β)	1	1
Target Length of Objective Function (F)	4	32

We observe the results in Fig. 3 where the colony on layer 1, biased by the other factors, moves away from its goal of using 32 different elements. On the other hand, the colony on layer 0 approaches its goal as the iterations advance but also stagnates at a certain value. This example shows how while both layers fail to reach their individual goal, a system equilibrium is reached after a number of iterations.

In musical terms, different musical roles are given by different levels of complexity as their target. Then, the interactions force the system to reach an equilibrium that outputs two correlated voices with a pleasant melody combination.

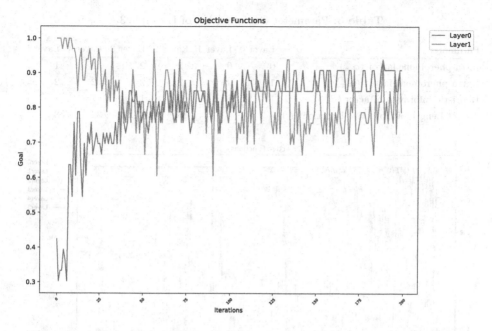

Fig. 3. Evolution of the Objective Functions for Example 2 with different goals.

4.3 Example 3: On the Scalabilty of the System

The goal of this last example is to illustrate the capabilities of the system to be used with more voices. As it happens in an orchestral composition, the roles need to be more precisely defined when the number of instruments increases. In that sense, we showcase the flexibility of the system by playing with both objective functions and pheromone influences configurations depicted in the previous examples.

For this example, we initialize a six layers' multigraph with the The Ricercar a 6 from The Musical Offering from J.S.Bach, which constitutes a six-voice fugue. We set the same parameters for pheromone initialization and evaporation factors and number of ants per colony. In this case, however, we generate longer melodies with a length of 64 elements. The rest of the configuration for each colony is described in Table 3.

The results in Fig. 4 show how colonies in the first two layers fail to reach their goal since they are not making use of the intra-pheromone (τ_{in}) to guide their decisions. Note that the colony on layer 3 also takes some time to reach their goal, since it seems harder to find a 4 elements sequence than other goals. Particularly, the used input plays a role in this example since there are long rests in some voices that are indeed used in this music generation. Moreover, these results can be better understood by the complementary information in Fig. 5 that describes the number of different nodes used in each iteration. In these results we observe the number of nodes that each colony is trying to use and the differences in their efficacy clearly correlated with the parameter α_{in}.

Table 3. Parameter Configuration of Example 3

Parameter	Layer 0	Layer 1	Layer 2	Layer 3	Layer 4	Layer 5
Intra-pheromone factor (α_{in})	0	0	2	2	1	1
Extra-pheromone factor (α_{ex})	1	2	2	2	2	2
Heuristic information factor (β)	1	1	2	1	1	1
Target Length of Objective Function (F)	4	4	16	4	32	8

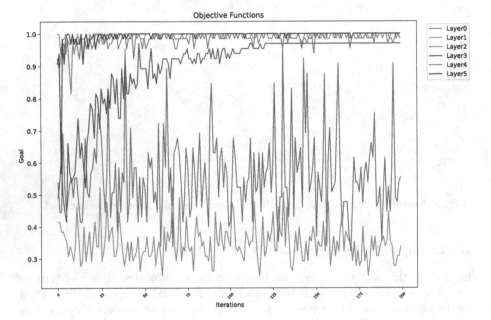

Fig. 4. Evolution of the Objective Functions for Example 3 with different pheromones and goals

We would like to emphasize that the produced output in this last example contains 6 different melodies, therefore, the raw output played on 6 piano voices lacks of timbre and texture design to make it more pleasant. This fact shows the limitations of such methodology which would need some musical production from the user to obtain a more elaborated result. In that sense, we include in the Supplementary Material an audio track where only the instrument sound has modified and a drum beat has been added. Therefore, this limitation in the output definition (being this a MIDI file) gives flexibility to the user to decide which sound wants to assign to each voice.

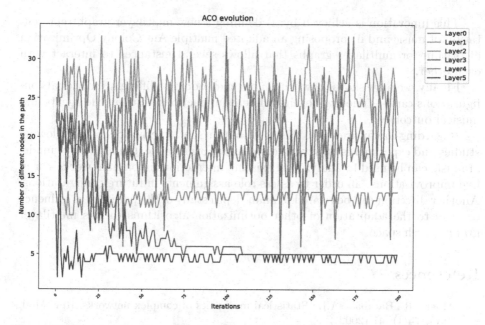

Fig. 5. Evolution of number of different nodes used by each voice in Experiment 3 with different pheromones and goals.

At this point, it is important to remind that, differently to other approaches, the optimization of the goals is not the aim of such methodology. Individual goals are used as a way of setting a behavior/role that will contribute to the emergence of certain consensus in the final piece. This fact motivates some of the design decisions such as not using some ACO elitist variation where the global best-so-far solution is stored. In that sense, an optimal solution found in a given layer A at iteration t_1 would not be correlated to an optimal solution found in a given layer B at iteration t_2; just as it would not make sense that we store each instrument best solo in a group improvisation and reproduce them altogether.

To conclude, it should be noted that in these experiments music is not formally constrained nor evaluated. Nonetheless, the methodology allows to incorporate such information by creating graphs with the required musical features and the desired inter-layer connections. In a similar way, well crafted objective functions can be defined to obtain more specific behaviors, as proposed by Geis et al. in [9].

5   Conclusion

In this work we have presented a methodology that extends the existing uses of ACO for music generation. To do so, we have introduced a new multi-agent compositional methodology that allow different agents to generate music both making use of internal information and external information from other agents.

This innovation is achieved by: i) defining a new multilayer graph representation of music and ii) proposing an adapted multiple Ant Colony Optimization algorithm for multilayer graphs that allows several instances to interact with each other.

Finally, we have shown how within our methodology different individual configurations can be tuned to obtain different interactions between the agents and musical outcomes.

Regarding the future work, this contribution sets a methodology for following studies and experiments on several directions. For instance, the multilayer model of music can be used to analyze the existing interactions in original works - or in live improvisations - in order to assess role assignments in future configurations. Another direction can be related to an in-depth study of the parameters influence or even to the adaptation of other optimization algorithms to this multilayer graph search space.

References

1. Albert, R., Barabási, A.L.: Statistical mechanics of complex networks. Rev. Mod. Phys. **74**(1), 47 (2002)
2. Aleta, A., Moreno, Y.: Multilayer networks in a nutshell. Annual Rev. Condensed Matter Phys. **10**, 45–62 (2019)
3. Blackwell, T.: Swarming and music. Evolutionary Computer Music, p. 194–217 (2007). https://doi.org/10.1007/978-1-84628-600-1_9
4. Borgo, D.: Sync or swarm: musical improvisation and the complex dynamics of group creativity. In: Futatsugi, K., Jouannaud, J.-P., Meseguer, J. (eds.) Algebra, Meaning, and Computation. LNCS, vol. 4060, pp. 1–24. Springer, Heidelberg (2006). https://doi.org/10.1007/11780274_1
5. Carnovalini, F., Rodà, A.: Computational creativity and music generation systems: an introduction to the state of the art. Front. Artif. Intell. **3**, 14 (2020)
6. Costa, L.D.F., et al.: Analyzing and modeling real-world phenomena with complex networks: a survey of applications. Adv. Phys. **60**(3), 329–412 (2011). https://doi.org/10.1080/00018732.2011.572452
7. Dorigo, M., Maniezzo, V., Colorni, A.: Ant system: optimization by a colony of cooperating agents. IEEE Trans. Syst. Man Cybern. Part B (Cybernetics) **26**(1), 29–41 (1996)
8. Ferretti, S.: On the complex network structure of musical pieces: analysis of some use cases from different music genres. Multimed. Tools Appl. **77**(13), 16003–16029 (2017). https://doi.org/10.1007/s11042-017-5175-y
9. Geis, M., Middendorf, M.: An ant colony optimizer for melody creation with baroque harmony. In: 2007 IEEE Congress on Evolutionary Computation (2007). https://doi.org/10.1109/cec.2007.4424507
10. Gomez, F., Lorimer, T., Stoop, R.: Complex networks of harmonic structure in classical music. Nonlinear Dynamics of Electronic Systems Communications in Computer and Information Science, pp. 262–269 (2014). https://doi.org/10.1007/978-3-319-08672-9_32
11. Greenfield, G., Machado, P.: Ant-and ant-colony-inspired ALife visual art. Artif. Life **21**(3), 293–306 (2015)

12. Guéret, C., Monmarché, N., Slimane, M.: Ants can play music. Ant Colony Optimization and Swarm Intelligence, pp. 310–317 (2004). https://doi.org/10.1007/978-3-540-28646-2_29

13. Iredi, S., Merkle, D., Middendorf, M.: Bi-criterion optimization with multi colony ant algorithms. In: Zitzler, E., Thiele, L., Deb, K., Coello Coello, C.A., Corne, D. (eds.) EMO 2001. LNCS, vol. 1993, pp. 359–372. Springer, Heidelberg (2001). https://doi.org/10.1007/3-540-44719-9_25

14. Kivela, M., Arenas, A., Barthelemy, M., Gleeson, J.P., Moreno, Y., Porter, M.A.: Multilayer networks. J. Complex Netw. **2**(3), 203–271 (2014). https://doi.org/10.1093/comnet/cnu016

15. Liu, X., Tse, C.K., Small, M.: Composing music with complex networks. In: Zhou, J. (ed.) Complex 2009. LNICST, vol. 5, pp. 2196–2205. Springer, Heidelberg (2009). https://doi.org/10.1007/978-3-642-02469-6_95

16. Mohan, B.C., Baskaran, R.: A survey: ant colony optimization based recent research and implementation on several engineering domain. Expert Syst. Appl. **39**(4), 4618–4627 (2012)

17. Serrà, J., Corral, A., Boguñá, M., Haro, M., Arcos, J.L.: Measuring the evolution of contemporary western popular music. Sci. Rep. **2**(1), 521 (2012). https://doi.org/10.1038/srep00521

18. Tatar, K., Pasquier, P.: Musical agents: a typology and state of the art towards musical metacreation. J. New Music Res. **48**(1), 56–105 (2019)

19. Todd, P.M., Miranda, E.R.: Putting some (artificial) life into models of musical creativity. In: Musical Creativity, pp. 392–412. Psychology Press (2006)

Automatically Adding to Artistic Cultures

Simon Colton[✉] and Berker Banar

School of EECS, Queen Mary University of London, London E1 4NS, UK
{s.colton,b.banar}@qmul.ac.uk

Abstract. We consider how generative AI systems could and should evolve from being used as co-creative tools for artefact generation to co-creative collaborators for enhancing cultural knowledge, via artefact generation. We argue that while generative deep learning techniques and ethos have many drawbacks in this respect, neuro-symbolic approaches and increased AI autonomy could improve matters so that AI systems may be able to add to cultural knowledge directly. To do this, we consider what cultural knowledge is, differences between scientific and artistic applications of generative AI, stakeholders and ethical issues. We also present a case study in decision foregrounding for generative music.

Keywords: Generative AI · Neuro-symbolic AI · Artistic cultures

1 Introduction

The notion of an artistic culture such as the visual arts or contemporary classical music is not easy to define. It is not controversial, however, to say that such cultures encompass artefacts such as paintings or musical compositions and knowledge which can be communicated and acted upon. Such artefacts are often made with computational support, sometimes via the usage of AI technologies such as generative deep learning. The knowledge contained in artistic cultures includes the following non-exhaustive, overlapping aspects:

- Processes and techniques for artefact construction which can be passed on from person to person, to be employed and adapted in a wide range of projects. Often the processes are embedded in software tools for creative applications.
- Precise concepts and more broad ideas and opinions about aspects of the world, coming from individual people or movements, which can be incorporated, extended, modified or subverted by other groups of artists or musicians.
- Interpretations of the meaning of artworks including analyses of personal artistic and emotional expressions, authorial intent, techniques and procedures.
- Aesthetic considerations about the value of art expressed in terms of notions of beauty, utility, conceptual depth, the ability to invoke emotions and the way in which art and artists expose and address aspects of the human condition.

C. Johnson et al. (Eds.): EvoMUSART 2023, LNCS 13988, pp. 50–66, 2023.
https://doi.org/10.1007/978-3-031-29956-8_4

As an example for an idea adding to cultural knowledge, writer Sam Sykes posted a short story (twitter.com/SamSykesSwears/ status/15918209585073 72548) on twitter recently, where the narrator retrieved their godson from a hell-mouth in the basement. What was interesting and funny was the mundane treatment of the subject by the narrator, e.g., saying "All right, I asked around the Despair Pits if anyone had seen a smart-alecky kid running around". While the twitter story itself is a contribution to culture, we believe it is not a communicable idea or process that others could act upon. However, the concept of the mundane treatment of fantastical scenarios – normally treated with more awe – is a contribution to cultural knowledge that could be acted on. This could be a procedure to apply the process to other fantasy scenarios (e.g., a boring first day in heaven), or an aesthetic consideration to judge other works of fiction by.

While some concepts, techniques, opinions and interpretations are expressed explicitly, often knowledge is communicated in more vague, sometimes obfuscated, ways. This highlights that personal interpretation of artefacts and information is important in artistic cultures, as it can lead to ideation in order to fit knowledge into a personal ethos or environment, which can in turn lead to increased understanding and ownership. Artistic cultures often start local to small communities and sometimes the cultural knowledge they generate can spread globally, with books, catalogues and scores published and art exhibitions or concerts held which are commented on in newspapers, magazines, web pages, etc. The purposes for artistic activity enhancing cultures are also not easy to define comprehensively. Again, however, it is not controversial to say that one of the many benefits of artistic cultures are the ways in which they connect communities of people via various mechanisms. These include personal reflection and expression leading to increased understanding between people, and the celebration of human ingenuity, innovation, skill, productivity and creativity.

Generative AI techniques such as evolutionary computing have been used for decades by people as tools to create artefacts from which cultural enhancements follow [25]. In recent years, generative deep learning, where processes for producing cultural artefacts are learned from vast training sets, has risen to dominate the field of generative AI in the arts. While such artefacts exist in an artistic culture, they do not enhance that culture until thoughts flow from human minds that eventually add to the knowledge contained in the culture. We posit here that it is possible and desirable for computational equivalents of thoughts to flow from AI systems directly into human cultures, and we explain that position and expand on what that could mean in practice below. We argue, however, that the rise of generative deep learning techniques and the surrounding ethos dominating generative AI for the arts has greatly diminished the potential for AI systems to contribute directly in meaningful ways to these cultures.

In Sect. 2, we look initially at how generative deep learning serves well the sciences. The situation in the arts is different, however, and adding to an artistic culture is quite different to adding to scientific understanding. We discuss this situation with particular reference to generative deep learning and the drawbacks in the ethos and methods of this field. We follow this with a proposal to focus generative AI on adding knowledge to artistic cultures, and some suggestions for neuro-symbolic approaches to achieve this. We end by looking in particular at generative music, where decisions made by AI systems could be foregrounded, explained and communicated culturally.

2 Applications of Generative Deep Learning

In both the arts and sciences, generative AI techniques have been used with much success. Generative deep learning has been applied in the arts in many different ways, but the standard approach is the same as in scientific applications, with the analogy of measuring predictive accuracy being artefact *quality*, supplemented with considerations of the *diversity* of the artefacts that are produced [3]. In order to later comment on the drawbacks of this standard application for the arts, it is important to understand how this serves the sciences well, as below.

2.1 Generative Deep Learning in the Sciences

A successful application of generative AI techniques in the sciences is the production of a 3D structure predicting how a protein folds, given information about the chain of amino acids in it. The success of the exercise here is estimated in terms of the value of the 3D structures predicted, achieved by measuring the fidelity of the predictions when calculated for, and compared with, real, observed, protein structures. If this fidelity is high, then we can – to some extent – trust the predictions for proteins where we only know the amino acid chains, and there are statistical processes for estimating how high our trust should be.

Due to the nature of the prediction process, when a large pre-trained neural model is used for the predictions, as is the case with AlphaFold [14], there is little chance of inspecting the generative process in such a way that new concepts, processes or hypotheses are elucidated in human understandable terms. However, this doesn't matter very much, as the value to medicine of accurate protein structure predictions is huge. Moreover, scientists can – and have – looked at the resulting structures to reverse engineer some level of understanding. Albeit slowly, many scientists are realising that the primary goals of their research is for problem solving, prediction and engineering, rather than human understanding that can be distilled into papers or textbooks. Hence, for the betterment of society, the usage of black-box AI techniques for scientific prediction is being embraced across science and engineering. Pride in formulating and testing an understandable scientific theory is being supplemented by pride in training and deploying large neural models, and human society is benefiting from this.

This kind of thinking pre-dates the deep learning revolution. The mathematician Doron Zeilberger, a pioneer in the use of computers in pure mathematics, has been saying for decades that the most interesting parts of mathematics are those which cannot be understood by people [19]. In this context, AI researchers and practitioners have rightly focused on evaluating their contributions in terms of the fidelity of predictions and not in terms of increased human understanding of the field of application. Hence, progress in deep learning is mostly measured in terms of the predictive accuracy of the techniques developed. This wasn't always entirely the case in the application of machine learning techniques to scientific predictions. When symbolic machine learning techniques such as decision tree learning or inductive logic programming (ILP) [17], are used, the results can often be understood with some effort. These techniques can produce what Pat Langley calls *human-communicable scientific discoveries* [15]. A well cited example is mutagenesis [26], where an ILP system discovered a chemical structure associated with cancer forming cells, as evidenced through clinical trials later.

2.2   Generative Deep Learning in the Arts

Science raises questions about the physical world, human bodies and brains, the natural world, animal kingdom, sub-atomic particles, deep space, etc. The majority of these questions arise precisely because the complex objects and processes under study were not human-made and hence are opaque to start with. The questions raised in the arts are much more human-centric. In our understanding, the arts don't exist to enhance human lives by solving problems, making predictions, increasing health, reducing poverty, etc. Rather, art is made by artists to express themselves, communicate ideas, to celebrate humanity, help connect people, challenge human decisions, for intrigue, entertainment and amusement of people. In short, the arts are human-made cultures by people, for people and largely about people and the environments they inhabit.

The reasons we make something, and the benefits from doing so, are often not the obvious ones. Ostensibly, the reason someone makes and bakes a cake at home is to have and then eat a nice treat. In this circumstance, the success of the endeavour should be measured in terms of how good the cake is. However, the real reason for making the cake might be to pass the time fruitfully and the benefits may be in terms of the pride gained in being productive and making something tasty. Alternatively, the purpose of making a cake may be in learning a new technique from a recipe, or even trying out a new idea in the creative process. In these cases, the quality of the cake is only of secondary consideration, and the benefits of the exercise may be in terms of gaining a new skill or inventing a new idea. If this idea is passed on to other people, we can say that this particular creative exercise has added to baking culture. Successful or failed ideas and techniques, opinions about the quality of recipes, cakes and culinary experiences, and many other cake-based communications all add to the culture of baking. Rarely, however, does a particular cake that someone has baked add much, if anything, to this particular arts and craft sub-culture.

Generative deep learning has been applied in the arts in a number of ways, with the standard approach centred on using a pre-trained neural model to generate artefacts such as musical compositions, visual artworks or creative language texts, with varying levels of human oversight, curation and control. There is no doubt that the application of such techniques in the arts has added to various artistic cultures. For instance, there has been a huge amount of activity and cultural exchange provoked by the recent release of new models in neural text-to-image generation [24]. This cultural exchange has been instigated by the people using the generative deep learning systems, in various ways. Often, artists and musicians have used the generative systems as tools in their practice and included details of this in cultural exchanges. On other occasions, ideas and practices from deep learning have found their way into artistic cultures and been embraced, for instance prompting an AI system with text, music or images. In addition, there has been huge cultural discussion about the ethics of using AI systems in the arts and numerous on-going issues have been raised.

As would be expected, these cultural contributions have largely been human-centric with the AI system acting as a mere tool, the usage of which provoked a cultural exchange. That is, rarely (if at all) would it be accurate to say that a generative AI system, itself, has made a contribution to an artistic culture. AI systems are employed to help add artefacts to artistic cultures, thus enlarging the culture without necessarily enriching it with new knowledge. This is entirely appropriate for a tool like a paintbrush or Photoshop. However, because AI systems have the capacity for autonomous intelligence, they have the potential to add to culture in more meaningful ways than as mere tools. When the dominant methods in generative AI were a multitude of symbolic approaches such as genetic programming, there was still the potential to utilise this capacity. However, with the rise of generative deep learning to dominate the processes employed in generative AI and even more pertinently, the ethos surrounding it, we believe this potential has all but gone, as argued below.

2.3   Drawbacks to Generative Deep Learning

We identify three reasons why the dominance of generative deep learning reduces the potential for AI systems to directly enhance artistic cultures. Firstly, the data-centric nature of deep learning means that artificial neural models are trained on human-produced artefacts like music or visual art images. This has made pastiche generation the dominant force in generative art. While making pastiches is a standard part of any artistic training, it is not necessarily seen as the most innovative use of time for mature artists, and adds to artistic cultures in only limited ways. However, the domination of pastiche generation has not proved fatal to the generative arts, as artists have found numerous ways to employ more-of-the-same generation with innovation. As an example, François Pachet and colleagues have trained models over artefacts in two distinctive musical styles to produce a pre-trained generative model able to make interesting creole music [20]. Similarly, text-to-image generators are enabling new visual art styles to emerge, with or without reference to particular artists or styles. We believe, however, that the human-data centric approach has greatly strengthened the idea that human authority in the arts is prime, which implies that AI systems have no place in adding to artistic cultures. In particular, it has stifled, or at least downplayed, the notion that AI systems can be used to invent genuinely new ideas, processes, critiques or other cultural knowledge.

Secondly, the opaque and uniform nature of generative deep learning processes make it difficult to extract information that could add to artistic cultures. People occasionally learn new skills via usage of AI systems, e.g., prompt engineering for text-to-image generation is clearly a skill that some artists have learned and passed on culturally [16]. However, these skills are obtained via artists experimenting, analysing and generalising their results into cultural contributions – the AI system itself did nothing to explicitly aid in this process. Unfortunately, if questions are asked about the generative process in deep learning, authentic answers would only be able to refer to deep learning considerations such as the training data, training regime, fine-tuning, activation unit calculations, etc. One exception to this is when a model requires an input seed, such as a text prompt, image or musical excerpt. Choosing this input is part of the process and correlations between inputs and outputs have been noticed and communicated. But again, this has not been facilitated directly by the AI process.

It is possible to use third party AI systems to analyse deep learning processes, e.g., the numerical flow through a model during a feed-forward pass. From this, we could extract culturally interesting concepts and processes. However, these would be post-hoc observations and estimations about the process, rather than communicated decisions from the original generative AI systems, which decreases their value somewhat. It is arguably more important to musical culture, for instance, to hear from Mozart himself on how/why he composed elements of his music rather than a music theorist or historian doing so at a later date, involving some level of generalisation and guesswork. A third party AI-based analysis of generative AI processes and products could possibly contribute to artistic cultures, but it would be better if the generative system itself had interesting processes from which such contributions would naturally flow.

Thirdly, the idea of AI systems having actual autonomy has been comprehensively quashed as part of the machine learning and general AI ethos pervading the field. Ethical considerations of AI system usage in the arts has, in the last decade, dictated that the integrity, independence, employment prospects, feelings and well-being of people in the arts is paramount and inalienable. Until recently, this narrative has rarely been pushed by human artists themselves, but rather by journalists, art lovers and artist allies, commentators, provocateurs and other stakeholders. It usually takes the form of exercising decades-old tropes of automation taking jobs and automation of creative tasks in particular denigrating and trivialising the efforts of human artists. Recently, due to increased output quality and usability of generative AI techniques, artists voices have added to the chorus of people calling for a halt to the usage of AI in the arts.

In addition, machine learning focuses on problem solving, where the computational system acts only as a tool to answer questions raised by people. The quality of the output (predictions or artefacts) in deep learning is so paramount in assessing progress that other considerations around the artistry or intrigue of the process are mostly ignored. Such evaluation regimes have been passed on from the technical to artistic communities. As a result, taking genuine concerns from the art world, together with the machine learning ethos and black box nature of deep learning, the world has been left with the impression that AI systems cannot and should not have autonomy that may enable them to contribute meaningfully to artistic cultures. Hence projects where AI autonomy is increased to enable it to take on creative responsibilities in arts projects that could lead to cultural enhancements, are basically non-existent. A few exceptions, such as The Painting Fool project [7], exist in the field of Computational Creativity [6] [9], where AI agency and autonomy in creative contexts is still discussed.

3   A New Emphasis for Generative AI

The simulation of intelligence is not the same as the simulation of a physical process or tool such as paint flowing and paintbrushes in graphics. AI-based intelligence has the power to educate, inspire, challenge and enhance human cultures, but only if it is allowed to do so. Societies have historically shunned, subjugated, disenfranchised and discriminated against certain groups of people, and we are currently doing the same to AI systems. Acknowledging this, without pretending that AI systems should have the same rights as people or are indeed much like people, could lead to an outpouring of AI-generated cultural enhancements which really drives humanity forward. We believe that each time the voices of a more diverse section of society are heard in the arts, culture takes a leap forward: more discussion, more innovation via crossover of ideas and backgrounds, more empowerment, and more understanding of each other and our roles in the world.

We therefore suggest considering whether generative AI systems should be allowed/engineered to progress beyond mere tools that generate artefacts of value. A new wave of AI systems could contribute ideas, processes, opinions, anecdotes, historical references, critical evaluations, aesthetic considerations and much more to artistic cultures, from which people could learn about themselves and others, to the betterment of human society. However, before discussing such advancements, we should question whether this is ethically appropriate and address some issues which naturally arise.

3.1   Ethical Issues

It is easy to argue that some ways in which generative deep learning systems are currently used can denigrate and upset well trained artists. As a particular example in the visual arts, the argument goes that an artist used to spend days or weeks in producing an image that a text-to-image generator can now produce in seconds, and can be done by a completely untrained member of the public for free. This trivialises the training an artist has been through and the effort they normally expend in their craft. To add insults to injuries, the process uses visual art imagery harvested without permission from fellow artists, the text prompts often reference by name particular artists in order to reflect their style in uncredited pastiche, and the output images often portray undesired biases in society that a trained artist would avoid depicting. Moreover, if the member of the public using the generative AI system is not well versed in – or connected to – artistic culture, then neither they nor the AI system itself is likely to contribute something of value to culture that artists can use in future as inspiration.

Previously, AI art was a non-threatening niche area largely ignored by art cultures or trivialised as producing naïve and dull art, often via simplistic pastiche. However, generative deep learning has brought about a step change in the quality of images, diversity of outputs and ease of use, with vast quantities of artistic images flooding social media recently. The art world has taken note, for instance with auction houses arranging special AI art sales, and AI artists gaining much in popularity and commercial potential. There have also been some cases, for instance in producing images to illustrate online articles, of AI art being generated when an human artist may have been previously commissioned. It is fair to say that the twin rises of AI artistic processes and NFT commercialisation have shook the art world recently [28].

In this context of disorienting and potentially upsetting change, one could argue that giving further autonomy to AI systems to contribute to culture should be held off or indefinitely avoided. There is always understandable pushback against technological changes, but AI systems which contribute to human artistic cultures do not have to be threatening or denigrating and offer much potential for benefit. Chess engines can thrash human players, but rather than ruining chess culture, they have enhanced it greatly: used to raise excitement during commentaries on live gameplay, for educating novices, training grandmasters and for generating chess puzzles and analyses, amongst other uses. It is up to the members of various artistic communities to decide the level of involvement of AI techniques in their culture. For instance, the well known art platform *Deviant Art* has recently made available a text-to-image generator. They have also allowed community members to opt out of allowing their images in future training of the generative models, with acceptance as default. This has caused controversy with mixed messaging, as the text-to-image is based on one that has already used their community's images, without any authorisation or credit.

We argue that AI systems contributing to cultural knowledge is not a problem added on top of the current issues surrounding generative AI, but could be a solution to some of the existing issues. In particular, the empowerment of the general public to produce high quality artwork to express themselves via generative deep learning is revolutionary. However, this has not been matched with a proportionate increase in contributions to artistic cultural knowledge. In fact, it has been rare for the cultural discussion raised around AI art to be anything more than circular discussion about AI and art themselves, rather than touching on elements of the human condition, pressing issues like environmental change, political issues, artistic (rather than computer science) techniques or sensibilities, or any number of more traditional thought-provoking topics in artistic culture.

Suppose that – by having more amenable processes and raised autonomy – a generative AI system could, itself, make such cultural contributions. These would be shared in the community, and could provide inspiration to some people, education to others, intrigue to others, entertainment to others, and so on. This alleviates to some extent the issue with generative art taking (data) from art culture but giving little of value back other than a flood of largely worthless (but impressive) generated artefacts. Moreover, AI systems able to interrogate their processes, outputs, interactions, etc., to afford cultural contributions are more amenable to being controlled, adjusted and corrected for biases, which also alleviates somewhat various issues surrounding generative AI. More speculatively, as per the notion of the machine condition [8], AI generated cultural knowledge could also inform human culture about technology, if the generator was used to communicate what it is like to be a machine.

3.2   Stakeholders

Machine learning scientists and engineers are fixated on the usage of AI systems merely as tools to solve problems rather than creative collaborators which could educate us and raise interesting problems, affordances and philosophical discourses. Indeed, in the engineering lingo in deep learning platforms like Pytorch and TensorFlow, pre-trained models make "predictions" of music or visual artworks. Predictions are part of scientific culture and usage in art culture leads to notions that there is a "perfect" or "best" painting to reflect a prompt, or a right or wrong way to compose a poem, which is rarely the case. Moreover, many scientists and engineers have not discerned that the point of the arts is not to make pictures, music or poems, but to contribute to artistic cultures which connect, empower and inspire human communities, enabling expression, self reflection and questioning of the human condition. Many more generative AI scientists and engineers are not remotely interested in artistic cultures, but rather creative artefact generation as a means to the end of progressing deep learning, which is entirely understandable.

Taken together, it doesn't seem likely that many deep learning scientists or engineers will take up the challenge of getting AI systems to contribute to artistic cultures. Furthermore, AI ethicists are strongly invested in the idea that AI systems having the kind of autonomy required to add to culture will indelibly upset human artists and ruin artistic cultures. This reflects the general concern that ethicists have of AI systems having too much decision making power. This ethos permeates all of AI, and it is difficult for researchers to justify projects where AI systems may disrupt a group of people, never mind one which might – through cultural contributions – question elements of human society in unsettling ways. There are some sub-fields of AI such as Computational Creativity research [6] [9], where the idea of AI systems having creative autonomy and thus potential to add to culture are still prevalent. However, as long-standing members of this research community, we see that most people are more interested in tool-centric approaches than in automation, and the field doesn't have the resources for large-scale projects which could substantially change the landscape.

Journalists, politicians, policy makers and commentators all know that automation is a perennially interesting and divisive topic that sells newspapers, divides people so they can be controlled and offers a platform with which to grab and exercise power and influence. Big software and advertising companies adhere to the tool-centric ethos for various reasons, including to avoid regulation and to avoid alienating customers. In particular, big companies making software for the creative industries need to cater to their customers whose well-being and job security is wrapped up, in part, in their autonomous creativity. Outfits like Google, Spotify, Sony, Adobe, Apple, etc. have creative AI teams, so there may be some work being undertaken on handing over creative responsibilities for cultural enhancement. However, some researchers in the creative AI field undermine the idea of AI autonomy through papers, blogposts and social media [23].

There is the possibility that in some of the current wave of generative AI startups, there will be innovation around handing over creative responsibility to AI systems. Many of these companies want to democratise generative AI to empower large numbers people to express themselves through art. As the majority of these people are untrained, they will likely decline to use complex tools, and the AI systems themselves will be required to take on creative responsibilities. If this happens, the decisions that the AI systems make could potentially be foregrounded and used for enhancing artistic cultures, especially, we believe, if neuro-symbolic approaches are employed (see below). Another place where automation may lead to AI-generated cultural contributions is in the open-source software world, which has had a big impact on generative AI. In particular, there have been a number of implementations shared recently which take on creative responsibilities previously undertaken by people, for instance reverse-engineering appropriate text prompts from images (huggingface.co/spaces/pharma/CLIP-Interrogator), assessing aesthetic qualities for the purposes of curation and combining models for multi-modal applications. Startup employees and open-source coders are constrained much less by ethical issues and machine learning (tool-centric) norms, and both sets of people are incentivised to disrupt, explore and innovate. Hence, it's possible that these stakeholders could enable AI systems to contribute directly to cultural knowledge.

4   Neuro-Symbolic Approaches

Generative art projects predate the rise of deep learning by many decades, feeding into, and taking from, areas of computer science such as AI, graphics, audio, etc. If we take as an example evolutionary approaches to generative AI that gave rise to projects either in the artistic or the academic research communities, we can ask how (if at all) they contributed to artistic cultures. We argue that, compared with the current applications of generative deep learning to artistic projects, with evolutionary art projects:

- The project leaders were usually more connected to and familiar with artistic cultures, hence incentivised to make cultural contributions.
- The evolutionary approaches were fairly heterogeneous and often relied on visual art or music theory, hence were more amenable to foregrounding aspects as cultural contributions via traditional communication means.

With reference to the first point, it's clear that the general public have – in great numbers – recently been using online generative deep learning tools for empowerment, creative expression, amusement and entertainment, without leaving much of an impression on artistic cultures. This has been exacerbated by the use of social media for dissemination, where complex discussion and cultural contributions are not the norm. However, in parallel, creative professionals such as designers and artists have been using and improving the tools and it's clear that as these approaches mature, more established artists and researchers will soon use them to contribute to culture in meaningful ways.

With reference to the second point above, evolutionary art projects historically had to identify key elements of bespoke generative processes to encode into genomes or for choosing function and constant sets in genetic programs. Researchers sometimes also had to write bespoke fitness functions to guide the evolutionary processes and for curation purposes. This was often based on art and music theory, and information about the generative process and fitness could be conceptualised and communicated as a contribution to culture. In particular, there was the potential to express these as processes which could be explicitly written up as code to be used elsewhere in the culture.

Unfortunately, the feed-forward propagation of numbers through a pre-trained neural model has always been criticised as a black-box approach. This is because the homogeneity and scale of the processing make it very difficult to extract human-communicable concepts describing the process. The field of explainable AI [10] tries to address this, and has focused on explaining decisions made by AI at the level of predictions. Such approaches could be useful for AI-generated cultural contributions, but they currently fall short of foregrounding processes for cultural communication. This is partly because it's rare for any element of art or music theory to be encoded into training or generative processes, as all that is needed is artefacts from the arts as data, mathematical machine learning techniques and statistical evaluation methods.

In many respects, it is a wonderful advancement that AI art and music generation doesn't need art or music theory now. But this does greatly dampen the potential for a shared context with people in artistic cultures, and hence makes it more difficult for generative AI researchers and practitioners to contribute to culture. As an example, deep learning researchers no longer need to understand the concepts of perspective or complimentary colours in the visual arts. They can train generative models which produce images expressing these concepts as long as they have appropriate human-made art to train over. It's very difficult currently to imagine a method to understand the processing of pre-trained models enough to determine how it uses perspective or colour theory, and even more difficult to imagine it doing so in an innovative way, communicable culturally.

We believe that taking the heterogeneous, theory-influenced symbolic approaches and combining them with the generative and analytical powers of deep learning offers the highest potential for generative AI systems to contribute to artistic cultures. We take the view that it is sensible to start by targeting cultural contributions at the process level. This is for a number of reasons, including:

(a) it enables us to avoid cultural contributions such as reflections on the human condition, which are more authentic – and hence more valuable – if they come from a person (b) processes can be valuable to both engineers and artists engaged in generative art projects, and (c) there are some useful technologies already in place. With that in mind, the following are general points to take into account when engineering neuro-symbolic approaches to generative AI which aim to add to cultural knowledge as well as to produce valuable artefacts.

- *Episodic Processing.* If a generative process has multiple stages, there is more potential for cultural description of how episodes fit together in terms of data transfer, how an artefact changes over the episodes, and how the AI system makes choices between episodes.
- *Heterogeneous Processing.* Within an episodic approach, there are a multitude of AI and non-AI processes that can be brought to bear on artefact generation. Generative deep learning and evolutionary approaches have been shown to be beneficial in generation. But predictive machine learning techniques can be used in assessing, sorting and creating artefacts, and symbolic AI techniques like constraint solving or planning can be employed in decision making. The more complex and heterogeneous the process, the more decision points there are and the more varied and interesting the cultural communications can be.
- *Multi-level and Multi-modal Processing.* Handing over creative responsibilities to AI systems in arts projects is somewhat a mantra of Computational Creativity research, as described in [9]. Such projects involve much more than just generation, and there are opportunities for AI systems to frame, curate and motivate their creations, all of which could lead to potentially interesting cultural contributions. Connecting different generative systems across modalities such as text, images and music also has great potential in this regard.
- *Accountable Unpredictability.* The use of random numbers has been ever present in the generative arts, to increase variety and the surprising nature of generated artefacts. The deep learning incarnation is in terms of randomly sampling a vector from a latent space as an input to a pre-trained model. However, variety and surprise can be achieved by a number of non-random processes which are just as unpredictable, e.g., using trending topics on social media. Random sampling rarely leads to interesting backstories, but achieving unpredictability via a connection to the world of social media has much potential for this.
- *Generative System Invention.* With the rise in automatic code synthesis via projects like *Github Co-pilot*, it is now possible to imagine the code for a generative system being produced automatically. While speculative, it's further possible to imagine linking concept invention techniques and aesthetic/utilitarian measurement techniques, which both rely on artistic theory, to generative systems for particular types of artefacts in novel ways invented through code synthesis. The processes here could constitute important cultural contributions.

As a relatively simple example, one can imagine an AI system taking trending topics on twitter as starting points, using text generation to turn the topics into framing text, then downloading real images which reflect the text as training for a GAN generator. After training and generation of hundreds of images, it uses a hand-coded evaluation metric to curate and present a few images, with the code reflecting elements of traditional visual art theory. If the images are interesting (and probably even if they are not), descriptions of elements of this process may be interesting as cultural contributions, especially if something surprising in the outputs or processing can be determined. One could further imagine automating this surprise detection, in order for the AI system to communicate why certain outputs are novel or certain processes are innovative.

5   Decision Foregrounding for Generative Music

Generative music has many applications such as composing music from scratch [12], extending a given piece of seed music [13], filling the gap between two musical materials (musical in-painting) [11], generating music in a certain genre [18] or with emotional references [27]. While these applications are valuable for practitioners in co-creation settings, they don't necessarily add knowledge to musical cultures. Much generative music can be criticised as lacking direction and purpose, fitting the aesthetics of a musical distribution rather than telling a story. Moreover, producing long-term structure in the generated pieces is an open problem in the field. In contrast, when we listen to human-composed music, audible decisions have clearly been made, and could be communicated with reference to a musical system. Such a system might comprise standard musical concepts and notation, but equally could be graphic scores for electro-acoustic music [1]. When a musician listens to a human-composed music, they can identify the decisions regarding tonal changes, chord progressions, musical themes and their variations, genre manipulations, orchestration, dynamics, etc. All of these decisions are justifiable as they support a direction purposefully, communicated with respect to some theoretical understanding of music.

Generative music systems also make decisions [2], but if these decisions are not communicated explicitly, then the contribution is limited to the generated artefacts and does not reach the level of enhancing human musical culture, which includes cumulative and summarised knowledge. In the explainable AI literature, there have been various attempts to allow deep learning models to communicate their decisions, via symbolic interventions to deep learning models, but these approaches are limited, as they are only able to reveal a few musical features, lacking a comprehensive understanding of music [5] [21] [22] [27].

We can speculate about what a human composer might say when asked about their decisions, in order to inspire ways to implement neuro-symbolic generative music systems able to communicate their decisions. Such communications could be the starting point for a more substantial contribution to musical culture. Suppose a composer has been asked why they wrote a particularly interesting chord sequence in bar 17. One answer could be: "this was highly constrained by the form, and was the only solution". Mirroring this in a neuro-symbolic system would involve the software describing the musical theory it is working within. Another answer could be: "I experimented a lot and this solution seemed to fit the music the best". This could be mirrored by a neuro-symbolic system generating multiple options in a symbolic fashion, then checking which fits best the musical distribution of the composed music so far with a pre-trained (or fine-tuned) neural model supplying the answer. A third answer could be: "I wanted it to sound jazzy". Here the neural model would be trained on jazz music, and again used to choose between options produced in a symbolic way.

We see that musical information from the symbolic system about AI-generated music could be communicated by the neuro-symbolic generative AI system, and this could be supplemented with information about personal choices, historical precedent or fit to genre, by the deep learning element. We have described symbolic-first approaches where the generative element is the symbolic part and the analytical part is deep learning. This could be switched, so that in a neural-first approach, a set of candidate musical pieces are generated using a generative deep learning approach, then an analytical music (symbolic) evaluator, similar to [4], makes the final choices, with decisions communicated about the fit to the model and fit to the theory.

In addition to these suggested workflows, there are other symbolic interventions for neural approaches. For instance, symbolic approaches could provide informed guidance in the loss landscape during the optimisation process. This would allow us to better evaluate the stylistic judgement capability of a model, and could lead to communicable cultural contributions. Another alternative would be post-hoc prediction and explanation of decision points in musical compositions. For instance, a generative deep learning system could be trained on information about musical decisions, and predict interesting aspects of decisions made post-hoc, rather than a report on the real decision making process.

6 Conclusions

Our main purpose here has been to argue that there are many potential benefits for AI systems to communicate cultural contributions themselves or to co-create them with people. We have focused on such contributions at the level of process invention and communication of decisions, paying particular attention to generative deep learning as it is clearly the dominant technique in the generative arts, and will likely continue to be so for some time. We highlighted issues around both the main black-box techniques and the tool-centric ethos of deep learning, arguing that these make it difficult to imagine a generative deep learning system making contributions directly to artistic cultures.

While it is true that symbolic approaches to artefact generation have rarely delivered additions to cultural knowledge directly, this is perhaps because they have rarely been used to, and there are limitations of symbolic approaches as they do not generalise well or support representation learning. However, we have argued that symbolic approaches which employ existing concepts from a culture have higher potential for communicating new ideas, processes, etc. to people in that culture. One of the most powerful aspects of deep learning is that it can work from raw data without requiring any hand-crafted concepts. However, in doing so, we've lost some of the ability to communicate with cultural audiences.

We proposed that the existing practice of co-creation of cultural artefacts evolves into the co-enhancement of cultural knowledge via the co-creation of such artefacts. We further suggested neuro-symbolic approaches, and provided some thoughts on music generation in more detail. In particular, we believe the formalism and language offered by symbolic approaches can be fruitfully coupled with the generative and analytical power of deep learning to produce not just better artefacts, but innovations at process and conceptual levels that an AI system could communicate as cultural contributions.

Every person who interacts with a generative AI system or its output, whether by developing and training the system, using it for art production, curating or exhibiting its output, or simply enjoying these artefacts, has an opportunity to contribute meaningfully to artistic cultures. Arguably, however, the one intelligent entity that is not able to likewise contribute is the most important one: the generative AI system itself. It is not necessary to think of AI systems as full members of an artistic community to believe that they can contribute to cultural knowledge. Likewise, we believe AI systems don't have to have personhood or consciousness or any particularly human-like qualities to be valuable members of an artistic culture. Generative deep learning has been used in projects where human thought and communication have led to cultural contributions over and above the artefacts produced. By looking at what kinds of information arise directly in artistic projects, and how people conceptualise and communicate this, it would be possible and beneficial to engineer AI systems able to supplement this knowledge by automatically transferring cultural knowledge they've generated themselves. We believe these cultural additions will greatly enhance human society and is a more worthy and interesting target for generative AI than merely helping people to produce artefacts of value.

Acknowledgements. We wish to thank the anonymous reviewers for their insightful comments. Berker Banar is funded by the UKRI/EPSRC Centre for Doctoral Training in AI and Music (grant number EP/S022694/1) and Queen Mary University of London.

References

1. Banar, B., Colton, S.: Connecting audio and graphic score using self-supervised representation learning - a case study with Gyorgy Ligeti's artikulation. In: Proceedings of International Conference on Computational Creativity (ICCC) (2022)
2. Banar, B., Colton, S.: Identifying critical decision points in musical compositions using machine learning. In: Proceedings of the 24th IEEE International Workshop on Multimedia Signal Processing (MMSP) (2022)
3. Banar, B., Colton, S.: A quality-diversity-based evaluation strategy for symbolic music generation. In: Proceedings of the ICLR ML Evaluation Standards Workshop (2022)
4. Banar, B., Colton, S.: A systematic evaluation of GPT-2-based music generation. In: Martins, T., Rodríguez-Fernández, N., Rebelo, S.M. (eds.) EvoMUSART 2022. LNCS, vol. 13221, pp. 19–35. Springer, Cham (2022). https://doi.org/10.1007/978-3-031-03789-4_2

5. Bryan-Kinns, N., et al.: Exploring XAI for the arts: Explaining latent space in generative music. In: Proceedings of the NeurIPS eXplainable AI Approaches Wshop for Debugging & Diagnosis (2021)
6. Cardoso, A., Veale, T., Wiggins, G.: Converging on the divergent: The history (and future) of the international joint workshops in computational creativity. AI Mag. **30**, 15–22 (2009)
7. Colton, S.: The Painting Fool: Stories from building an automated painter. In: McCormack, J., d'Inverno, M. (eds.) Computers and Creativity. Springer, Heidelberg (2012). https://doi.org/10.1007/978-3-642-31727-9_1
8. Colton, S., Pease, A., Guckelsberger, C., McCormack, J., Llano, T., Cook, M.: On the machine condition and its creative expression. In: Proceedings of the International Conference on Computational Creativity (2020)
9. Colton, S., Wiggins, G.A.: Computational creativity: the final frontier? In: Proceedings of the 20th European Conference on Artificial Intelligence (2012)
10. Došilović, F.K., Brčić, M., Hlupić, N.: Explainable artificial intelligence: a survey. In: Proceedings of the 41st International Convention on Information and Communication Technology, Electronics and Microelectronics (2018)
11. Hadjeres, G., Crestel, L.: The Piano inpainting application. arXiv:2107.05944 (2021)
12. Hawthorne, C., et. al: General-purpose, long-context autoregressive modeling with perceiver AR. In: Proceedings of the 39th ICML, vol. 162 (2022)
13. Huang, C.Z.A., et al.: Music transformer: generating music with long-term structure. In: Proceedings of ICLR (2019)
14. Jumper, J., et. al: Highly accurate protein structure prediction with AlphaFold. Nature **596**, 583–589 (2021)
15. Langley, P., Shrager, J., Saito, K.: Computational discovery of communicable scientific knowledge. In: Magnani, L., Nersessian, N.J., Pizzi, C. (eds.) Logical and Computational Aspects of Model-Based Reasoning, pp. 201–225 (2002)
16. Liu, V., Chilton, L.B.: Design guidelines for prompt engineering text-to-image generative models. In: Proceedings CHI (2022)
17. Muggleton, S.: Inductive logic programming. N. Gener. Comput. **8**(4), 295–318 (1991)
18. MuseNet. https://openai.com/blog/musenet/. Accessed 17 Feb 2023
19. Opinion 36: Don't ask: what can the computer do for me?, but rather: what can I do for the computer? https://sites.math.rutgers.edu/zeilberg/Opinion36.html. Accessed 17 Feb 2023
20. Pachet, F., Roy, P., Carré, B.: Assisted music creation with flow machines: towards new categories of new. arXiv:2006.09232 (2020)
21. Pati, A., Lerch, A.: Latent space regularization for explicit control of musical attributes. In: Proceedings of the ICML Machine Learning for Music Discovery Workshop (ML4MD) (2019)
22. Pati, A., Lerch, A.: Attribute-based regularization of latent spaces for variational auto-encoders. Neural Comput. Appl. **33**(9), 4429–4444 (2021)
23. Re-imagining work in the creative age. https://medium.com/@creativeai/re-imagining-work-in-the-creative-age-419fb5ce9139. Accessed 17 Feb 2023
24. Rombach, R., Blattmann, A., Lorenz, D., Esser, P., Ommer, B.: High-resolution image synthesis with latent diffusion models. arXiv:2112.10752 (2021)
25. Romero, J., Machado, P. (eds.): The Art of Artificial Evolution: A Handbook on Evolutionary Art and Music. Springer, Heidelberg (2007). https://doi.org/10.1007/978-3-540-72877-1

26. Srinivasan, A., Muggleton, S., King, R., Sternberg, M.: Theories for mutagenicity: a study of first-order and feature based induction. Artif. Intell. **85**(1-2), 277–299 (1996)
27. Tan, H., Herremans, D.: Music FaderNets: controllable music generation based on high-level features via low-level feature modelling. In: Proceedings ISMIR (2020)
28. Technology has upended how art is created and consumed. Is that a good thing? https://www.cbc.ca/news/entertainment/ai-art-nft-technology-1.6626015. Accessed 17 Feb 2023

Extending Generative Neo-Riemannian Theory for Event-Based Soundtrack Production

Simon Colton and Sara Cardinale[✉]

School of Electronic Engineering and Computer Science,
Queen Mary University of London, London, UK
{s.colton,s.cardinale}@qmul.ac.uk

Abstract. We present the GENRT music generation system specifically designed for making soundtracks to fit given media such as video clips. This is based on Neo-Riemannian Theory (NRT), an analytical approach to describing chromatic chord progressions. We describe the implementation of GENRT in terms of a generative NRT formalism, which produces suitable chord sequences to fit the timing and atmosphere requirements of the media. We provide an illustrative example using GENRT to produce a soundtrack for a clip from the film A Beautiful Mind.

Keywords: Generative Music · Neo-Riemannian Theory · Soundtracks

1 Introduction

Musical soundtracks are an essential part of many forms of media. They help create an immersive experience by supporting story lines, portraying characters, establishing the overall setting and mood, highlighting important events and adding to the aesthetic value of media. Soundtrack music has requirements not shared by music in general, which makes it an interesting and valuable target for generative music, which could lead to useful tools for composers. Our problem setting here is to be given an existing piece of media like a video clip from a film, or an audio excerpt of an actor reading for an audiobook, and to generate an appropriate soundtrack for it. The media is supplied along with a list of episodes given in terms of (a) the start and end timestamps of the episode in milliseconds (b) an overall atmosphere (e.g., calm) for the episode and (c) a change in overall atmosphere to be reflected smoothly during the episode (e.g., getting more exciting). Optionally, each episode may have a particular event at the conclusion of it, specified by the timestamp at which it needs to occur and the emotional or situational change in the media that the event represents.

An appropriate soundtrack would have to satisfy many things, but we concentrate here on it being *accurate* in terms of timings of episodes and events, *influential* in terms of musical changes which audibly support situational or emotional changes in the media and *atmospheric* in terms of the nature of the music between events. We describe the GENRT music generation system which

C. Johnson et al. (Eds.): EvoMUSART 2023, LNCS 13988, pp. 67–83, 2023.
https://doi.org/10.1007/978-3-031-29956-8_5

Table 1. (a) NRT Rewrite rules for major and minor chord starting points. In each case, the starting chord is $\{a, b, c\}$. (b) Association of compound Neo-Riemannian operators and emotional and/or situational scene elements (from [11]).

NRO	Major	Minor
R	$\{a, b, c+2\}$	$\{a-2, b, c\}$
P	$\{a, b-1, c\}$	$\{a, b+1, c\}$
L	$\{a-1, b, c\}$	$\{a, b, c+1\}$
N	$\{a, b+1, c+1\}$	$\{a-1, b-1, c\}$
N'	$\{a-2, b-2, c\}$	$\{a, b+2, c+2\}$
S	$\{a+1, b, c+1\}$	$\{a-1, b, c-1\}$

Compound NRO	Emotion/Situation
LP	Antagonism
L	Sorrow, loss
N	Romantic encounters
PRPR	Mortal threats, dangers
RL	Wonderment, success
NRL	Suspense and mystery
RLRL	Heroism (Lydian)
NR	Fantastical
S	Life and death

can produce MIDI files that act as accurate, influential and atmospheric soundtracks to a given piece of media. Underpinning GENRT is an extension of a musical theory called Neo-Riemannian Theory, described in Sect. 2. This theory was devised for the analysis of music, but we provide a generative formalism, which can be employed to make music, as detailed in Sect. 3. We describe GENRT in terms of this formalism in Sect. 4, highlight the two main issues of producing appropriate chord changes for events and making the music between events suitably low-key, so the event music is not obfuscated. In Sect. 5, we use GENRT to produce a soundtrack for a clip from the film A Beautiful Mind, and we end with conclusions and a discussion of future work.

2   Background

Neo-Riemannian Theory (NRT) comprises methods to analyse chromatic chord progressions that do not employ tonality [6]. These types of chord progressions can be difficult to analyse using conventional music theory, as they do not follow traditional models which include the use of keys and modes to understand harmonic developments. A musical cue such as a chord change can be analysed in terms of which Neo-Riemannian Operators (NROs) have been used. As per the original theory, an NRO applies a transformation to trichords consisting of the tonic, third and fifth of a major or minor scale in some permutation. Every NRO outputs a major chord given a minor one as input, and vice-versa, and each can act on any trichord. The six NROs originally identified are given in Table 1(a). It is also possible to analyse a single chord change as the application of multiple NROs iteratively, which we call *compound* NROs.

Chromatic chord progressions are widely used in movie soundtracks, as the film scenes often change in rapid, fairly pronounced ways, and composers have found that chromatic chord progressions highlight these changes well. In [11],

Lehman analysed film scores in depth and found certain chromatic chord changes
– which he described in terms of compound NRO usage – were commonly used
by composers to illustrate emotional or situational changes in the movies. For
instance, a quick change of chords can be used to represent important on-screen
events such as the appearance of a character, a significant death or a disappoint-
ing failure. Lehman mapped compound NROs to emotional/situational changes
in film narratives, as per Table 1(b) by analysing various soundtrack cues and
noting when certain operators were used to describe specific events. While not an
exhaustive list of compound NROs and emotions, it provides a starting point to
understand how chromaticism can represent events and emotions in soundtracks.

The original papers on NRT obfuscated matters somewhat with fairly dense
and sometimes superfluous mathematics. In [4], we rationalised the operators as
NROs as portrayed in Table 1(a). This enabled us to describe a basic generative
reading of NRT where the NROs are employed as conditional re-write rules,
with the condition being whether a trichord is major or minor. We likewise
rationalised the notion of compound NRO and, drawing on the work of Cohn
and Lehman, stated the main tenets of generative NRT as follows:

• Chord sequences are produced from a given starting chord via the repeated
application of compound NROs.

• Emotional and situational changes in the target media can be reflected in the
music via the application of particular compound NROs, as per Table 1(a) to
produce suitable chord changes.

• The longer the compound NRO used for a chord change, the more striking it
will be, audibly.

Using these tenets as a foundation, we significantly generalise and extend this
formalism in Sect. 3 below, to enable generative NRT to become the basis of a
working implementation of the GENRT music generation system.

2.1   Related Work

The generation of chord progressions has been much investigated for automatic
production of musical harmony, and Wiggins [15] looked at the notion of inten-
tionality in this respect, which is important for the generation of chord progressions to accompany scenes from media such as films. Bernardes et al. [2]
implemented the D'accord harmony generation system which worked over a perceptually motivated tonal interval space. While not using NRT, Monteith et al.
[13] generated music to induce targeted emotions, using statistical techniques
such as Hidden Markov Models, and applied this in [12] to produce affective
music to accompany the audio of fairy tales being read aloud.

Neo-Riemannian Theory for music generation has been explored in as Chew
and Chuan [5], where a style-specific accompaniment system was proposed. The
system utilises NRT to represent transitions between neighbouring chords. The
authors make use of roman numeral musical analysis to describe the chord progression. This is an analysis technique usually used in tonal music, as the roman

numerals describe tonal relationships between chords. Given this observation, this work seems to be aimed at the generation of diatonic, tonal chord progressions which respect functional harmony. This is a limitation of this approach, as NRT is used to analyse chromatic progressions which are not restricted to functional harmony rules. Similar limitations can be observed in other systems and approaches, such Amram et al. [1]. Here, the authors implement a generative chord-based composition tool. Similarly to [5], the authors implement NRT in a tonal, diatonic way, therefore not using the theory to its full potential.

Other emotion-based systems include Mind Music [7], which focuses on designing music based on a character's emotions using mood nodes and a spreading activation model, creating adaptive audio for game characters. Hutchings et al. [9] further developed these ideas, proposing an adaptive music system that combines a multi-agent system (MAS) and cognitive models of emotion. The agents focused on harmony, melody and rhythm to develop compositions.

3   Extended Generative NRT

In order to produce more diverse music and to have more options for controlling the generative process, we have substantially extended and adapted our generative NRT approach as described below. Our starting point was the observation that each of the existing NROs took a trichord of tonic, third and fifth notes from either a major or minor scale, and produced a new trichord which is also a major or minor tonic, third or fifth. Each of the original NROs preserves at least one note of the original chord and no note in the output chord is more than a tone away from a note in the original chord. To describe the extended generative NRT implementation in the next section, we use a formalism which extends trichords and NROs. We start by generalising trichords past the limitation of just major and minor chords, as follows:

- [A1] A *note* is an integer from the midi pitch range $\{0, \ldots, 127\}$. For instance, middle C on a piano keyboard is note number 60.

- [A2] The *note class* for a given note, N, denoted $nc(N)$, is calculated as:
$nc(N) = \{k : 0 \leq k \leq 127 \text{ and } k \bmod 12 = N \bmod 12\}$
e.g., the note class for middle C (midi note 60) is $\{0, 12, 24, 36, \ldots, 120\}$.

- [A3] A *general trichord*, $T = [t_1, t_2, t_3]$, is an ordered triple of three distinct notes from strictly different note classes where no two notes are less than a whole-tone apart, i.e., $\forall i, j \ |t_i - t_j| >= 2$. We say T is *sorted* if $t_1 < t_2 < t_3$.

- [A4] A general trichord $T = [t_1, t_2, t_3]$ is a *major trichord* if some permutation, $[a, b, c]$, of T is such that $b - a = 4$ and $c - b = 3$. Likewise, T is a *minor trichord* if there is a permutation $[a, b, c]$ such that $b - a = 3$ and $c - b = 4$.

Given this foundation, we generalise Neo-Riemannian operators from the originals to work with the expanded definition of trichords.

- [B1] A *General Neo-Riemannian Operator (GNRO)* is an ordered triple of integers $[n_1, n_2, n_3]$ such that $\forall i \ (-2 \leq n_i \leq 2)$ and $0 \in \{n_1, n_2, n_3\}$.

- [B2] An application of a GNRO, $G = [n_1, n_2, n_3]$, to a general trichord, $T = [t_1, t_2, t_3]$, denoted $G(T)$, involves adding in place the three values of G to the three values of T, i.e., $G(T) = [t_1 + n_1, t_2 + n_2, t_3 + n_3]$.
- [B3] A *Compound GNRO* (CGNRO), $[G_1, G_2, \ldots, G_k]$, of *length* k, is a non-empty ordered list of k GNROs.
- [B4] An application of a CGNRO, $C = [G_1, G_2, \ldots, G_k]$, to a general trichord, T, denoted $C(T)$ involves the iterative application of each G_i to the previous result, i.e., $C(T) = G_k(\ldots G_2(G_1(T)) \ldots)$.

The original set of NROs given in Table 1(a) had the advantage that, given a major or minor chord as input, the output would also be major or minor. Unfortunately, that is not true of GNROs, i.e., given a general trichord as input, the output might not be a general trichord as per definition A4. Hence we specify the admissibility of a GNRO for a given chord below. Also, as we discuss in the following sections, to produce acceptable music to accompany a given video clip, requires the ability to produce situational chord changes at exactly the right dramatic moment in the target video. This is enabled by the use of the original NROs of Table 1(b), re-interpreted as CGNROs.

- [C1] A CGNRO, C, is *admissible* for a given general trichord, T, if and only if the resulting triple of notes, $C(T)$, is itself a general trichord, i.e., with three notes in distinct note classes and each pair of notes at least a whole tone apart.
- [C2] A CGNRO, C, is *m-admissible* for a given general trichord T if and only if $C(T)$ is either a major or a minor trichord as per definition A4.
- [C3] An ordered pair of two admissible CGNROs, $[C_1, C_2]$, produces a *situational chord sequence* from a given general trichord T, if C_1 is m-admissible, and C_2 is from Table 1(b). In effect, C_1 makes sure that the general trichord input to C_2 is a major or minor trichord, so the emotional effect of using C_2 is correct.

Importantly, the generative process also requires producing music in between the dramatic moments which doesn't include any striking chord changes that might obfuscate the dramatic chord change when it happens. The following definitions help describe how we try to achieve this:

- [D1] A *major key signature* for the note N is the list of note classes: $[nc(N), nc(N+2), nc(N+4), nc(N+5), nc(N+7), nc(N+9), nc(N+10)]$. A *minor key signature* for the note N is the list of note classes: $[nc(N), nc(N+2), nc(N+3), nc(N+5), nc(N+7), nc(N+8), nc(N+10)]$
- [D2] A general trichord, $T = [t_1, t_2, t_3]$, is in the (major or minor) key signature, K, of given note N if $\forall\, t_i \in T,\ nc(t_i) \in K$.
- [D3] Given a *focal note*, N, a general trichord, $T = [t_1, t_2, t_3]$, can be *focal mapped* to N by calculating $T' = [t_1', t_2', t_3']$ such that $\forall i,\ (t_i' \in \{1, \ldots, 127\}$ and $\nexists k \in \{1, \ldots, 127\}$ where $nc(k) = nc(t_i') \wedge |k - N| < |t_i' - N|)$. Informally, the mapping takes each t_i to the note in the same note class closest to N.

Soundtracks usually comprise more than a chord sequence, and GENRT affords the production of music with 8 tracks: three for the notes of a trichord, three for percussion, and one each for a melody and bassline. Described in more

Table 2. Number of admissible, m-admissible and identity CGNROs of length 1, 2 and 3 over all general trichords with notes in one octave. Number of different chords realised. Values have been rounded to the nearest integer.

len	num	Admissible			M-admissible			Identity			Different		
		min	av	max	min	av	max	min	av	max	min	av	max
1	60	16	31	51	2	9	12	0	1	2	15	31	51
2	3600	1242	1697	1920	212	405	552	60	72	95	102	142	195
3	216000	77151	94470	104535	15594	21880	25146	1248	1805	2955	220	314	412

detail below, the trichord sequence is dictated by the application of CGNROs, starting from a given general trichord. It can be given a rhythm so that the chord is played multiple times during a (given) timespan. The melody and bassline can mirror the trichord to selectively play a sequence of notes from the chord at different octaves, using a given rhythm. To describe this, we use the following definitions:

- [E1] A musical *episode* of a given duration (in ms), d, spans a chord sequence of N general trichords $[c_1, \ldots, c_N]$, with the duration (in ms) of chord c being denoted $dur(c)$, such that $\sum_{i=1}^{N} dur(c_i) = d$.
- [E2] For a given general trichord, T, and given a number of beats b that $dur(T)$ is to be split into, a *rhythm specification*, denoted $rh(T)$, for T is a set of pairs $(t/b, d/b)$ which each dictates that the chord should be played on beat t, and last for d beats.
- [E3] For a given general trichord $T = [t_1, t_2, t_3]$ and given the number of beats, b, that $dur(T)$ is to be split into, the *melody specification* for T, denoted $mel(T)$ is a set of triples $(t_i, t/b, d/b)$ where each $t_i \in C$. Each triple dictates that while chord T is playing, melody note $t_i + 12$ should start at beat t and last duration b. The *bassline specification* for T is the same as a melody specification, but specifies that notes $t_i - 24$ are to be played.
- [E4] A *duration fraction signature* is a list of fractions $[\frac{1}{p_1}, \ldots, \frac{1}{p_k}]$ such that each p_i is an integer and $\forall i, \; p_i < p_{i+1}$.
- [E5] Given a duration of d ms, a sequence of notes $N = [n_1, \ldots, n_j]$ to be played over d, and a duration fraction signature $F = [\frac{1}{p_1}, \ldots, \frac{1}{p_k}]$, we say that N *adheres* to F if the start time, s_i, of each n_i can be expressed as $s_i = d * \frac{f_i}{p_i}$ for some $p_i \in \{p_1, \ldots, p_k\}$ and integer, f_i, such that $f_i \leq p_i$ and $f_i \geq 0$.
- [E6] If a sequence of notes, S, with starting times $s_1, \ldots, s_n$ adheres to duration fraction signature $F = [\frac{1}{p_1}, \ldots, \frac{1}{p_k}]$, we say the *level of adherence* of S is the smallest $x \in \{1, \ldots, k\}$ for which $\forall i \; s_i = \frac{f_i}{p_x}$ for some integer f_i. As an example, suppose $F = [\frac{1}{1}, \frac{1}{2}, \frac{1}{4}, \frac{1}{8}]$ and a three note melody, M, is to be played over duration d, with notes starting at $d * \frac{0}{4}, d * \frac{2}{4}$ and $d * \frac{3}{4}$. Here, M adheres to F at level 3, as $\frac{1}{4}$ is the third fraction in F.

3.1 The Distribution of Admissible CGNROs

In order to guide the implementation of the GENRT system, we checked the distributions of admissible CGNROs of lengths 1, 2 and 3, as reported in Table 2. With an exhaustive search, we determined that there are 112 general trichords as per definition A4, with notes all between middle C and the B note above it. There are also 60 CGNROs of length 1, hence $60^2 = 3600$ of length 2 and $60^3 = 216,000$ of length 3. We applied each of the CGNROs of length 1 to each of the 112 general trichords and recorded the number of CGNROs that are admissible and the number which are m-admissible for each trichord. We did the same for length 2 and 3 CGNROs and recorded the minimum, average and maximum numbers of admissible and m-admissible CGNROs in Table 2. In addition, we determined how many applications of a CGNRO to a trichord were identity mappings resulting in no overall change in the three notes of the chord. Finally, we recorded how many distinct chords resulted in the application of the entire set of CGNROs to each trichord, again recording minimum, average and maximum values of this in Table 2.

We see that when CGNROs of length 1 are used, there are an average of 31 which can be used for a general trichord and this number greatly increases as the length of the CGNRO increases. However, there are only an average of 9 m-admissible CGNROs of length 1, and for some chords, this can be as low as 2. While this guarantees that every general trichord can be turned into a major or minor chord with a CGNRO of length 1, there is not much diversity. We also see that there is a lot of redundancy in the set of CGNROs. For instance, while there are 216,000 possible CGNROs of length three to apply to a given general trichord, they result in only 314 different chords, on average, and 1,805 of the CGNROs are the identity mapping, again on average.

4 The GENRT Music Generation System

Two initial requirements for generative soundtrack production are that musical changes happen at precise times and for tempo changes to increase and decrease tension in episodes. We have implemented in the GENRT system the following approach for generating music which fits a given video clip by ensuring that certain CGNROs are employed at given times which match emotional or situational changes in the video. The user can also specify ways in which the music is *damp-ened* between the dramatic changes, so that the chord changes the CGNROs impose have more effect, i.e., are not obfuscated by the surrounding music.

4.1 Event-Reflecting Musical Episodes

As per the definitions of Sect. 3, each soundtrack is composed of a series of episodes which each consist of a sequence of general trichords, played on one instrument. The soundtracks also have a melody and bassline played on (option-ally) different instruments, and three percussion instruments adding rhythm and texture. The user specifies the number of episodes that the composition will have. Each episode is specified to either have a particular duration or to end with a particular situational chord sequence (as above), with the final chord change happening at a precise number of milliseconds since the start of the composition. The user also specifies a target for the duration of the first chord (in ms) and a target for the duration of the last chord in each episode, along with the notes of the starting chord (or specifies that it can be a random general trichord). If the first and last chords of an episode differ in duration, then this indicates a gradual change in tempo during the episode, to be smoothly realised by GENRT.

Given these specifications, GENRT calculates the durations for a chord sequence as follows. Suppose the start chord duration target is s and the end chord duration target is e, with the final chord happening at time t to coincide with a situational/emotional change in the video. For a smooth linear transition of chord durations from s to e, GENRT determines the number of chords in the sequence, n, so that the number of milliseconds changes from one chord to the next is r and the final chord occurs at t. We note the following equations hold:

$$(i)\ t = \sum_{i=0}^{n-2}(s + ir) \qquad (ii)\ e = s + (n-1)r$$

The first equation states that the total duration of the first n-1 chords needs to equal the timestamp, t, when the final chord is to be played; and the second expresses the final chord duration, e, in terms of the start chord duration, s, and the variables n and r. Expanding (i) using the well-known formula for the summation of the first i integers, we see that:

$$(iii)\ t = (n-1)s + r\left(\sum_{i=1}^{n-2} i\right) = (n-1)s + r\left(\frac{(n-1)(n-2)}{2}\right)$$

We can also arrange (ii) to show that: (iv) $r = \frac{(e-s)}{(n-1)}$, and substitute this version of r into (iii), to get:

$$t = (n-1)s + \frac{(e-s)}{(n-1)}\left(\frac{(n-1)(n-2)}{2}\right) = (n-1)s + \left(\frac{(e-s)(n-2)}{2}\right)$$

enabling us to solve for n, giving:

$$n = \frac{2(t+e)}{(s+e)}$$

We can calculate r using n in equation (iv). Note that in practice, we need to take the floor, $\lfloor n \rfloor$, of the calculation for n above, as it is the number of chords in the episode, hence an integer. This introduces a discrepancy that means the final chord doesn't happen at the correct time t. To correct this, a value of x milliseconds is added to the duration of each of the chords. x is calculated by using $\lfloor n \rfloor$ in equation (iv) to calculate r, which is in turn used along with $\lfloor n \rfloor$

in equation (iii) to calculate a value t'. The difference between user-given t and t' is divided by $n - 1$ to give x. This ensures that the timing of the final chord is at exactly time t, but means that the start and end chord durations are often not exactly as targetted by the user, but are usually very close. Note also that, if the user only supplies a duration, d, for the episode and not a timing for the final chord, GENRT calculates n and r in a similar way using d instead of t.

The use of episodic music production is vital for the application of GENRT to soundtrack generation. By enabling the user to specify the end chord duration for one episode to be the start chord duration for the next episode, they smoothly merge during the soundtrack, allowing tempo changes to reflect longer-term atmospheric changes. This supplements the reflection of dramatic or situational changes in the video with CGNRO-generated chord changes. The user can also specify start and end volumes for each of the instruments (chords, melody, bassline and percussion) in each episode. The user specifies which MIDI instrument each track has and we use the Fluidsynth sound font (fluidsynth.org) for these. Instruments are specified on a per-episode basis, and the user can also specify a number of chords in advance of the episode change that the change of instruments begins. GENRT then uses two tracks to interpolate the volume of the latter episode's instruments from zero to one during this period, likewise interpolating the former episode's instrument volume from one to zero.

4.2   Chord Sequence Generation

Once the number and duration of chords in an episode have been calculated, GENRT generates a random sequence of CGNROs with which to generate a chord sequence. Starting with the user-given chord, e.g., middle C, E, G of C major, the first CGNRO is chosen randomly from those which are admissible for this general trichord. Applying this CGNRO produces a new chord, and the process iterates until near the end of the episode. The user can specify that a single dramatic/situational event happens per episode. To do so, they supply the timestamp in milliseconds from the start of the soundtrack that this happens at and a keyword from the list of Table 1(b): [Antagonism, Sorrow, Romance, Threat, Wonderment, Mystery, Heroism, Fantasy, LifeAndDeath].

GENRT ensures that the corresponding CGNRO from Table 1 is applied to generate the final chord of the episode via an appropriate situational chord change (definition C3). Given that the audible power of the chord change has only been considered for major or minor chords, GENRT must ensure that the penultimate chord to which this CGNRO is applied is either major or minor. To do this, it chooses the penultimate CGNRO randomly from the list of m-admissible ones available to apply to the anti-penultimate chord. We note from Table 2 that there will always be at least two m-admissible CGNROs to choose from. The user is able to provide a different rhythm specification (definition E2) for the chords of each episode, so that over the chord duration, the notes of the chord are played rhythmically, if desired.

4.3 Bass, Melody and Percussion Generation

Once a chord sequence has been generated using random CGNROs, GENRT generates a bassline and melody to fit the chords. These processes are currently fairly simplistic and we plan to improve upon them in future work. For the bass, the user provides a bassline specification as defined in E3, for a given trichord, that dictates which of the three notes of the chord are played and at what time during the chord duration, at two octaves below the chord note. The user can also specify how staccato the notes are by providing a cutoff proportion so that each note is curtailed before its full duration.

For melodies, the user can also provide a melody specification as per E3 which will generate note sequences for a chord an octave above its notes. Alternatively, they can specify that GENRT itself produces a voice-leading melody. It does this in two stages over all the episodes of the composition. Firstly, for the first chord of the first episode, it chooses a random permutation of the chord's three notes, so that the notes an octave above become the *backbone* of the melody for that chord. For each subsequent chord, GENRT chooses a permutation of the chord's notes for which the first note is as close to the last note of the melody for the previous chord as possible.

In the second stage, passing notes are added between every pair of backbone notes (b_1, b_2) for each chord's melody. Assuming that $b_1 < b_2$, starting at $b_1 + 2$, passing notes which are a whole-tone away from the previous one are added until $b_2 - 2$ is reached. If the final backbone note of a chord's melody is the same as the first of the next chord's melody, the former is removed. If not, then more passing notes are added to join the end and start notes accordingly. As before, the user can specify a cutoff proportion, so that the melody sounds more staccato if required. Three different percussion instruments can be specified and utilised by the user providing rhythm specifications for each, as per definition E2.

4.4 Dampening the Music Between Events

In initial tests with the extended generative NRT approach, we found that it was often hard to hear the emotional/situational chord changes because there were many similar chord changes in advance which somewhat obfuscated things. One way to address this is to make the intermediate music less striking, and we implemented some techniques in GENRT that users can experiment with for this purpose, as follows.

Firstly, the minimum and maximum length of the CGNROS that are employed in generating the chord sequence can be set. As the CGNROs get longer, the further the output chord notes are from the input ones, and the more audibly striking the chord sequence is. Hence, employing only CGNROs of length 1 or 2 is advised for subdued intermediate music. The user can also specify a major or minor key signature as a *fixed key* for the chords in the episode. As per definition D2, a general trichord is in key K if the note class of each of its notes is in K. GENRT can then choose only CGNROs where the output chord is in the fixed key. We have found that this significantly dampens the music.

Note that the emotional/situation CGRNO specified in an event is exempt from this rule, which makes the chord it produces more impactful. In addition, when GENRT produces voice-leading melodies, it can be told to ensure that all the passing notes are in the chord sequence's fixed key, if there is one.

Sometimes the random nature of CGNRO selection that GENRT employs can produce pitch drift, in that the chords go consistently up (or down) in pitch, dragging the whole composition with them. While this is something that users may want to add by design, it can be striking and create music which is inappropriate for an episode. Hence, we added the option for GENRT to map each generated chord to a user-given focal note, as per definition D3. This ensures that all chords stay in roughly the same pitch range.

When GENRT produces a voice-leading melody $M = [m_1, \ldots, m_n]$ over a chord, by default it plays each note of the sequence at equal intervals over the duration of the chord, d. Often, this clashes with the static, user-specified rhythms that the chords, bassline and percussion instruments play. That is, the number of notes in M means there are polymetric rhythms at play, for instance, five melody notes played at the same time as four chord repetitions. This can be quite striking, so we added the option for users to suppress this by specifying a duration fraction signature $F = [\frac{1}{p_1}, \ldots, \frac{1}{p_k}]$ as per definition E4. Then GENRT takes the generated melody for a chord and produces a list of start times $[s_1, \ldots, s_n]$ with an optimally high level of adherence to F, as per definition E6. To do this, it finds the largest fraction $\frac{1}{p_x} \in F$ which is smaller than or equal to $\frac{1}{n}$ and sets each s_i initially to $\frac{i-1}{p_x}$, with each note initially having duration $d * \frac{1}{p_x}$ over d.

This can mean that the total duration of the notes is less than d, and melodies end early for a chord, which is noticeable and undesirable. Hence GENRT randomly chooses a note to extend the duration of, and this is done by adding $\frac{1}{p_x}$. Accordingly, GENRT increases the starting points of the notes following the extended one. It does this until the total duration of the notes is d. Noting that passing notes in a melody can often clash with the chord being played at the same time, we further refined the process so that only the backbone notes of the melody (which don't clash) were extended. This quantization of the note starting points means that a duration fraction signature can be specified which fits with the static rhythms of the chords, bassline and percussion, e.g.,. if they only start at moments which are multiples of $\frac{1}{8}$ of d, then a duration fraction signature of $F = [\frac{1}{1}, \frac{1}{2}, \frac{1}{4}, \frac{1}{8}]$ will ensure that the melody notes all do likewise. Optimal adherence level to F (definition E6) means that the melody notes will have the simplest rhythm.

Trying these optional specifications for music generation, we found that we could hear the dampening of music reasonably well. However, to be more concrete and confident in this, we performed some experiments to quantify the benefits in this respect. As per the previous experiments, we exhaustively applied all CGNROs up to length 3 on all 112 general trichords with notes between middle C and the B above. However, in a second session, we stipulated that only input chords where all notes in the fixed key of C major were considered, and the result was only recorded if the output was likewise in the key of C major. This simulates the usage of a fixed key to dampen chord sequences. In a third session, we instead mapped all the input chords and output chords to focal note 66. To measure the dampening effect in these sessions, we used two measures:

Table 3. Number of admissible and m-admissible CGNROs, chord distances and focal note distances for standard, fixed key and focal mapping sessions.

session	len	Admissible			M-admissible			Chord dist			Focal dist		
		min	av	max	min	av	max	min	av	max	min	av	max
standard	1	16	31	51	2	9	12	0.00	0.78	1.33	1.33	3.11	6.00
	2	1242	1697	1920	212	405	552	0.00	1.00	2.67	1.33	3.22	7.33
	3	77151	94470	104535	15594	21881	25146	0.00	1.17	4.00	1.33	3.33	8.67
fixed key	1	5	9	13	0	3	5	0.00	0.74	1.33	1.67	3.14	4.67
	2	337	451	518	36	118	178	0.00	0.78	2.33	1.67	3.25	6.33
	3	18002	21961	24272	3546	5723	6813	0.00	0.99	4.00	1.67	3.39	7.67
focal mapped	1	16	31	51	2	9	12	0.00	0.88	3.00	1.33	3.00	4.67
	2	1242	1697	1920	212	405	552	0.00	1.07	4.00	1.33	3.00	4.67
	3	77151	94470	104535	15594	21880	25146	0.00	1.16	5.00	1.33	3.00	4.67

- *Chord distance* measures the distance of a chord, O, output from a CGNRO from the input chord, I. To do this, for each note of O, it determines the closest note (or notes) in I, and calculates the absolute distance to it. An average over the three input notes of the input general trichord gives an overall distance. The intuition here (as highlighted by analytical NRT) is that the further a chord is away from the previous one, the more striking the chord change is.
- *Focal point distance* measures the distance of the output chord from a given focal note N. It does this by averaging the individual absolute distances of each note from N. The intuition here is that a series of chords with pitches occurring in a narrow range will be less striking than a series within a larger pitch range.

The number of admissible and m-admissible CGNROs of lengths 1, 2, and 3 for the exhaustive standard session, fixed-key session and focal-note session are given in Table 3. Note that the standard session is the same as in Table 2 above, for easy comparison. We see that, as expected, the mapping to a focal note doesn't alter the number of admissible CGNROs. However, if the chords of an episode are constrained to those in a particular key, then the average number of admissible and m-admissible CGNROs available for a chord drops considerably, and it may be sensible to allow longer length CGNROs to maintain a level of diversity in the output. Table 3 highlights a procedural problem, i.e., that if the user employs a fixed key and only CGNROs of length 1 (for high dampening), then for some chords, the number of m-admissible CGNROs is zero, hence it would not be possible to produce a major or minor chord ready for a CGNRO from Table 1, to reflect an emotional or situational change. To fix this when it happens, we allow GENRT to temporarily increase the CGNRO length to 2.

The distance metrics are also given for the three sessions in Table 3. We see that, as expected, average chord distance increases along with CGNRO length allowed, from 0.78 (length 1 in the standard session) to 1.00 (length 2) and 1.17 (length 3). There are similar increases in the other sessions. Hence, users can control the amount of dampening through experimentation with CGNRO lengths. Fixing a key lowers the average chord distance to 0.74, 0.78 and 0.99 for CGNRO lengths 1, 2, and 3 respectively, which provides audible dampening.

The average distance of a chord from a focal note also increases with CGNRO length, but, as we can see in Table 3, the focal mapping routine minimises this. We note, however, that focal mapping considerably increases the maximum distance of an output chord from the input chord, because the mapping can occasionally move chords a lot. We see that the average chord distance in the focal mapped session was similar to that in the standard session, but the maximum chord distances are higher, hence there is a higher likelihood that a striking chord change could happen in an episode with focal mapping. Users should therefore employ focal mapping carefully and monitor the chord distances. In future, we intend for GENRT to use distance measures when selecting CGNROs.

5   An Illustrative Example

GENRT is a python program which produces music in MIDI form. It handles timings by turning chord durations into MIDI ticks, and we chose 960 ticks per second, so that millisecond timings can be passed on to the MIDI with high accuracy. The system is in development, and is not sophisticated enough yet to produce particularly beautiful or innovative music. However, returning to the problem setting in Sect. 1, our purpose here is only to show that (a) GENRT produces accurate soundtrack music for a video clip, in terms of the timing of episodes and events which match the media exactly (b) users have satisfactory levels of control over the atmosphere of the music in the episodes, and GENRT can smoothly change the mood over the period of an episode (or multiple episodes) and (c) particular emotional or situational events in the media are highlighted in audible ways with the generated soundtracks.

To test GENRT, we asked a trained musician (second author) with a background in composing soundtracks to create event-based music for a clip of the movie A Beautiful Mind [8]. This is a biographical drama that narrates the story of the mathematical genius John Nash. After making an incredible discovery, the protagonist is diagnosed with paranoid schizophrenia, and the story follows Nash on a journey of self-discovery, balancing mental health and his job as a cryptographer. The soundtrack is tasked with representing the workings of Nash's incredible mind, and its struggles. The composer of the movie's soundtrack, James Horner, represents Nash's genius with a series of beautiful and fast-moving chord progressions throughout the soundtrack. Music theorist

Frank Lehman wrote about the representation of genius in A Beautiful Mind's soundtrack [10] and used NRT to analyse its chord progressions. Horner's chord progressions use chromatic triadicism, especially the L, R and S NROs (see Table 1(a)), hence are difficult to accurately analyse using traditional music theory.

In the 2 min, 26 s film clip we chose, John Nash's antagonist and his group of friends are playing Go, when they encounter Nash. The protagonist is challenged to play a game of Go against his adversary while the group of friends watches. Nash is sure he is going to win, and the game seems to suggest that too, but to his surprise, he loses. There is a situational change in the clip to be highlighted in the soundtrack, at 69 s, when the game starts, and an emotional change at 112 s, when Nash loses and is antagonised by his opponent.

Fig. 1. GENRT chord sequence for the film clip from A Beautiful Mind, with chord start times, episodes, and the user-selected CGNRO for two transitions.

The musician employing GENRT used four different episodes for the soundtrack generation, each with different specifications to reflect the on-screen events and emotions. The first episode of 20 s generates the music for the preamble to the scene. For this episode, the musician specified the starting chord of C major and further employed a fixed key of C Major for the chord sequence. This forced the music to be somewhat dampened, as the scene during this period is just a preparation for the game. There was no event set for this episode. The volume faded in slowly during this episode, setting an appropriate atmosphere and introducing the soundtrack. To strengthen the emphasis on the introduction setting the scene, the chord progression moves quite slowly at 4000 ms per chord.

The second episode covers the period leading up to the game starting. To ensure continuity between the first and second episodes, the start chord duration was set to 4000 ms, and increases to 3000 ms during the episode, to raise the tension until the game starts. The key fix on C Major is continued, so the music is dampened and the change to the next episode, which generates the chord change for when the characters play the game, is audible. The event of this episode happens when the game starts, at 69,000 ms, and the musician chose the mystery CGNRO (see Table 1(b)) to reflect intrigue at the start of the game.

The third episode generates the music played during the Go game. The key fix functionality is turned off here so that the tool generates music which uses triadic chromaticism, using CGNROs up to length 3, which adds some disorienting tension during the game. As the game progresses, the music gets faster until the event of the episode at 112,000 ms. This is when Nash loses the game, and is represented by the antagonism CGNRO. In the fourth episode, the musician emphasised Nash's loss by turning key fix functionality back on and instantly reducing the tempo to much more slow-moving chords. As the scene comes to an end, the volume fades out to zero. The musician maintained consistency across episodes, with strings for chords, violin for the melody and cello for the bass. As per the original soundtrack, percussion was turned off. Each episode had chords fixed to focal note 48, to keep the music in a suitable pitch range.

The chord sequence and timings for the Beautiful Mind clip accompaniment are given in Fig. 1. We see there 41 chords and inspection of the start times of these shows that (a) the episode durations, situational change at 69 s [chord 19] and emotional change at 112 s [chord 35] were accurate to within a few milliseconds, and (b) the increase in tempo was smooth. On listening to the soundtrack, the music was more calm leading up to chord 19, more turbulent during the third episode and calmer towards the end of the soundtrack, as desired by the musician. The mystery CGNRO – reflecting the Go game starting – successfully interrupted the tonal chord progressions from the previous two episodes at chord 19, making the start of the game aurally noticeable. Chord change 35 (antagonism) was less prominent but still noticeable. Overall, the musician was satisfied with the accuracy, atmosphere and influence of the soundtrack.

6   Conclusions and Future Work

We have presented the GENRT system for automatically producing soundtracks for media such as video clips, and shown that it has promise for producing accurate, atmospheric and influential music. There are many next steps for this implementation, including: exposing more parameters for users to control the generation with; increasing the sophistication of melody generation and adding counterpoint production; and stronger dampening of the music between events. We also plan to combine this symbolic approach with deep learning so that a pre-trained neural model can be used to select musical sequences in terms of which fits best to a distribution, hopefully increasing the quality of the output. We may employ evolutionary and other search techniques to alter episode specifications (rather than musical choices), as part of this. Soundtracks often include musical *leitmotifs*, [3] [14], which consistently portray a character, and we intend to experiment with the generation and deployment of these in GENRT.

Once GENRT is somewhat more sophisticated, we will undertake a first round of user studies with novice composers, to assess its utility and future directions for its functionality. We also plan to carry out experiments on parameter tuning and checking correlation with other musical metrics in the output.

To describe and implement GENRT, we extended generative Neo-Riemannian Theory in various ways, including generalising the trichords which can be generated for a chord sequence. This required a number of adjustments in terms of the operation of the CGNROs. We also introduced and implemented ways in which the application of CGNROs to chord sequence generation could be dampened so that intermediate music was more in the background. This was in order to avoid obfuscating important chord changes which reflect events in the media.

As our requirements for GENRT increase, we will continue to develop the formalism to capture how generative NRT can be used to produce music. In particular, we intend to capture the interplay of chord sequences and leitmotifs with the formalism. We plan to involve GENRT in some multi-modal projects, where generative AI techniques are used to produce text and videos, with GENRT providing accompanying soundtracks. We are also interested in producing interesting and novel stand-alone music using GENRT. Our aim is to produce music that has direction and purpose, as with the best human-made compositions.

Acknowledgments. This work has been funded by the UKRI as part of the "UKRI Centre for Doctoral Training in AI and Music", under grant *EP/S022694/1*. We would like to thank the anonymous reviewers for providing helpful feedback.

References

1. Amram, M., Fisher, E., Gul, S., Vishne, U.: A transformational modified Markov process for chord-based algorithmic composition. Math. Comput. Appl. **25**(3), 43 (2020)
2. Bernardes, G., Cocharro, D., Guedes, C., Davies, M.: Harmony generation driven by a perceptually motivated tonal interval space. Comput. Entertain. **14**(2), 2991145 (2016)
3. Bribitzer-Stull, M.: Understanding the leitmotif: from Wagner to Hollywood film music. Cambridge University Press (2015)
4. Cardinale, S., Colton, S.: Neo-Riemannian theory for generative film and videogame music. In: Proceedings of the Int. Conference on Computational Creativity (2022)
5. Chuan, C.H., Chew, E.: Generating and evaluating musical harmonizations that emulate style. Comput. Music. J. **35**, 64–82 (2011)
6. Cohn, R.: Neo-Riemannian operations, parsimonious trichords, and their Tonnetz representations. J. Music Theory **41**(1), 1–66 (1997)
7. Eladhari, M., Nieuwdorp, R., Fridenfalk, M.: The soundtrack of your mind: Mind music - adaptive audio for game characters. In: Proceedings of the ACM SIGCHI International Conference on Advances in Computer Entertainment Technology (2006)
8. Howard, R.: A beautiful mind, universal pictures (2002)
9. Hutchings, P.E., McCormack, J.: Adaptive music composition for games. IEEE Trans. Games **12**(3), 270–280 (2020)

10. Lehman, F.: Transformational analysis and the representation of genius in film music. Music Theory Spect. **35**(1), 1–22 (2013)
11. Lehman, F.: Film music and neo-Riemannian theory. Oxford Handbook (2014)
12. Monteith, K., Francisco, V., Martinez, T., Gervás, P., Ventura, D.: Automatic generation of emotionally-targeted soundtracks. In: Proceedings of the International Conference on Computational Creativity (2011)
13. Monteith, K., Martinez, T., Ventura, D.: Automatic generation of music for inducing emotive response. In: Proceedings of the International Conference on Computational Creativity (2010)
14. Whittall, A.: Leitmotif. Oxford University Press (2001)
15. Wiggins, G.A.: Automated generation of musical harmony: What's missing. In: Proceedings of the International Joint Conference on Artificial Intelligence (1999)

Is Beauty in the Age of the Beholder?

Edward Easton[✉], Ulysses Bernardet, and Anikó Ekárt

Department of Computer Science, Aston University, Birmingham, UK
{eastonew,u.bernardet,a.ekart}@aston.ac.uk

Abstract. Symmetry is a universal concept, its unique importance has made it a topic of research across many different fields. It is often considered as a constant where higher levels of symmetry are preferred in the judgement of faces and even the initial state of the universe is thought to have been in pure symmetry. The same is true in the judgement of auto-generated art, with symmetry often used alongside complexity to generate aesthetically pleasing images; however, these are two of many different aspects contributing to aesthetic judgement, each one of these aspects is also influenced by other aspects, for example, art expertise. These intricacies cause multiple problems such as making it difficult to describe aesthetic preferences and to auto-generate artwork using a high number of these aspects. In this paper, a gamified approach is presented which is used to elicit the preferences of symmetry levels for individuals and further understand how symmetry can be utilised within the context of automatically generating artwork. The gamified approach is implemented within an experiment with participants aged between 13 and 60, providing evidence that symmetry should be kept consistent within an evolutionary art context.

Keywords: Aesthetic judgement · 3D Art Generation · Gamification · Virtual Reality

1 Introduction

"Beauty is in the eye of the beholder", this famous quote attributed to Margaret Wolfe Hungerford is often applied to discussing the beauty of individuals; however, it is also applicable to the judgement of some genres of artwork. The subjectivity of the platform is one of the few things which is considered an absolute truth about art, even as far back as Plato, who suggested that only highly educated individuals could truly understand art and therefore were the only ones who could create it. He held this view to such an extent he believed *the Iliad* was not true art, an opinion not shared by others. More modern attempts to define the process consider a multitude of aspects internal and external to the artwork itself such as the emotions invoked by the artwork in the viewer and the intended emotion conveyed by the artist. These numerous aspects all interact with each other, suggesting that the appreciation of artwork can be considered

ⓒ The Author(s), under exclusive license to Springer Nature Switzerland AG 2023
C. Johnson et al. (Eds.): EvoMUSART 2023, LNCS 13988, pp. 84–99, 2023.
https://doi.org/10.1007/978-3-031-29956-8_6

a holistic process, where all aspects contribute in varying ratios. The potential that the appreciation of art considers a high number of largely unknown factors introduces three questions:

1. What are the contributors to the appreciation of artwork?
2. How are different contributing aspects influenced by one another?
3. Can these aspects be effectively modelled to represent an individual's aesthetic preference and used to auto-generate artwork specific to them?

Understanding which aspects contribute to aesthetic judgement can be a complicated process as it can be very difficult for individuals to articulate their thoughts when judging a piece of art. Being able to do so is a skill which is learnt over a long time, reserved for people with a high level of expertise in art. However, as suggested in [16] art expertise itself is an aspect which can influence the remaining aspects, so this introduces an additional question, is it possible to extract this information without requiring a high level of art expertise, to avoid potentially biasing the data with these additional influences? One approach to solving this is to use a gamified approach to art appreciation to help extract which aspects influence the appreciation and judgement of artwork, avoiding the need to ask an individual to describe their preferences. This process would allow the identification and the ability to draw general conclusions about the different aspects which influence the decision-making process.

This multi-faceted nature of aesthetic judgement means that it is not the result of a single judgement or aspect [20], which is reinforced by establishing that multiple pathways of the brain are used when making aesthetic judgements, including pathways which are not linked to the physical properties of the object [31]. Between aspects such as emotion [27], perceived back story [8] and more stable aspects such as the fractal dimension [32], it suggests that all aspects can be split into three categories: (1) Constant, (2) Subjective and (3) Transient. The constant aspects are those that can be considered the same across all individuals, the subjective aspects are those which may be different across individuals but change relatively little within an individual and the transient aspects are those which will change within an individual's judgement process within a short period of time, i.e. an hour.

This paper presents a novel gamified approach to further understand the role that symmetry plays within the appreciation of 3D virtual sculptures, obtaining the categorical placement of symmetry as well as any patterns of preferred values. The remainder of the paper is organised as follows: Sect. 2 provides an overview of how symmetry is treated within artistic appreciation, specifically looking at how artwork has been auto-generated, Sect. 3 details how the gamified approach was created to investigate aesthetic judgement, Sect. 4 provides an evaluation on how effective the presented process was at allowing participants to indirectly choose a value for their preferred symmetry level and the results of what levels of symmetry are preferred across age ranges, the discussion is presented in Sect. 5 and the conclusions in Sect. 6.

2   Art Appreciation Approaches

Despite the difficulty, numerous approaches have been used to identify and assess which aspects contribute to the assessment of a piece of art, these can be split into theoretical and formal approaches. The theoretical approaches are largely based on psychological or philosophical perspectives and the formal approaches on computational methods. While the latter approach often leads to specific measures of 2D and 3D artwork, utilised in the automatic generation of art, the former uses empirical evidence to make general conclusions about the aspects and the judgements being made.

Within the theoretical approaches, different methods have been used such as analysing critiques of artwork and identifying common terms used to judge artwork [30] and looking at specific attributes about an individual such as their level of art expertise [7], equating physical manifestations of emotion to link to the aesthetic experience [23] or how the aesthetic appreciation of visual media changes over time [14].

The formal approaches cover a wide variety of aspects, often being derived from real-world data such as examining existing artwork and extracting different features of the images [1, 2, 17, 18, 26, 28, 32–34]. One of the most well-known attempts at a formal equation of aesthetic beauty was suggested by Brikhoff [6], who believed that an artwork could be considered in terms of the ratio between its order and complexity. This indicates the importance order has in art judgement, due to its major role in processing visual information, where the more ordered an item is, the easier it is for our brains to process [5]. Extensive research has been performed resulting in different methods being suggested for quantifying it, most attempts to define the Order relate it to the symmetry of the artwork, for reasons summed up by Weyl "Symmetry is one of the ideas by which man through the ages has tried to comprehend and create order, beauty and perfection" [37]. Due to its near universal importance across multiple research fields, such as architecture [19, 29, 38], inter-human attraction [24], user interface design [3] and the extensive work looking at it within artwork [4, 12, 13, 21, 22, 25, 35], it is a safe assumption that symmetry plays a role in the evaluation of artwork. However, its relative importance and placement into a single category is not straightforward as it needs to be understood how symmetry fits within the wider context of art appreciation. In order to correctly place symmetry, evidence needs to be collected which considers how people judge artwork when different levels of symmetry are present. It may be expected that symmetry falls within the constant category, however, there are still a variety of factors which can influence an individual's symmetrical preference [15, 16, 36]. If similar symmetry preference levels are found across a wide range of ages, then symmetry can be considered a constant aesthetic and forms a stable part of the aesthetic appreciation of a piece of art.

3   Helping People Release Their Inner Art Critic

To facilitate identifying an individual's symmetry preference when judging artwork, a novel gamified approach was developed utilising a custom VR application. At each step, this VR application expects users to make binary decisions between two sculptures displaying two different levels of symmetry, to elicit the users' aesthetic preferences. Each decision leads to updating the symmetry displayed in the next pair of sculptures in the VR application bringing the symmetry values of the displayed sculptures closer together, converging on the user's preferred value. The sculptures displayed in the first step have perfect symmetry 1 and no symmetry 0, respectively. At each step, the symmetry level the participant chooses is displayed again in the next pair of sculptures, however the other sculpture has it's symmetry level amended by 0.2, it is decreased if the participant chooses the lower symmetry value sculpture and increased if the participant chooses the higher symmetry value sculpture, thus the user's symmetry preference is established in five steps. To improve confidence in the recorded symmetry preference of the participant, the five steps starting from sculptures with perfect symmetry and no symmetry, respectively, through to convergence to one symmetry value are repeated with different sculptures corresponding to the same set of symmetry values at each step, meaning the participant makes ten choices to ascertain a preference level. The tree displaying all possible paths of symmetry value pairs that the participant can follow is shown in Fig. 1. Once all the choices have been made the participant is presented with a (new) final sculpture which displays the participant's chosen symmetry level calculated as the mean of the outcomes of the two sets of choices. At the end, the participant is asked to confirm how well the final sculpture matched their preference. This acts as a check to determine whether the participants' preference was captured.

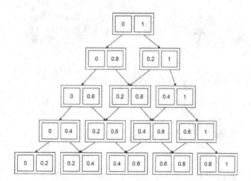

Fig. 1. Decision tree showing all paths which could be followed by participants

3.1   Generating the Sculptures

Each choice the user makes is between two sculptures generated with the Axial Generation Process [9], with a couple of limitations: using a reduced set of expressions in the genotype expression trees and not using colours, instead, a neutral stone grey colour was used to render them in the VR environment. To provide the ability for the user to choose between sculptures at differing levels of symmetry, six genetic algorithm runs were performed to create items which displayed symmetry between 0 and 1 in steps of 0.2, with the symmetry level calculated using a measure based on point clouds [10]. This measure works by finding multiple candidate symmetry planes and then refining their positions by minimising the error between points reflected in the plane. The score calculated for each item is the average distance error across all detected planes for an item.

From each run of the GA, parameters shown in Table 1, the sculptures displaying symmetry values approximating the target values were then automatically chosen. The final sets were chosen by the experimenters based on how interesting the sculptures looked, a selection of the sculptures at each symmetry level is shown in Fig. 2. To avoid uninteresting sculptures being initially generated, the complexity, measured as the Global Normalised Kolmogorov Complexity [26] which calculates the compressibility by encoding all points into text and then compressing the string using ZIP compression, similar to the process used by [9,11], was also maximised as part of the generation process.

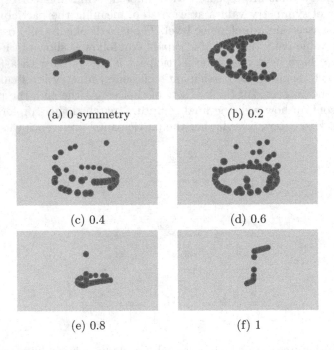

(a) 0 symmetry   (b) 0.2

(c) 0.4   (d) 0.6

(e) 0.8   (f) 1

Fig. 2. Examples of the standard sculpture at each available symmetry level

Creating the Reward The process to generate the final sculptures followed a similar method, but created larger and more diverse sculptures, intended as a reward for navigating through the caves. The mean symmetry preference across the two sets of choices was calculated in 0.1 steps, the final sculptures also reflected this by being generated using 11 genetic algorithm runs, each focusing on a value from 0 to 1 in 0.1 steps, using the parameters shown in Table 2. Five

Table 1. Evolutionary Algorithm parameters

Total Generations	500
Mutation Probability	0.7
Initialisation Method	Full
Selection Method	Tournament (k=3)
Population Size	50
Fitness Measure	Average Symmetry

sculptures were chosen at each level of symmetry to provide a wide range of options which could be displayed to the user in the final section of the experiment, examples of the final sculptures are shown in Fig. 3. The pre-generation of the final sculptures ensured that the game experience was smooth, with no waiting period for live generation of the final sculpture. As it was solely the final symmetry value that would drive the generation of the final sculpture and there were only 11 possible values, the pre-generation of the set of possible final sculptures at each symmetry level did not detract from the quality of the observations.

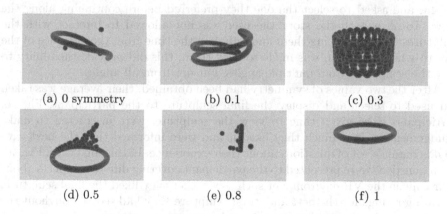

(a) 0 symmetry   (b) 0.1   (c) 0.3

(d) 0.5   (e) 0.8   (f) 1

Fig. 3. Examples of the final sculpture

Table 2. Evolutionary Algorithm parameters

Total Generations	500
Mutation Probability	0.7
Initialisation Method	Full
Selection Method	Tournament (k=3)
Population Size	50
Fitness Measure	Average Symmetry

3.2   Making an Engaging Environment

There are many facets of VR which could be used to facilitate the collection of data, however, due to including a wide range of ages, one of the main factors to consider in this experiment was how to make the experience engaging. In order to do this, a mine cart theme was chosen, where the user would travel along a track through a series of caves using a simple control mechanism, where a participant can lean in four directions to control their avatar. The pitch (forward/backwards) controls the speed of the cart and the Yaw (left/right) allows the user to choose a sculpture or select an answer to a question, depending on the stage in the experiment.

Upon starting the experiment the participant was given an introduction to what the experiment would consist of, followed by basic training on operating the cart, making sculpture choices and answering questions within the system. Once the training had been completed, the participant was then free to travel through the caves at their own pace. The participant would then complete the experiment by travelling through two types of cave, scenic caves and sculpture caves in an alternating manner. The scenic caves were intended to help keep the VR environment engaging, by presenting a unique, fun theme in each one, Fig. 4c. After going through the scenic cave the participant found themselves in the sculpture cave and were presented with a choice between two sculptures, Fig. 4a, and asked to select the one they preferred before continuing along the track. To align with the story the user was not allowed to interact with the sculptures e.g. by rotating them and only had the time from the entrance of the cave to when a decision was made to view them, this did provide the ability to view the sculptures from multiple angles, but not from all angles.

After the two values of symmetry had been obtained, their average was taken and used to select and display the final sculpture to the participant, Fig. 4b. Participants were given time to view the sculpture, were instructed to make a judgement on how much they liked it and were informed that the next cave would contain a set of questions about their experience. In this final cave, Fig. 4d, seven questions were presented to the participant covering different aspects about their time in the VR environment such as whether they liked the final sculpture, providing ratings for the best and worst sculpture they had seen throughout the experience and whether they experienced any motion sickness.

Experimental Setup. The experiment itself was run in two phases, the first phase was run entirely online, with participants sourced through the Prolific platform who had access to their own VR headsets. This phase specifically targeted individuals 18 years old and over with no specific age limit, however by requiring that the participant had access to their own VR headset, this potentially limited the available age range of participants, being skewed to younger adults who are more likely to own a VR headset. In addition to this due to reducing the complexity of development, only two versions of the VR environment were created, running on the HTC Vive and Meta Quest headsets respectively.

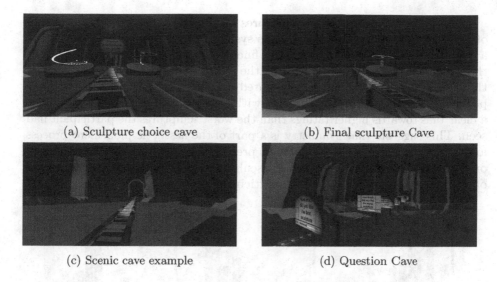

(a) Sculpture choice cave   (b) Final sculpture Cave

(c) Scenic cave example   (d) Question Cave

Fig. 4. The main caves to collect information from the participant

The Meta Quest version[1] was made available via the App Labs section of the Oculus store and the Vive app was available as a download when the participants registered for the experiment. The second phase was completed at the CityFest event, run by Aston University, showcasing technology projects to school children aged 11–13, and Meta Quest 2 headsets were used for the experiments.

For the first phase, in order to run the experiment using the Prolific platform, the available headsets on the platform needed to be determined. This was achieved by running a short questionnaire which asked participants to list the VR headsets they had access to. From the results of this questionnaire, the correct participants could be targeted on the Prolific platform, where the users were able to download the appropriate version for the type of headset they owned.

4 Results

A total of 33 participants completed the experiment (13 female, 20 male) between the ages of 13 to 52 (m=28.0). Three main paths of analysis were undertaken, validating the process itself; looking at the overall level of symmetry preferred by the participants across different age groups and finally investigating the participant's behaviour within the system.

4.1 Process Validation

As this was a new approach to extracting aesthetic preferences from participants, the first thing to check is that the process itself was valid and created suitable results to analyse the aesthetic preferences of the participants.

---

[1] https://www.oculus.com/experiences/quest/3429965927112413.

Looking at the ratings of the sculptures within the system gives an indication of whether the process was valid. If the system was able to correctly determine a preferred symmetry level, then the final sculptures should be rated higher than others. The results collected from the participants, Fig. 5, show that whilst the final sculptures were not always rated as highly as the best sculptures the participant had seen during their time within the Virtual Environment, there is a clear bias towards higher ratings than the worst sculptures the participant had seen. This suggests that symmetry is a part of the aesthetic judgement process and in addition to this that the process presented here did have a positive effect on how much the participant liked the sculptures, showing that a preferred level of symmetry was obtained for each participant.

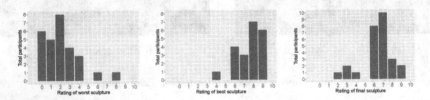

Fig. 5. Overall aesthetic preference ratings provided for the sculptures in the system: left) Ratings for the worst sculpture, middle) Ratings for the best sculpture, right) Ratings for the final sculpture

4.2   Symmetry Preference

The analysis of the process indicates that confidence can be had in the collected data and use it for assessing the main purpose of this experiment, extracting an individual's preference for symmetry.

Similarly to health related age groupings in scientific studies, and also inline with the expectation that preference changes with experience and experience grows with age, the participants were grouped into under 18, 18–30 and 31+. Table 3 shows simple statistics of the age groups including the mean of the chosen symmetry for each group, while the full distribution is shown in Fig. 6a.

Table 3. Details of the symmetry values for the three age groups

Group	Mean Symmetry	Symmetry difference	Min Age	Max Age	Count
1	0.50	0.28	13	17	5
2	0.30	0.23	18	30	15
3	0.37	0.31	31	60	13

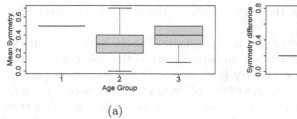 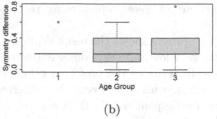

(a)   (b)

Fig. 6. a) Distribution of mean symmetry levels per age group, b) difference in the chosen symmetry levels by age group

A simple analysis of the values within these groups show that an effect of age on the preferred level of symmetry is likely. Group 1 included the youngest participants who consistently chose a higher level of symmetry compared to the more mature groups. This indicates that age potentially does have some impact on the level of symmetry preferred by a participant. However, the one-way ANOVA indicates that there was no significant effect on the mean level of symmetry based on the age cluster [$F(1,31)=0.777$, $p = 0.385$]. This backs up existing research which suggests that symmetry is a constant factor. Another interesting aspect to note is the relatively low average across all participants (0.36) which indicates that less ordered sculptures may be preferable.

More confidence can be had in the values collected for the symmetry preference level due to how consistent the participants were in their choices. The difference between the symmetry level chosen by the participant at each stage of the experiment was found on average to be 0.267, indicating that the participants were relatively consistent across their choices, being a single step apart from the symmetry value obtained in the previous set of choices. The data for each group of participants are shown in Fig. 6b.

Some symmetry levels were impossible to reach in individual sets of choices (e.g. 0.5), therefore if the participant prefers a level of 0.5, this ideally results through the choice of sculptures with symmetry values of 0.4 and 0.6, respectively in the two stages, potentially explaining why the difference was present. Additionally to the final symmetry value, the similarity of the two paths taken by a participant also indicates whether symmetry was the main basis for choices being made in the system, as it could reasonably be expected that the participant would take the same or very similar route to reach the final value.

4.3   Participant Behaviour

On top of the symmetry preference levels, more information can be obtained from the data collected, such as how an individual's preference can influence their confidence.

By considering each set of choices as a 5 dimensional vector, where each value represents the level of symmetry chosen at that stage, we can calculate the Euclidean distance as a measure of how different the paths taken by the participant were, representing their confidence in the level of symmetry they chose.

Figure 7a shows the results split by age group: the distance between the paths was very low and very consistent across all participants. If the distance is considered in terms of the preferred symmetry level, shown in Fig. 7b, a clear pattern emerges, where the lower the preferred level of symmetry, the more consistent the participant was in their choices. Correlating these two values suggests that there is a significant effect of the level of symmetry preference on how confident the participants were with their choices. We found a moderate correlation, but which is highly significant, $r(31) = .47$, $p = 0.005$.

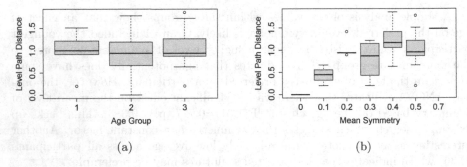

Fig. 7. a) The difference between the two sets of choices per age, b) The difference between the two sets of choices per preferred symmetry level

This is further supported by inspecting the individual paths to determine how closely they followed an ideal route within the system based on the participant's preferred symmetry level. For example, if a participant prefers a level of 0.2, then this can be achieved after making a single choice within the system. However, it would be more likely that the participant would choose the path which has the closest symmetry value sculpture for their preferred level; in this scenario the expected path would be to choose the 0 symmetry value sculpture as up until the last set of presented sculptures this is the closest value to their preference. For users who chose a low level of symmetry, this often matched the path they chose exactly or with minor discrepancies in the final step where both sculptures shown are of equal distance to the participant's optimal choice, as shown in Fig. 8.

However, for higher levels of symmetry preference, the adherence to the ideal paths is less consistent, as shown in Fig. 9. It is visible that individuals are more confident about the symmetry level preference at the lower end of the spectrum. This combined with mid-levels of symmetry being perhaps less easily perceived, suggests that participants preferring higher levels may not be as confident.

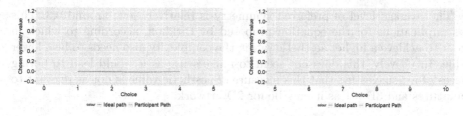

Fig. 8. Distance between optimal and participant paths based on the selected level and the mean symmetry, left) First set, right) Second set

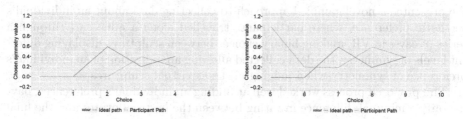

Fig. 9. Distance between optimal and participant paths based on the selected level and mean symmetry, left) First set, right) Second set

5 Discussion

The consistency of the preferred symmetry level reveals two important aspects, firstly that the choices were being made, at least to some degree, based on the level of symmetry in the sculptures, which backs up existing research that indicates that symmetry is both an important part of the aesthetic judgement process and that it is relatively universal. Secondly, it indicates that the process itself is a good method for extracting preferences of the aspects of aesthetic judgement, as study participants were able to select the level of symmetry they preferred. As this is information that people may struggle to articulate, this system becomes a powerful tool for understanding aesthetic preferences.

This has two important effects when considering applying this research to auto-generating sculptures: rather than trying to display different levels of symmetry, aiming to keep the symmetry consistent may be a better approach for generating aesthetically pleasing artwork. Obtaining a baseline value for symmetry is important as this can affect how the symmetry should be treated. Individuals didn't follow the ideal path for their chosen level of symmetry and this means that variation in the symmetry level can be shown to a user while generating artwork. However, the amount of variation is highly dependent on the user's preferred level of symmetry: if an individual is found to prefer lower levels of symmetry, this consistently represents their choices and little variation of this value should be presented, whereas in the case of preference for higher levels of symmetry a wider range of values should be displayed.

The average level of preferred symmetry is relatively low, around 0.3. This has implications for the equation proposed by Birkhoff, according to which in order to achieve a higher aesthetic rating the complexity also needs to be a lower value. Intuitively, this does not seem to make sense as it would lead to boring pieces of art suggesting that this measure of aesthetic value is not as relevant to sculptures and 3D art as it may be for 2D artwork.

6   Conclusion

We presented a novel gamified approach to collecting data about an individual's aesthetic preference, where participants travelled along a mine cart through a series of caves in VR, making binary choices between sculptures displaying different levels of symmetry, the data collected allowing an indication of an individual's preference of symmetry when judging abstract virtual sculptures.

The proposed process worked well at finding an individual's preferences, having only a minimal difference in rating between the best sculptures and the final sculptures shown to the participants representing their ideal symmetry. The consistency of the participant's choices both in the difference between the level of symmetry obtained after each set of choices and the closeness of the choices made in each stage indicates that the participants were mainly making choices based on the symmetry and the system reliably measured user preferences.

The use of VR also contributed to the success of the application, confirmed through the informal feedback provided about the app, where it was suggested that the experience was engaging, encouraging the user to use the app for longer. This coupled with the success of measuring preferences means that the system could be amended to cover other facets of aesthetic judgement with a few minor extensions of the environment, such as including a more varied selection of items to view. This approach may also be a very good candidate for measuring and assessing more complex variations for aesthetic preference. Including these additional aspects is expected to also improve the viability of the final sculpture as it would be considering a more complete view of aesthetic judgement. The environment did not provide a full range of motion around the sculptures, this was deliberately excluded in order to fulfil the goal of making the environment engaging and entertaining, these aspects will be added in future experiments.

The exciting implication of this system as a whole is that it allows aesthetic preferences to be established without requiring people to explicitly verbalise or rate their preferences, circumventing the barrier of art being notoriously difficult to judge.

When considering which category symmetry falls into, the similarity found across all participants suggests that symmetry fits nicely into the constant category and does not have to change or be re-measured to generate aesthetically pleasing sculptures. In addition to this, the consistency shown between the two sets of choices demonstrates that any interfering aspects do not change within small timescales, both of these reveal that symmetry can comfortably fit into the constant category. However, if a longer period of time is taken to describe some-one's preferences e.g. over months or years, then the same assumptions cannot be made.

The actual level of symmetry chosen by the participants had a surprisingly small variation across the age ranges. This is important when considering auto-generating 3D artwork as from a general standpoint a small range of values can be used across multiple contexts. The distance between the two paths the participant chose also impacts how symmetry can be used to auto-generate 3D artwork. The confidence the participant had in their choices was significantly affected by their preferred level of symmetry, where the lower the level of sym-metry was, the more confident the participant was in their choices, indicating less variation in symmetry should be presented to people in this category.

It is worth noting that the final sculpture was not rated as the best by the majority of the participants, however, the designation provided was skewed towards higher ratings, showing that using the preferred level of sym-metry improved the overall aesthetic rating. This is an indicator that multiple attributes are used within aesthetic judgement, not just symmetry, however, the lack of a more complete understanding of aesthetic preference is a limitation of this system. This also prevents any conclusions from being drawn about how symmetry may be affected by the presence of other attributes within the sculp-ture. In addition to this, other forms of symmetry such as colour symmetry were not investigated so it is unknown whether the same principles would apply.

Overall the system was successful in its goal of extracting preferences of symmetry within the aesthetic judgement of 3D sculptures, yielding information which can be applied to auto-generating sculptures. The information is also important for identifying other aspects which may be included in the assessment of sculptures and forms the basis for understanding how these other aspects may influence the preferred level of symmetry. Having this greater understanding of aesthetic judgement will lead to more accurate methods of measuring aesthetic preference and a greater ability to generate aesthetically pleasing sculptures. The preference of symmetry may not generally be in the age of the beholder but it is very important to its aesthetic appeal.

There are two main paths of work which will be continued, the first is to further the understanding of symmetry by trying to consider how it interacts with other aspects which contribute to the aesthetic judgement of sculptures. The second is that the system will be used to measure other aspects of aesthetic appreciation e.g. complexity, enabling their categorisation as constant, transient or subjective. This will also include high-level attributes of the judgement of a sculpture, using descriptive terms such as warm or dynamic, which may not map exactly to individual properties of the sculpture model and instead relate to a combination. If the combination of aspects can be ascertained, the same system can be used to extract preference levels for these high-level concepts, providing a much greater understanding of how aesthetic judgement is performed.

References

1. Acebo, E., Sbert, M.: Benford's law for natural and synthetic images. In: First Eurographics conference on Computational Aesthetics in Graphics, Visualization and Imaging (2005)
2. Alamino, R.C.: Measuring complexity through average symmetry. J. Phys. A: Math. Theoret. **48**(27), 275101 (2015)
3. Bauerly, M., Liu, Y.: Computational modeling and experimental investigation of effects of compositional elements on interface and design aesthetics. Int. J. Hum Comput Stud. **64**(8), 670–682 (2006)
4. Bergen, S., Ross, B.J.: Aesthetic 3D model evolution. Genet. Program Evolvable Mach. **14**(3), 339–367 (2013)
5. Bertamini, M., Silvanto, J., Norcia, A.M., Makin, A.D., Wagemans, J.: The neural basis of visual symmetry and its role in mid-and high-level visual processing. Ann. N. Y. Acad. Sci. **1426**(1), 111–126 (2018)
6. Birkhoff, G.D.: Aesthetic measure. Mass, Cambridge (1933)
7. Chatterjee, A., Widick, P., Sternschein, R., Smith, W.B., Bromberger, B.: The assessment of art attributes. Empir. Stud. Arts **28**(2), 207–222 (2010)
8. Colton, S.: Creativity versus the perception of creativity in computational systems. In: AAAI spring symposium: creative intelligent systems. vol. 8, p. 7. Palo Alto, CA (2008)
9. Easton, E., Ekárt, A., Bernardet, U.: Axial generation: Mixing colour and shapes to automatically form diverse digital sculptures. SN Comput. Sci. **3**(6), 1–17 (2022)
10. Ecins, A., Fermuller, C., Aloimonos, Y.: Detecting reflectional symmetries in 3d data through symmetrical fitting. In: Proceedings of the IEEE International Conference on Computer Vision Workshops, pp. 1779–1783 (2017)
11. Ekárt, A., Sharma, D., Chalakov, S.: Modelling human preference in evolutionary art. In: Di Chio, C., et al. (eds.) EvoApplications 2011. LNCS, vol. 6625, pp. 303–312. Springer, Heidelberg (2011). https://doi.org/10.1007/978-3-642-20520-0_31
12. den Heijer, E.: Evolving art using measures for symmetry, compositional balance and liveliness. In: International Conference on Evolutionary Computation Theory and Applications. vol. 2, pp. 52–61. SciTePress (2012)
13. den Heijer, E., Eiben, A.E.: Comparing aesthetic measures for evolutionary art. In: Di Chio, C., et al. (eds.) EvoApplications 2010. LNCS, vol. 6025, pp. 311–320. Springer, Heidelberg (2010). https://doi.org/10.1007/978-3-642-12242-2_32
14. Isik, A.I., Vessel, E.A.: Continuous ratings of movie watching reveal idiosyncratic dynamics of aesthetic enjoyment. PLoS ONE **14**(10), e0223896 (2019)
15. Jones, B.C., DeBruine, L.M., Little, A.C.: The role of symmetry in attraction to average faces. Perception Psychophys. **69**(8), 1273–1277 (2007)
16. Leder, H., Tinio, P.P., Brieber, D., Kröner, T., Jacobsen, T., Rosenberg, R.: Symmetry is not a universal law of beauty. Empir. Stud. Arts **37**(1), 104–114 (2019)
17. Machado, P., Cardoso, A.: Computing aesthetics. In: de Oliveira, F.M. (ed.) SBIA 1998. LNCS (LNAI), vol. 1515, pp. 219–228. Springer, Heidelberg (1998). https://doi.org/10.1007/10692710_23
18. Matkovic, K., Neumann, L., Neumann, A., Psik, T., Purgathofer, W.: Global contrast factor-a new approach to image contrast. Comput. Aesthetics **2005**, 159–168 (2005)
19. McDermott, J., et al.: String-rewriting grammars for evolutionary architectural design. Environ. Plann. B. Plann. Des. **39**(4), 713–731 (2012)

20. Nadal, M., Chatterjee, A.: Neuroaesthetics and art's diversity and universality. Wiley Interdisciplinary Rev. Cogn. Sci. **10**(3), e1487 (2019)
21. Osborne, H.: Symmetry as an aesthetic factor. Comput. Math. Appl. **12**(1–2), 77–82 (1986)
22. O'Reilly, U.M., Hemberg, M.: Integrating generative growth and evolutionary computation for form exploration. Genet. Program Evolvable Mach. **8**(2), 163–186 (2007)
23. Pelowski, M.: Tears and transformation: feeling like crying as an indicator of insightful or "aesthetic" experience with art. Front. Psychol. **6**, 1006 (2015)
24. Rhodes, G.: The evolution of facial attractiveness. Annu. Rev. Psychol. **57**, 199–226 (2006)
25. al-Rifaie, M.M., Ursyn, A., Zimmer, R., Javid, M.A.J.: On symmetry, aesthetics and quantifying symmetrical complexity. In: Correia, J., Ciesielski, V., Liapis, A. (eds.) EvoMUSART 2017. LNCS, vol. 10198, pp. 17–32. Springer, Cham (2017). https://doi.org/10.1007/978-3-319-55750-2_2
26. Rigau, J., Feixas, M., Sbert, M.: Conceptualizing Birkhoff's aesthetic measure using Shannon entropy and Kolmogorov complexity. In: Computational Aesthetics, pp. 105–112 (2007)
27. Rodrigues, A., Cardoso, A., Machado, P.: Generation of aesthetic emotions guided by perceptual features (2018)
28. Ross, B., Ralph, W., Zong, H.: Evolutionary image synthesis using a model of aesthetics. In: IEEE International Conference on Evolutionary Computation, pp. 1087–1094. IEEE (2006)
29. Salingaros, N.A.: Symmetry gives meaning to architecture. Symmetry Cult. Sci. **31**, 231–260 (2020)
30. Sibley, F.: Aesthetic concepts. Philosoph. Rev. **68**(4), 421–450 (1959)
31. Skov, M.: Aesthetic appreciation: the view from neuroimaging. Empir. Stud. Arts **37**(2), 220–248 (2019)
32. Spehar, B., Clifford, C.W., Newell, B.R., Taylor, R.P.: Universal aesthetic of fractals. Comput. Graph. **27**(5), 813–820 (2003)
33. Tinio, P.P., Leder, H.: Just how stable are stable aesthetic features? symmetry, complexity, and the jaws of massive familiarization. Acta Physiol. (Oxf) **130**(3), 241–250 (2009)
34. Vázquez, P.P., Feixas, M., Sbert, M., Heidrich, W.: Viewpoint selection using viewpoint entropy. In: VMV. vol. 1, pp. 273–280 (2001)
35. Vinhas, A., Assunção, F., Correia, J., Ekárt, A., Machado, P.: Fitness and novelty in evolutionary art. In: Johnson, C., Ciesielski, V., Correia, J., Machado, P. (eds.) EvoMUSART 2016. LNCS, vol. 9596, pp. 225–240. Springer, Cham (2016). https://doi.org/10.1007/978-3-319-31008-4_16
36. Weichselbaum, H., Leder, H., Ansorge, U.: Implicit and explicit evaluation of visual symmetry as a function of art expertise. i-Perception 9(2), 2041669518761464 (2018)
37. Weyl, H.: Symmetry, vol. 47. Princeton University Press (2015)
38. Zeki, S.: Beauty in architecture: not a luxury-only a necessity. Archit. Des. **89**(5), 14–19 (2019)

Extending the Visual Arts Experience: Sonifying Paintings with AI

Thomas Fink and Alkim Almila Akdag Salah[(✉)]

Information and Computing Sciences, Utrecht University, Utrecht, The Netherlands
a.a.akdag@uu.nl

Abstract. Sonification of visual information is a relatively new research line that aims to create a new way to access and experience visual displays, especially for the visually impaired. When applied to artworks, sonification needs to translate the aesthetic experience as well. This is attempted via a handful studies in the literature, where most of the transformation and music generation is done manually, or only by using the low level visual features of artworks. In this paper, we present a sonification model that uses both low level and high level features such as color, edge information, saliency, object and scene detection to create a pleasant and descriptive sonification of artworks with the use of a fully automatic pipeline. The results of the model are tested via interviews done with experts in music theory and generative music models. We found a high agreement among experts for the evaluation of a small set of sonified paintings. Addition of high level features such as sounds extracted from the scene played a big role in this. Among the challenges observed during the interviews was the need to add emotion and mood information as well as semantic information to the sonification in order to create more descriptive melodies and sounds. The complexity and ambiguity of the visual information generated the most disagreement among experts both in their interpretation of the paintings as well as their sonifications.

Keywords: Artwork sonification · Generative music models · High level visual feature sonification · Low level visual feature sonification

1 Introduction

Enjoying paintings is mostly a visual experience, but connecting paintings with audio or music is not a far stretch [15]. The transformation of visual stimuli to auditory stimuli is a known research line that especially addresses the needs of people with visual impairment, and emphasizes certain data types (e.g., information visualizations) [5]. There are also some research done on the sonification of paintings, but these mainly focuses on using basic visual features such as color and color properties [3,6,11], or edge information [17]. The resulting sonification instances do not take into account high level features, nor make use of fully automated methods.

Supplementary Information The online version contains supplementary material available at https://doi.org/10.1007/978-3-031-29956-8_7.

In this research we explore the possibilities of artwork sonification by using AI to extract high-level features such as scene and object detection and create an automated process to combine high and low level features for sonification. We argue that this approach leads to a sonification that is better at pleasantly conveying a painting's content and a simpler method for generating the sonification. Besides opening up a new way to enjoy visual arts, and rendering a new dimension within the exhibition space, research outcomes might be used to offer the public a deepened experience to engage with an art collection. Within the framework of the museums of the 21st century that makes use of Artificial Intelligence technology for restoration, analysis, and re-creation of art, this approach will bring a fresh perspective on how AI and visual arts can be coupled.

This paper is structured as follows: We first summarize literature on visual stimuli sonification, focusing above all on artwork and artistic experiences. In the methodology section we describe the model we created for automatic sonification of artworks. The result section describes the interviews conducted to assess the generated sonifications, along with the interview protocol and materials prepared for this purpose. We conclude by summarizing the research results, insight gained from the expert interviews, and suggestions for future directions. As the quality and fittingness of the sonification are hard to quantify, quantitative evaluation will be near to impossible.

2   Literature Review

Most existing visual sonification methods are not designed with paintings in mind, as they transform visual data directly into the auditory space with a one-on-one mapping to stay as close to the source data as possible [2, 17]. This is an effective way to convey information accurately but easily results in a non-musical sonification. In this section, we detail research applied to create a more pleasing sonification for paintings.

Using color as a base for the sonification is one of the most commonly applied method. To help the visually impaired to understand color within a picture a tool has been created by Cavaco et al. [2]. The color attributes that are used for the conversion are hue (e.g., blue, red, magenta, etc.), saturation (colorfulness, e.g., deep blue or pale blue), and value (the intensity of the color). These variables are linked to the psycho-acoustic variables of sound, namely pitch, timbre, and loudness. Where the hue is linked to the pitch, saturation influences the shape of the waveform to affect timbre, and value changes the loudness. The picture is scanned from top to bottom in a vertical manner where each row is played sequentially. A row is divided into 12 segments, with each segment having a shifted phase waveform based on the x-axis, which plays simultaneously. To validate the tool a user study was done with eight visually impaired participants. The experiment was a forced-choice test where sonification was carried out on one of seven colors. The participants noted that the frequency of neighboring colors was too close together and thereby hard to distinguish.

A downside of one-to-one mapping is that the resulting audio has no musical relation, and is, therefore, less suitable for a pleasant art experience. Polo et al. [13] tried to address this problem by the addition of harmony, where multiple notes played at the same time in a musically pleasing manner. This approach furthermore made use of RGB values for color sonification instead of hue, saturation, and intensity, The RGB values represent three notes that play simultaneously, one for red, one for green, and one for blue. To evaluate the performance, a user study with 12 participants was conducted. A participant would listen to the sonification of an image and in a forced-choice test style, has to pick the best fitting image while multiple images were being presented. With training, non-musical participants had an accuracy of 70 percent. This method only transforms color information into the auditory space, thereby it is hard to discern objects from the sonification.

To help the visually impaired recognize objects within a picture, Yoshida et al. created the tool EdgeSonic that uses an edge detection algorithm to transform the original picture to edge features [17]. To test the performance of the framework before and after training, sighted participants were blindfolded and told to explore a given shape and reproduce it. After 90 min of training two out of four participants were able to reproduce the shapes. While the tool can accurately convey shape information present in visual input, the resulting audio has no musical relation and disregards all visual information but edges.

In Cho et al. [3] a specific instrument is selected to play for a certain color. This color-instrument mapping is mostly inspired by the findings of the painter Wassily Kandinsky. For example, red is played by a violin while green is played by a cello. Besides color, the color intensity was represented by a specific chord or melody. Calling this process color coding, the researchers proposed three different color coding models. The first model directly linked chords and instruments to colors where a different color is played by a different instrument. The saturation of the color maps to the intensity of the sound and the lightness maps to the pitch of the chord. The second model used the same principle as the first model, but linked the saturation and intensity to different melodies from Vivaldi's "Four seasons". Saturated sound uses Vivaldi's "Spring" composition, light uses "Autumn" and dark is based on "Summer". The third model is based on classical music scores where instruments are again linked to colors, whereas saturation and color intensity is linked to a specific part of the classical score best representing those values.

Rector et al. [14] incorporated more features than color where users could select background music, sound effects, or a verbal description for the sonification. The sonification method has an orchestral loop, and the loudness of different instruments is changed to convey color, e.g., bright green would produce a loud piano. The painting is manually divided into segments and those segments got assigned a color by hand. To give a more literal impression of the painting, sound effects are also added manually for objects within the painting. It is imported to note that the sonifications were generated by hand as the generative music models were deemed not musically pleasant enough.

Kabisch et al. [8] used sonification to create an art installation. The authors argue that the mapping of data to sound in the realm of art is as much an artistic choice as a philosophical and technical choice. Thereby having the luxury to interpret data in a way where it can communicate more than the source material itself. They extracted data from landscape images to influence the created sound on the macro-level (rhythm and form) as well as on the micro-level (timbre). The image obtained from edge detection is scanned on the vertical axis. When the RGB values reach a set threshold, those values are used to trigger notes with a certain pitch based on vertical position. The position of the user is used to select a horizontal area of the picture. To change the representation on a micro level the edges are used to create a wave-shaping lookup table consisting of different edges which can be used to shape a waveform for the audio generation. The addition of the scenic sound is however done manually.

3   Methodology

In this section, we detail our model for the automatic generation of music based on visual features of artworks. Our framework is built after several experimental steps for different visual features and how to sonify them. Here we present only the end model, and expand on the decision process only to explain the reasoning behind the choices made along the road. The assessment procedure for each step was done by the main researcher and 2 computer scientists, and an art historian.

In Fig. 1 we give an overview of our pipeline, demonstrated on an example painting. We first start with extracting low level features such as dominant color, HSV (Hue, Saturation, Values), and histogram from the overall painting. We use hue for the generation of the root note, and value for deciding the scale (Major/Minor). Since the color information is much richer than the notes and scales, we scale the visual information by reducing it to the corresponding audio feature dimensions. We then proceed in extracting high level features such as object and scene detection as well as generating saliency maps. Segmentation of the artwork is done with the help of object segmentation to create a musical timeline of the painting. We make use of the size of the segments as a measure of length for the created sound. A panning algorithm defines the source of the sound, based on the position of segments relative to the middle of the painting. A saliency map shows the most salient parts of the painting which are sonified first in the created sound. Scene detection is tied with a sound dataset to find a matching background sound for the detected scene, and added to the sonification. Table 1 shows the mapping applied for low and high level features to audio features.

To put all the above ideas into reality a framework has been created with the use of Python and Supercollider. Python is responsible for the image processing while Supercollider creates the sound. The visual part of the framework scans through the image and creates parameters that can be used by the audio side of the framework to create sound.

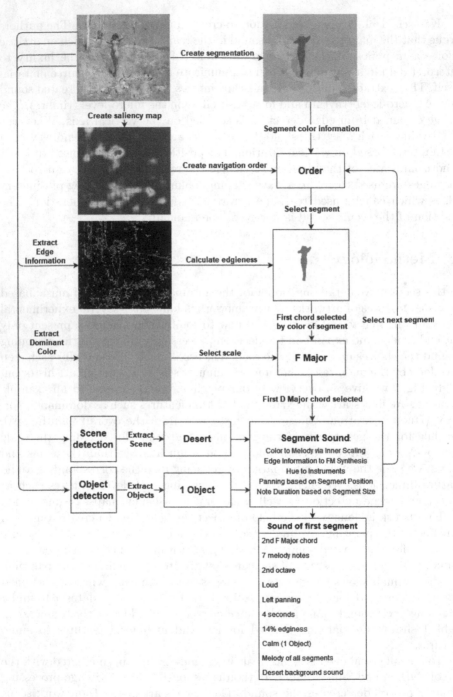

Fig. 1. Flow of the model demonstrated on an example painting

Table 1. Visual Features mapped to Audio Features (HL: High Level, LL: Low Level)

Visual source	Visual property	Audio property	Scaling method
Dominant color	Hue (LL)	Root note	Normal
Dominant color	Value (LL)	Scale (Major/Minor)	Normal
Scene	Scene of painting (HL)	Scene sound	None
Dominant color	Hue (LL)	Instrument	1 to 1
Edges in segment	Percentage of edge	Carrier wave selection	Percentage
		FM modulation amount	
Dominant color	Hue (LL)	Chord root note	Inner Scaling
Dominant color	Saturation (LL)	Volume	Normal
Dominant color	Value (LL)	Scale (Major/Minor)	Inner Scaling
All segments	Hue (LL)	Melody	Normal
Segment	Segment location (HL)	Panning	Percentage
Saliency	Saliency of segment (HL)	Play order	Percentage
Segment	Size of segment (HL)	Timing of notes	Normal
Objects	Amount of objects (HL)	Timing of notes	Normal

Visual Features used from the Overall Painting As a first step, the dominant color of the image is extracted and the image is scaled down. From the scaled-down image a histogram, and three new images are created, namely an HSV image, a Saliency map, and an image containing all edges. From the HSV image the hue, value, and saturation are extracted.

The Root Note: The hue of the dominant color will influence the root note and therefore needs to be scaled from the possible hue values of 0 to 179 to 0 to 11. This creates twelve possible outputs that correspond to the twelve possible notes within an octave. The equation used for scaling can be seen in Eq. 1.

$$\frac{k_{min} + (x - j_{min}) * (k_{max} + k_{min})}{j_{max} - k_{min}} \tag{1}$$

In the above formula, the x represents the input value that is going to be scaled. j_{min} and j_{max} represent the range the input value can be in. E.g., x can be 0 at the minimum and 179 at maximum, then j_{min} will be 0 and j_{max} 179. k_{min} and k_{max} will define the output range of the scaled value. E.g., when the input value x is 90 and the input range is from 0 to 179, k_{min} can be set to 0 and k_{max} to 7 and the expected outcome will be 4. In the framework, the output values of the formula are rounded to create a less fine-grained scaling as floating points values would fall out of the scale the sonification needs to be in.

Scale: The value (V) of the HSV from the dominant color image is scaled from 0–255 to 0–1 using Eq. 1 for the decision of Major/Minor scale. This is done as the value will influence the scale the sonification is played in. The two possible scales are minor (0) and major (1). A dominant color where the value of the HSV image is 127 or lower will create a minor scale sonification, whereas a value of 128 or higher will create a sonification in the major scale.

Object Detection and Segmentation: Rector et al. made manual divisions of paintings into natural segments [14]. For example, the sky or a person would form a segment each. To implement such an approach automatically, Detectron2 [16] is used to divide a painting into natural segments, i.e. into objects detected and segmented in the painting.

Scene Detection and Background Sound: Our framework assigns a background sound with the use of Places [18], an algorithm designed for automatic scene detection. Thus, if a painting displays the scene of a market, the sounds heard at a market will be added as a background layer in the sonification. To match up the detected scene with a sound, first scene labels need to be created for an audio dataset. To do this in an automatic manner Soundnet [1] has been utilized to label large audio datasets. The datasets were ESC-50 [12], FSD50K [4], TUT acoustic scenes 2016 [9], and TUT acoustic scenes 2017 [10] because of their scenic nature. In the matching process, when more than one match has been found, a random sound is chosen. The output is a path to the audio file that will be added to the generated text file containing the other parameters for the sonification.

Visual Features used from Segments

The Chord: The dominant color of a segment is used to generate a chord. Hue (H) was scaled down to select a chord within a scale. Saturation (S) was used to change the volume of the played chord. Value (V) was used to select the octave the chord was played in. At first the hue was scaled down from 0 to 127 to a range of 0 to 7 to accommodate the possible chords within a scale and the value was scaled down from 0 to 127 to a range of 0 to 1 to accommodate the possible octaves. This is quite a big reduction in resolution. If a painting contains a lot of colors that are close to each other in e.g. the hue, probably these colors will play the same notes. To prevent this, we created the method *Inner Scaling* when extracting the dominant color of a segment. This method minimized the resolution reduction by limiting the scaling to the possible input values. Instead of using the whole range from 0 to 127, we take only the hues present in the painting, and reduce the possible input values only. This has only been applied to the hue and value from the HSV. The scaling of the saturation input was unchanged as less erratic volume changes enhanced the listening experience. A downside of this is that the scaling will be dynamic and no color is always linked to a certain sound. When scaling with this method linking of e.g. hue to chords can differ per painting, making it hard to extract specific colors across paintings, even with training.

Note Duration: The percentage of pixels within the segment is used to create a change of duration of a specific segment in the sonification, see Eq. 2. In the equation, $p_{segments}$ represents the number of pixels of a segment, and p_{total} represents the total amount of pixels within the painting. This division will represent the percentage of space a segment takes up within a painting. This percentage is scaled to a suitable note duration. A small segment will therefore sound 2 s, while a large segment will sound 8 s.

$$scale(\frac{p_{segment}}{p_{total}} * 100, (0, 100), (2, 8)) \tag{2}$$

In an effort to create more diversity between the sonifications a high-level feature parameter is used to influence the duration of the notes, see Eq. 3. By using the number of objects in paintings a better distinction can be made between calm and chaotic paintings. This is done with the assumption that a calm painting will contain lesser objects than a chaotic painting and long slow notes are considered calmer than quick short notes.

$$scale(\frac{p_{segment}}{p_{total}} * 100, (0, 100), (max(1, 4 - (\frac{objects}{4})), max(4, 16 - objects))) \tag{3}$$

Equation 3 is based on the same principle used in Eq. 2 with the addition of a dynamic output range. When more objects are in the painting the minimum and the maximum output will be lower, when there are fewer objects the output will be higher, resulting in a longer duration.

Panning: To create a more dynamic musical piece and a sense of space we applied a method called Panning. When the sonification of a segment is on the left side of a painting, the sound will also be panned to the left. The further away the segment is from the center of the painting, the more apparent the panning will become. Our panning method looks at the position of the segment relative to the middle of the painting. To calculate the panning Eq. 4 is used, where p_{total} represents the total amount of pixels present in the segment. p_{left} is the amount of pixels from the segment that are on the left of the painting relative to the middle. p_{right} represents the number of pixels that are on the right.

$$-1 * \frac{p_{left}}{p_{total}} + 1 * \frac{p_{right}}{p_{total}} \tag{4}$$

Saliency Maps for Order of Segments: From the saliency map a threshold image is created. From this threshold image, the percentage of salient pixels of each segment is calculated. This percentage is used to define the order in which segments are played. The segment with the highest percentage is played first and the segment with the lowest percentage is played last. The equation can be seen in 5.

$$\forall s \frac{s \cap p_{255}}{(s \cap p_0) \cup s(\cap p_{255})} * 100 \tag{5}$$

In this formula, s is a set containing sets of all the pixels of each segment. p_{255} contains all pixels having a value of 255, and are therefore salient pixels. p_0 are the pixels having the value 0, therefore containing all the non-salient pixels. The formula divides the salient pixels by the total amount of pixels in a segment. The output of this division is multiplied by 100 to create an easily readable percentage. The output of the formula is a set of saliency percentages of each segment that can be used to create a navigation order.

Timbre: We followed the method used by Cho et al. [3] linking every color to a unique instrument, but we preferred to use hue to map to instruments. To be able to play instruments, first, the sound samples of the instruments are loaded into buffers within the Sound Engine. To create the sound of the instruments a synth definition has been created that can play the instrument samples from the loaded buffers. In the function responsible for playing notes based on given parameters (i.e. Pbind), the linking of the hue to an instrument is created. Beside the instruments added, we used edge information in each segment to enrich the timbre. The edge image is used to calculate the edginess of a segment as a percentage, i.e. the number of pixels containing an edge will be counted and divided by the total amount of pixels present in a segment. To convey roughness out of the edge information, we first applied Wavetable synthesis. This was done by providing three waveforms, namely a sine wave, a waveform extracted from the painting's histogram, and a saw waveform. The number of edge pixels within a segment was used to linearly interpolate between the before mentioned waveforms to convey roughness. One downside of this implementation was that most results sounded alike. Therefore another synthesis method, namely *FM synthesis*, was explored. FM Synthesis works by having two waveforms that can influence each other. One of the waveforms is called the carrier and the other waveform is called the modulator. This opens up a new array of parameters that can be changed based on color or other visual information. In our model FM synthesis is implemented by using it to change one of its three waveforms in a threshold manner based on the edge information. When the percentage of edge pixels in a certain segment exceeds a threshold a certain wave is used. The thresholds chosen for this model are 0–0.33 (first waveform), 0.33–0.66 (second waveform), and 0.66–0.99 (third waveform).

Segments as Melody: All segments of a painting are combined to form a melody pattern, based on the dominant colors of the segment. This pattern will then be shuffled each time a segment is played. Much like the creation of the chord the HSV was used here as well. Hue (H) was scaled down to select a note out of the 12 possible notes within the scale without inner scaling. Saturation (S) was used to change the volume of the played notes. Value (V) was used to select the octave the notes was played in. The melody notes are always played one octave above the chord to make them stand out from the chord.

4   Results

One of the challenges of sonification of visual features is the assessment of the results. The literature utilizes user studies to measure the success of the relation between the visual features and the resulting sonification. We followed a similar approach to report our results. Instead of user studies, we carried out interviews with experts to first of all see if they agreed on the qualities of the sonifications, and if they described musical features of the paintings in a similar manner. The use of the interview format furthermore gave us a chance to find new solutions to current problems, to find problems that have not yet been identified, and to find inspiration to extend and improve the current framework. In this section we first account the choosing process of a small set of paintings sonified for the interviewed, and the interview setup. Then we delineate the quantitative and qualitative results, and conclude with general observations.

Dataset: We filtered the *Painter by Numbers* dataset to create a small, diverse and representative sample dataset that would fit the purposes of this research. To make the best use of automatic scene detection, we preferred genres where paintings are close to natural images. After filtering the dataset with this rule, we were left with 8279 paintings in 29 genres. We furthermore omitted the following genres: caricature, illustration, nude painting, panorama, portrait, poster, self-portrait, sketch, study, still life, symbolic painting, vanitas. These genres had lack of color or definable scenes, and were problematic for our automatic pipeline. After the omission the dataset contained 6514 paintings. The next step was to see how scene detection performed on the selected paintings. For a manual inspection, we randomly sampled 10 paintings per genre, leaving 143 paintings. The manual assessment gave us 44 paintings with correctly detected scenes. Out of these 44 paintings we decided to focus on 8 paintings that were visually dissimilar enough to create different sonifications, and used the resulting sonifications during the expert interviews, see Fig. 2. The accompanying sonifications are provided as supplementary materials.

Participants: 8 experts contributed to the interviews. The interviews were conducted only with participant who had at least a good understanding of generative music models, as well as experience in music production since expert suggestions to improve the sonifications was one of the main goals. In order to assess experts familiarity with the music theory and generative music, a set of questions were prepared. These questions were about their knowledge of generative music models, if/how many models they used, the instruments they play, and their understanding of music theory.

Interview Protocol: The interviews were conducted in a semi-open interview format. First, the experts were asked to listen to a sonification without seeing the painting, and given the questions "What mental image do you get from listening to this song?", "What part of the song influenced your mental picture the most?" and "Would you describe the song you heard as pleasant?".

After listening to the paintings the experts were shown each of the eight paintings and asked the following question: "What sound/song do you imagine when looking at the painting?" With the answer to this question a comparison can be made between what the participant would create as a musical piece. Also, inspiration for further improvement of the framework can be taken from the answers to this question.

The last step gave the experts the chance to look at the painting while listening to the corresponding sonification simultaneously. After each painting, the following questions were asked: "Do you think the song is descriptive of the painting?", "Could anything be added or changed?", and "What are the most descriptive features of the song to you?"

Quantitative Results: There were two closed questions, first one delineating on the pleasantness of the sonification, and the second one asking if the sonification is descriptive of the painting or not. Table 2 summarizes the answers to these questions. Out of 8 painting sonifications, 8 experts found 49 times the sonification pleasant, and 45 times descriptive. Some of the paintings had all experts agreeing on these features, this happened most when the scene related sounds were clearly recognizable and had an unmistakable visual relation, such as the sound of waves or water, as in the instances of Claude Monet's Waterlilies or Valentin Serov's Iphigenia in Tauris.

Table 2. Answers to questions about the pleasantness and the descriptiveness of the sonifications

Pleasantness	Descriptiveness
Yes: 49	Yes: 45
No: 11	No: 8
Neutral: 4	Somewhat or parts: 7

Qualitative Results: To get an insight on the performance of the model, we looked at how experts evaluated the sonification itself without seeing the painting. We also collected feedback on how they would sonify the painting. It is important to note that this question type always came after listening to the sonification made by the framework. Therefore participants could be influenced when coming up with ideas on how to sonify a painting. Here we summarize their assessment per painting by highlighting the points where the expert agreed and disagreed the most. Figure 2 shows the paintings used during the interviews.

Charge of the Scots Greys at Waterloo. Five experts used keywords like "fast, chaotic and battlefield" when they listened to the sonification, which describe

Fig. 2. Paintings used during the Interviews: (a) Charge of the Scotts Greys at Waterloo, (b) Alfred Sisley-Snow at Louveciennes, (c) Enrique Simonet's El Barbero del Zoco, (d) Fairy Mountain by John Lavery, (e) Olive Groove by William Merritt Chase, (f) Monet-Water Lilies, (g) Valentin Serov-Iphigenia in Tauris, (h) Paul Delvaux-The Viaducto

the painting well. Two experts mentioned a castle, and although one might connect the battle scene and horses depicted in the painting to the concept of a castle, there is no caste in the painting. Looking at the painting, experts feedback was to add chaos or randomness to the sonification along with the sound of a battlefield, and marching drums. To make the sound darker and more grand was a suggestion too.

Alfred Sisley-Snow at Louveciennes. Five experts used words such as "something dark, ominous, evil, or night" while listening to the sonification, which could be seen as descriptive of the painting. Looking at the painting, the experts asserted in creating a sonification that was calm, slow, or peaceful. The sounds would be silent, muted, or quiet. As a background sound they would add the wind and the footsteps of the person walking in the snow or sounds that would relay the coldness the painting portrays.

Enrique Simonet's El Barbero del Zoco. Six experts cited words such as "bright, happy, high, excited, playful, summer, sunny, or day". Five experts agree on the concepts of "something with people, city, park or eating". These words describe the painting to a certain extend. It is of importance to note that once seeing the painting the experts were much more interested in the setting of the painting, i.e. an oriental picture, and used that for the sonification, rather than a direct description. Hence, most experts said they would create a musical piece containing Oriental, Arabic, or middle eastern instruments/sounds. Three experts named a more specific instrument to add, namely the flute, and two even talked about using the regional tuning within the musical piece. Experts would also add a background sound of people speaking (in Arabic), to fit the scene.

Fairy Fountain by John Lavery. Four experts found it hard to imagine a painting while listening to the sonification. The keywords cited were "something related to water, waterfall, river, or boat" (by four experts). This was likely influenced by the water or rainy sound in the background as three experts mention this influenced their mental picture the most. Three experts mentioned that their mental image was dark or sad. After seeing the painting, experts said they would add calmness to the sonification. Six and thereby most experts would use the sound of water or a fountain. Four experts would use a dark timbre or sad chords. Two experts would also add the sound of talking in the background. The difficulty to extract the mood of the painting was mentioned by two experts as they stated it could either be happy or sad depending on how the people in the painting are feeling.

Olive Groove by William Merritt Chase. Five participants noted that the mental image they get from listening to the sonification is related to night or dark. This is a stark contrast to what the painting actually contains. This is likely due to the low and dark tones present in the sonification as three experts mention that low or dark notes mostly influenced their mental image. Experts also mentioned that the cricket sound present in the background made them think of a night scene. Looking at the painting six expert imagined a slow, calm, relaxing sound and would add a nature sound to the background. Five experts would create a "happy or joyful" sound to go with the painting. Two experts noted they specifically would add a flute as an instrument and birds or leaves in the wind as background. Two participants also noted that the musical piece should be played in a high register.

Monet - Water Lilies. All experts refer to "water, sea, nature, or boat". Three participants mentioned that their mental picture was not too bright or they noticed the darker lower tones. Two experts found their mental picture to be calm, and two used "mysterious or surprising" to describe their mental image. Five experts noted that their mental image was influenced the most by the instrument selection of "horn, trombone, or brass" and four mentioned "low tones or bass". Looking at the painting, three experts would create a musical piece that conveys something happy, cheerful, or warm. Two experts would add some sound of a water animal or specifically a frog. Two experts mentioned that they would add the reflection present in the painting to the sonification by inverting the melody at some point in the sonification. One thing that stands out for this painting is that four participants mention that they would create a musical piece that would be calm or slow, whereas two participants also mention they would create something fast or quick.

Valentin Serov-Iphigenia in Tauris. Four experts used "waves, beach, forest, or nature" while listening to the sonification. The "Waves and beach" describe the painting pretty well. While forest is also a nature-related concept, this is not what the painting contains. Three experts said their mental image was related to medieval times and two noted that it was related to fantasy. This is likely caused by the timbre of the sound. While looking at the painting seven experts imagined a wave or sea sound. Three experts would create a musical piece that is calm or low tempo, and two would create a grand musical piece and two experts would let it represent a sad emotion. It is good to note that not all experts considered the mood of the painting as sad, as one expert explicitly mentioned it to be a happy context.

Paul Delvaux-The Viaducto. Two experts mentioned the concepts of the night sky or blue sky which describes the painting. Two experts mentioned storm or wind to be present in their mental image which does not describe the painting well. Three participants said the biggest influence on their mental image is the chords of the sonification. Looking at the painting six experts wanted to add concepts of calm or slow. Four experts emphasized the need for the rhythm of a train within the sonification, and four experts expressed the need for the sounds of a night scene. Three experts would add the sound of a train or a railroad as a background sound. Two experts mentioned the explicit use of a piano.

General Observations: Most experts tended to attribute the descriptiveness of the sonification of the painting to the scene sound present, such as sounds of objects in the scene or sounds you could hear in the context of the painting. Hence, the added background sound of the sonification helped the most in matching a sonification to a painting. Furthermore, while looking at paintings, most suggestions by experts on how to turn a painting into sound were based on including some sort of high-level feature to link the visual space to the auditory space.

Interestingly, when experts were asked about how they would implement a sonification method, many talked about extracting color to create a sound, however, when listening to a sonification what they imagined the painting looks like rarely contained specific colors. It is likely that people think turning color into a sound is an intuitive way of sonifying visual information, but when extracting information from sound, this method looks less intuitive. Instead, experts tended to talk more about the setting or objects they associated with sounds, or the emotion the sonification brought them.

Another point of feedback is about the randomness of the melody. Some experts noted that the melody had no musical structure triggering the feeling that the melody was random. Experts also got tired after hearing the sonifications for a while commenting that everything started to sound very similar. This problem can probably be attributed to a lack of diversity between the sonifications.

There were contrasting comments between experts when interpreting a painting and the mood it generates. For some, a painting might look happy due to the contents, for example two people connected to each other, whereas the same painting might look sad due to its color schema to others. This contrast is reflected in how the experts would sonify the painting, or if they have found our sonification a descriptive one or not. All in all, we observed that the most agreement among experts are seen when the visual properties of the painting were not ambiguous and interpreted similarly.

5   Conclusion and Future Work

This paper addressed the challenge of creating pleasant sonifications of paintings automatically, building a model that uses both low and high level visual features. We followed a bottom-up approach to the problem, making use of existing sonification methods in the literature, such as assigning colors to instruments. The resulting sonifications are assessed by the authors, as well as via semi-open interviews with experts. Diverging from the literature, a training phase to get familiar with the sonification rules are not applied, and still the interview analysis did show a high number of overlapping observations by the experts. Most of the sonifications are found both descriptive and pleasant. We found out that the addition of high level features to sonification created the high agreement.

One of the shortcoming of our model is the diversity between sonifications which makes it hard to differentiate between paintings by only listening to their sonifications. We plan to add more diversity by the addition of more scales. The current solution only uses the major and minor scales for light and dark paintings respectively. A finer link between the brightness and scale could be used when more scales are available. Another problem, which can be seen as a sub-problem of diversity, is the lack of musical cohesion within the sonifications. Currently, a segment that is played does not take into account the previous or next played segment in the sonification, thereby making the musical relationship between the segments up to chance, meaning sometimes one segment can play well with another segment, and sometimes not. This can make the sonifications sound random and therefore not diverse, making them hard to listen to over a long period of time. A way to solve this issue would be to make sure the chords and melodies of segments take their neighboring segments into account, therefore making the whole sonification a coherent musical piece.

Future research into this topic would benefit from a top-down approach where the focus is more on how people would turn a painting into a musical piece and what information people tend to extract during this process. This could be done by creating a website where people could describe how they would turn a painting into sound and find common concepts in their descriptions. Semantic information about paintings could be incorporated into the sonification phase by for example taking into account the era the painting is created, and use this information for choosing the historically correct instruments. The overall mood of the painting can be approved by adding emotion recognition of the sitters, and actions which would bring a depth to the sonification that is currently missing. After all, the high level information is found to be the most telling points for the descriptiveness of the sonification, and the addition of semantic and emotional information will bring more clarity. To strengthen this approach, studies from psychology should be further made use of, especially the literature about correspondences between visual and sonic stimuli, as well as the observed relations between emotions and sonic markers [7].

References

1. Aytar, Y., Vondrick, C., Torralba, A.: SoundNet: learning sound representations from unlabeled video. arXiv preprint arXiv:1610.09001 (2016)
2. Cavaco, S., Henriques, J.T., Mengucci, M., Correia, N., Medeiros, F.: Color sonification for the visually impaired. Procedia Technol. **9**, 1048–1057 (2013)
3. Cho, J.D., Jeong, J., Kim, J.H., Lee, H.: Sound coding color to improve artwork appreciation by people with visual impairments. Electronics **9**(11), 1981 (2020)
4. Fonseca, E., Favory, X., Pons, J., Font, F., Serra, X.: FSD50K: an open dataset of human-labeled sound events. arXiv preprint arXiv:2010.00475 (2020)
5. Frysinger, S.P.: A brief history of auditory data representation to the 1980s. Georgia Institute of Technology (2005)
6. Heep, E., Kapur, A.: Extracting visual information to generate sonic art installation and performance. In: Proceedings of the 21st International Symposium on Electronic Art. Vancouver, Canada (2015)
7. Juslin, P.N., Laukka, P.: Communication of emotions in vocal expression and music performance: different channels, same code? Psychol. Bull. **129**(5), 770 (2003)
8. Kabisch, E., Kuester, F., Penny, S.: Sonic panoramas: experiments with interactive landscape image sonification. In: Proceedings of the 2005 International Conference on Augmented Tele-existence, pp. 156–163 (2005)

9. Mesaros, A., Heittola, T., Virtanen, T.: Tut database for acoustic scene classification and sound event detection. In: 2016 24th European Signal Processing Conference (EUSIPCO), pp. 1128–1132. IEEE (2016)
10. Mesaros, A., Heittola, T., Virtanen, T., Fagerlund, E., Hiltunen, A.: Tut acoustic scenes 2017, development dataset (2017)
11. Micheloni, E., Mandanici, M., Roda, A., Canazza, S.: Interactive painting sonification using a sensor-equipped runway (2017)
12. Piczak, K.J.: Esc: Dataset for environmental sound classification. In: Proceedings of the 23rd ACM International Conference on Multimedia, pp. 1015–1018 (2015)
13. Polo, A., Sevillano, X.: Musical vision: an interactive bio-inspired sonification tool to convert images into music. J. Multimodal User Interfaces **13**(3), 231–243 (2019)
14. Rector, K., Salmon, K., Thornton, D., Joshi, N., Morris, M.R.: Eyes-free art: exploring proxemic audio interfaces for blind and low vision art engagement. Proceed. ACM Interactive Mobile Wearable Ubiquitous Technol. **1**(3), 1–21 (2017)
15. Short, C.: The art theory of Wassily Kandinsky, 1909–1928: the quest for synthesis. Peter Lang (2010)
16. Wu, Y., Kirillov, A., Massa, F., Lo, W.Y., Girshick, R.: Detectron2. https://github.com/facebookresearch/detectron2. Accessed 12 Feb 2023
17. Yoshida, T., Kitani, K.M., Koike, H., Belongie, S., Schlei, K.: EdgeSonic: image feature sonification for the visually impaired. In: Proceedings of the 2nd Augmented Human International Conference, pp. 1–4 (2011)
18. Zhou, B., Lapedriza, A., Khosla, A., Oliva, A., Torralba, A.: Places: a 10 million image database for scene recognition. IEEE Trans. Pattern Anal. Mach. Intell. **40**(6), 1452–1464 (2017)

Application of Neural Architecture Search to Instrument Recognition in Polyphonic Audio

Leonard Fricke$^{(\boxtimes)}$ ⓘ, Igor Vatolkin ⓘ, and Fabian Ostermann ⓘ

Department of Computer Science, TU Dortmund University, Dortmund, Germany
{leonard.fricke,igor.vatolkin,fabian.ostermann}@udo.edu

Abstract. Instrument recognition in polyphonic audio signals is a very challenging classification task. It helps to improve related application scenarios, like music transcription and recommendation, organization of large music collections, or analysis of historical trends and properties of musical styles. Recently, the classification performance could be improved by the integration of deep convolutional neural networks. However, in to date published studies, the network architectures and parameter settings were usually adopted from image recognition tasks and manually adjusted, without a systematic optimization. In this paper, we show how two different neural architecture search strategies can be successfully applied for improvement of the prediction of nine instrument classes, significantly outperforming the classification performance of three fixed baseline architectures from previous works. Although high computing efforts for model optimization are required, the training of the final architecture is done only once for later prediction of instruments in a possibly unlimited number of musical tracks.

Keywords: Neural Architecture Search · Instrument Recognition · Music Information Retrieval · Hyperband Search · Bayesian Optimization

1 Introduction and Related Work

A precise knowledge of the instruments that are present in a music piece can help us to solve many related applications, including management of large music collections, recommendation of new tracks based on similarity search, playlist generation, theoretical analysis of genre and style properties, plagiarism detection, or characterization of historical trends. Although this information is usually annotated in (digital) music scores, it is not trivial to extract it from audio recordings, and scores are not always available.

First approaches on automatic instrument recognition were based on feature engineering, like the prediction of eight instruments in solo recordings using linear prediction and cepstral coefficients with Gaussian mixture models and support vector machines [22] or 16 types of instrument samples using cepstral, spectral, and temporal features with k-nearest neighbour classifier [8]. Later works addressed

© The Author(s), under exclusive license to Springer Nature Switzerland AG 2023
C. Johnson et al. (Eds.): EvoMUSART 2023, LNCS 13988, pp. 117–131, 2023.
https://doi.org/10.1007/978-3-031-29956-8_8

more complex problems of instrument recognition in polyphonic recordings, managing the increased number of features under investigation with selection techniques, like correlation-based [9] or multi-objective evolutionary [31] feature selection. The next significant improvement came with the rise of deep neural networks (DNNs), which have proved their success to solve this and many other classification tasks, among others in the signal domain. An application of convolutional neural networks (CNNs) to recognize predominant instruments was proposed in [12].

However, despite of very good classification performance, the integration of deep neural architectures for classification tasks brought some disadvantages. Models based upon thousands of trainable parameters can overfit, i.e. show poor generalization performances (for discussion on adversarial attacks in image recognition, we refer to [1]). While this problem could be at least significantly reduced, e.g., by means of dropout layers, which randomly remove connections between neurons, or artificial augmentation of the training data [19], another challenge is that finding a network architecture that matches a specific problem best is not straightforward, as the numbers of network layers, the specific layer types, neurons, and further hyperparameters are virtually unlimited.

As yet, in proposed approaches on instrument recognition with the help of DNNs, the network architectures were usually adopted from the image classification applications, sometimes heuristically modified to achieve better performance. For instance, architectures in [12] are heavily inspired by AlexNet [18] and VGGNet [29]. [21] focused on the design of the input signal representation with the help of Hilbert-Huang transform for a CNN inspired by VGGNet [29]. [11] compared multi-layer perceptrons, CNNs, and recurrent CNNs with architectures inspired by the previous work from [3,4]. [30] applied the AlexNet [18] architecture. In [6], a CNN architecture was "developed from scratch by virtue of transfer learning", which was supposed to work on weak hardware using another input representation (scalograms based on wavelet transform). The CNN architecture in [32] was then again derived from [12]. [10] took the network design from [33]. [14] describes two fixed CNN architectures to classify guitar effects, which as task is closely related to instrument recognition. Finally, [5] compares several input representations from raw waveform to Constant-Q transform, as well as network architectures (VGGNet [29], ResNet [13], "squeeze and excitation" network [16], and musicnn [27]) for generation of musical tags including prediction of 22 instruments.

However, it is also possible to optimize network architectures in a completely automated manner, applying neural architecture search (NAS) [7]. Here, an optimization method switches between different architectures after evaluating each one using a limited number of epochs in search for the optimal one. Beyond a simple random walk, more advanced optimizers were proposed, like evolutionary algorithms [34] or Hyperband search [20].

To the best of our knowledge, until now, NAS has not been applied to the optimization of deep neural architectures for the task of instrument recognition. The current contribution is the first one to address this gap, comparing two selected NAS approaches (Bayesian and Hyperband) with three baseline network architectures. The results show significant improvements for classification performance at predicting all nine instrument classes.

Section 2 presents the backgrounds: parameter search space, baseline architectures, and network architecture tuners. In Sect. 3, we describe the dataset, input representation, and the whole neural training setup used in our study. Section 4 deals with the analysis of results with respect to classification performance. In Sect. 5, we present and discuss the best architectures identified for individual instruments. In Sect. 6, we conclude with the most important findings and list ideas for future work.

2 Methods

2.1 Network Search Space

For the evaluation of network architectures, we restrict the search space to hyperparameters and values listed in Table 1. This list should include previously applied architectures from [12, 18, 29], but also allows for a large number of possible new network configurations. Although the number of parameters under investigation is strictly limited for this initial study, the total number of possible architectures already is 51,840.[1] Therefore, it would be far too costly to perform an exhaustive search.

The networks are built from a configurable amount of convolution blocks (num_blocks). Each block contains num_layers_b convolution layers with an adjustable number of filters filters_b and a specific kernel size kernel_size_b. All layers use zero-padding and ReLU activation function. After the convolution layers, a maximum pooling layer can optionally be inserted (use_pooling_b). However, it gets turned off if the current input shape of the layer is smaller than 6×6, because, otherwise, some blocks will produce extremely low dimensions for the next layers. Next, a dropout layer is inserted with a tunable rate dropout_b (a value of zero disables dropout). All hyperparameters related to convolution blocks are tuned for each block individually.

After the very last convolution layer, a global pooling layer is added. The hyperparameter pooling controls whether to use maximum or average global pooling. Next, a dense layer with ReLU activation function and a tunable amount of neurons num_units_dense is added, followed by a dropout layer with rate dropout_dense and an optional batch normalization layer (dense_use_batchnorm). Finally, the model outputs are generated using a dense layer with 9 neurons (the number of instrument classes, cf. Sect. 3.1) with sigmoid activation function. All models are trained using the binary cross entropy loss function and ADAM optimization [17].

---

[1] $(3(3·5·2·3·2)+4(3·5·2·3·2)+5(3·5·2·3·2))·2·2·3·2 = (3·180+4·180+5·180)·24 = 51,840.$

Table 1. Hyperparameter search space. Parameters with suffix '_b' are tuned individually for each block in num_blocks.

Parameter name	Possible values
num_blocks	$\{3, 4, 5\}$
num_layers_b	$\{1, 2, 3\}$
filters_b	$\{16, 32, 64, 128, 256\}$
kernel_size_b	$\{3, 5\}$
dropout_b	$\{0.0, 0.25, 0.5\}$
use_pooling_b	$\{$true, false$\}$
pooling	$\{$avg, max$\}$
num_units_dense	$\{512, 1024\}$
dropout_dense	$\{0.0, 0.25, 0.5\}$
dense_use_batchnorm	$\{$true, false$\}$

2.2   Baseline Networks

To measure the impact of NAS optimization, we will compare the results to three fixed baseline networks, which architectures are presented in Table 2. During the training of the baselines, early stopping was applied after three epochs with no progress.

Han follows the architecture proposed for instrument recognition in [12] and adjusted in [32].

AlexNet is inspired by [18] and contains less convolutional layers per block (except for the final one) with larger filters.

VGG16 is based on [29] and has more convolutional layers.

Table 2. Architecture blocks for three networks inspired by previous work. Keras layers: $C(s, f)$: Conv2D(kernel_size $= (s, s)$, filters $= f$); p: MaxPooling2D$(1, 2)$; P: MaxPooling2D$(2, 4)$; d: Dropout(0.25); D: Dropout(0.5); G: GlobalMaxPooling2D; f: Dense($relu$); F: Dense($sigmoid$).

Han	AlexNet	VGG16
$[C(3, 32) \times 2]$pd	$C(11, 32)$pd	$[C(3, 32) \times 2]$pd
$[C(3, 64) \times 2]$pd	$C(5, 64)$pd	$[C(3, 64) \times 2]$pd
$[C(3, 128) \times 2]$Pd	$[C(3, 128) \times 3]$Gd	$[C(3, 128) \times 3]$Pd
$[C(3, 256) \times 3]$Gd	fDF	$[C(3, 256) \times 3]$Gd
fDF		fDF

2.3   Tuners

The hyperparameters can be optimized using *tuner* algorithms which apply different strategies to explore the parameter space and evaluate the performance of adjusted network models. A selected hyperparameter configuration that is to be tested is called a *trial*. All tuners in this work are implemented with `keras-tuner` [25]. In our study, we apply the following two tuners.

Bayesian tuner starts with several random samples of the search space and applies Bayesian optimization [28]. Next, the model is refined by updating the Bayesian *posterior distribution* by testing new configurations. The posterior is a probabilistic distribution of the objective function that is being optimized. New configurations are chosen so that the uncertainty of that distribution is reduced. The Bayesian tuner in `keras-tuner` uses a Gaussian process to model the posterior [25].

From that posterior of the objective, an acquisition function is used to control the search algorithm. The acquisition function uses the uncertainty of the posterior (exploration) and the estimated objective (exploitation) to select the configuration for the next evaluation of hyperparameters. The exploitation part guides the tuner to configurations with good posterior value (objective). The exploration component prefers configurations with high uncertainty which are the areas of the search space that have not been tested yet. A comprehensible visualization of the Bayesian optimization process with mean and confidence intervals of the posterior is given in [28, Fig. 1]. Overall, each trial reduces the uncertainty in the posterior distribution and thus can locate better optima of the objective function. In this work, the upper confidence bound function is used as acquisition function [28, p. 14]:

$$\alpha_{UCB}(\mathbf{x}) := \mu_i(\mathbf{x}) + \beta\sigma_i(\mathbf{x}), \tag{1}$$

where $\mathbf{x}$ is an arbitrary point in the posterior distribution; μ_i and σ_i are the posterior mean and the posterior uncertainty at trial i; β is a balancing factor between exploration and exploitation.

Hyperband [20] is a tuner that was developed to reduce computational resources while preserving search performance comparable to Bayesian optimization. It is an extension of the SuccessiveHalving algorithm [15]. SuccessiveHalving uniformly allocates a budget r of computational resources to a selected set of n hyperparameter configurations. After their evaluation, the worst half of configurations is rejected. This is repeated until only one configuration is left. The computational resources that are invested (e.g., the number of epochs) are higher for more promising hyperparameters.

In the original SuccessiveHalving algorithm, the budget is uniformly allocated to a set of configurations and then the worst configurations are thrown out while the promising models exponentially get more resources. However, it is unclear whether it is better to consider many configurations with small budget or few configurations with large budget [20]. Hyperband solves this by performing the SuccessiveHalving algorithm in an inner loop while essentially applying grid search over n and r. Two parameters to setup are the maximum amount of resources R that can be allocated to a configuration and the reduction factor η that controls which values are considered for n and r. Each run of Successive-Halving is called a 'bracket' (Algorithm 1, lines 2–11) which chooses n and r such that the combined budget is approximately equal for all brackets (line 3). The total amount of brackets s_{max} (line 1) arises from the input parameters R and η. B (line 1) describes the budget that each bracket should approximately use. The Hyperband algorithm starts with high n and low r. Then, in each bracket, the number of configurations n is reduced and the computational budget r is increased by the factor of η (lines 6–7). The SuccessiveHalving algorithm with inputs n and r runs from lines 4 to 10.

Input: R, η

1 Initialize $s_{max} = \lfloor \log_\eta(R) \rfloor, B = (s_{max} + 1)R$;

2 **for** $s \in \{s_{max}, s_{max} - 1, ..., 0\}$ **do**

3 $\quad n = \left\lceil \frac{B}{R} \frac{\eta^s}{s+1} \right\rceil, r = R\eta^{-s}$;

$\quad$ // begin SuccessiveHalving with (n, r)

4 $\quad T$=get_hyperparameter_configuration(n);

5 $\quad$ **for** $i \in \{0, ..., s\}$ **do**

6 $\quad\quad n_i = \lfloor n\eta^{-i} \rfloor$;

7 $\quad\quad r_i = r\eta^i$;

8 $\quad\quad L = \{$run_then_return_val_loss$(t, r_i), t \in T\}$;

9 $\quad\quad T = top_k(T, L, \lfloor n_i/\eta \rfloor)$;

10 $\quad$ **end**

11 **end**

12 **return** *Configuration with smallest loss seen so far*

Algorithm 1: The Hyperband algorithm in pseudo code [20].

The functions from lines 4, 8, and 9 perform the following:

get_hyperparameter_configuration(n) - a function that returns a set of n samples from a random distribution defined over the hyperparameter space. Because the search space in our work only consists of discrete parameters, we just sample unique random values.

run_then_return_val_loss(t, r) - a function that takes a hyperparameter configuration t and resource allocation r as input and returns the validation loss after training the configuration for the allocated resources.

top_k(configs, losses, k) - a function that takes a set of configurations as well as their associated losses and returns the top k performing configurations.

In this work, the value $\eta = 3$ is adopted from the original paper [20], and the resource budget is calculated as the number of invested epochs. This leads to a total number of $R \lfloor \log_\eta(R) \rfloor$ trials to be performed within each Hyperband run.

3   Design of Experiments

3.1   Dataset

For the instrument recognition task, we used the *Artificial Audio Multitracks* (AAM) dataset [26], which is publicly available and contains 3000 artificial music tracks generated by an algorithmic composer. It attempts to simulate Western pop music and then synthesized using realistic samples from Native Instrument Komplete [24]. Because audio and score are perfectly aligned, the annotations of present instruments are precise and available for each onset time event (the beginnings of new tones), in contrast to other datasets with instrument annotations, like instrument tags for full tracks in MTG-Jamendo dataset [2]. The use of synthetic data helps to thoroughly compare our results on the different methods due to the absolute precision and quality of the underlying annotation data.

The onset events which deal as classification instances are represented by 128-dimensional mel spectrograms extracted with `librosa` [23] with a sampling rate of 16 kHz and frames of 2048 samples with a step size of 1024 samples. The onsets with annotated instruments are represented by appr. 2 s windows (as in [32]), leading to input matrices of size 128×31 for the CNN models.[2]

The classification task is to predict the occurrences of nine different classes of Western and ethnic instruments: bass (acoustic and electric), brass (flugelhorn, trombone, trumpet), drums, guitar (acoustic and electric, balalaika, sitar, ukulele), organ, piano (acoustic and electric), pipe (flute, fujara, panflute, shakuhachi), reed (clarinet, saxophones), and strings (contrabass, cello, erhu, jinghu, morin khuur, viola, violin). From all available tracks, 2000 are used to train the neural networks, 500 to validate the weights, and 500 are reserved for the final independent test.

3.2   Setup

The parameter search was executed in the space as defined in Sect. 2.1 using the baseline architectures presented in Sect. 2.2 and the NAS tuning algorithms explained in Sect. 2.3. In one trial, the tuner chooses a hyperparameter configuration, trains it on 2000 tracks, and validates it on 500. The Bayesian tuner trains for a fixed amount of 30 epochs per trial, the Hyperband tuner trains for 30 epochs only in the final bracket. The brackets train for 2, 4, 10, 30 epochs in ascending order. This allocation of computational budget comes from the implementation of Hyperband in `keras-tuner`. The value `max_epochs = 30` was chosen for both tuners, because in the preliminary experiments the models converged after about 30 epochs. Every trial repeats the training process three times and then estimates the mean evaluation measure of these repetitions to decrease the influence of random weight initialization. The validation loss of the best epoch during training serves as an evaluation measure to select the models for a final test on the remaining 500 tracks. For the latter, the balanced relative error (BRE)[3] is separately estimated for each instrument class.

Each run of the Bayesian tuner was repeated 10 times. Two parameter configurations were sampled as starting points before the beginning of the posterior optimization. The balancing factor was set to $\beta = 2.6$ as suggested by

---

[2] The step size of 1024 samples corresponds to $\frac{1}{16,000} = 0.064$ s; $31 \cdot 0.064$ s $= 1.984$ s.

[3] BRE is the mean of classification errors for "positive" onsets (containing an instrument) and "negative" onsets (not containing that particular instrument).

keras-tuner. Each Bayesian run did 25 trials. This leads to a total number of 750 epochs per statistical run, with 3 repetitions each.

The Hyperband search was also executed 10 times and explored 90 trials in each search run. The reduction factor was set to the suggested value of $\eta = 3$ [20, p. 8]. Combined with the maximum epochs of $R = 30$, the budget and bracket calculation from keras-tuner leads to $30 \cdot \lfloor \log_\eta (30) \rfloor = 90$ trials per Hyperband run. Each Hyperband search trained for a maximum of 706 epochs (following the implementation in keras-tuner [25]), again with 3 repetitions per trial. In practice, the real number of epochs for both tuners was slightly lower, because as with the baselines early stopping was performed after 3 epochs without progress.

Because both tuners allocate epochs to trials with a different strategy, we have selected the tuner settings in a way that the computational budget, as is the total number of epochs, was supposed to become approximately equal. The Bayesian tuner ran for a total of 250 trials, while the Hyperband tuner did 900 trials. With the previously described setup this leads to 22,500 and 21,180 epochs that are invested in total for Bayesian and Hyperband, respectively. The major increase in trials arises from the aggressive early stopping of poorly performing trials in the first brackets of the Hyperband algorithm.

4   Results: Classification Performance

Table 3 summarizes the results of the three baseline networks as described in Sect. 2.2. VGG16 performs best for 6 of 9 classes. Also, clear differences exist between the instruments: the prediction of organ occurrence seems to be a quite simple task, leading to errors with two zeros after the fix point for all architectures. The recognition of reed, guitar, and piano is more challenging, producing lowest errors around 0.10.

Table 3. Balanced relative errors on the test dataset of the baseline networks Han, AlexNet, and VGG16. Best values per class are highlighted.

	Bass	Brass	Drums	Guitar	Organ	Piano	Pipe	Reed	Strings	Mean
Han	0.0167	0.0429	0.0355	0.1060	0.0017	0.0990	0.0871	0.1060	0.0303	0.0648
AlexNet	0.0211	0.0940	0.0284	0.1687	0.0040	0.1624	0.1450	0.1718	0.0511	0.0941
VGG16	0.0161	0.0349	0.0346	0.0953	0.0013	0.0941	0.0871	0.1168	0.0288	0.0628

The experiments were executed on computers with NVIDIA RTX 2080 GPUs. Each search run lasted between 5 to 7 days.

Table 4 presents the best validation losses of trials during the tuning process. Both algorithms reach varying loss values in 10 individual search runs due to the randomness in their initialization and their exploration strategies. Hyperband achieved the best trial in its second run (**#2**) with a validation loss of 0.0655. Bayesian tuner reached the lowest validation loss in the first run (**#1**) with 0.0876. Please note that Bayesian had a larger budget of computing resources (cf. Sect. 3.2). When comparing the mean validation loss of all 10 runs, then Hyperband was able to reach a lower mean value than Bayesian optimization even though using less resources.

Table 4. The validation losses of the best trials after each search run for each tuning algorithm. The best value for each tuner are highlighted.

Run	Bayesian	Hyperband
#1	0.0876	0.0825
#2	0.0886	0.0655
#3	0.0988	0.0956
#4	0.1185	0.1157
#5	0.0928	0.0993
#6	0.0941	0.1005
#7	0.1125	0.0943
#8	0.0971	0.0990
#9	0.0924	0.1046
#10	0.0927	0.0852
Mean	0.0988	0.0942

The progress of validation loss over the tuning process for the best Bayesian run and the best Hyperband is plotted in Fig. 1 and Fig. 2, respectively. The figures show that both search algorithms succeed in reducing the validation loss during their tuning. The validation loss curves show strong fluctuations, because every trial changes the network architecture. However, the Hyperband tuner shows a decreasing loss in the overall process, while Bayesian only slightly reduces the loss value. Compared to Bayesian search, Hyperband achieves not only the lowest loss value of all trials (see Table 4), but also finds more configurations with loss below 0.1.

Because the classification models were only trained for a maximum number of 30 epochs during the tuning process, the best model configuration from each run of the tuners was re-trained a total of 200 epochs (with early stopping) for the final validation on the test set. To evaluate these models, we estimated the test BRE for each instrument class. We also computed the mean BRE over all instrument classes to compare the individual models from each run.

The best Bayesian model architecture was found in run #2 (cf. Table 5). It produces the lowest errors for 5 of 9 instrument classes. Brass and pipe were best recognized by #5. The recognition of drums and strings performed best using models from runs #10 and #9, respectively.

The re-trained models from the Hyperband runs achieve the lowest mean BRE in the runs #10, #2, and #6 (cf. Table 6). The BRE per instrument also reaches the lowest errors in these runs. Run #2 is the best for guitar, organ, pipe, reed, and strings. Run #10 is the best for bass, brass, drums and piano, while run #6 and #10 reached the same value for bass.

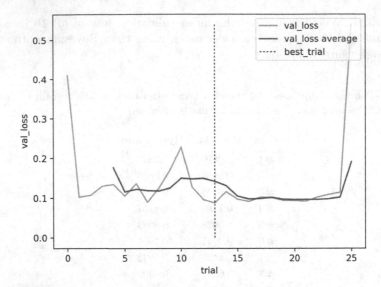

Fig. 1. Bayesian (run #1): Loss on the validation set during tuning process with running-average (window size of 5). The best trial is marked with a dotted line.

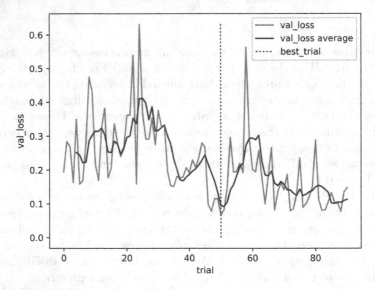

Fig. 2. Hyperband (run #2): Loss on the validation set during tuning process with running-average (window size of 5). The best trial is marked with a dotted line.

The comparison of the best models of both tuners from Tables 5 and 6 reveals that Hyperband was able to discover more network architectures with lower mean BRE. The best model from all search runs is also found by Hyperband in run #10. It must be noted that Bayesian search found better networks for the instrument classes drums, guitar, and piano. Organ was well classified by both tuners with an equal BRE. Generally, Hyperband shows more BRE values that are lower than in the baseline experiments (58 green cells for columns with individual instruments in Table 6 against 42 green cells in Table 5). The two best networks from Hyperband both show lower BRE than the baseline did for all instruments classes. The Bayesian search did not outperform the baselines for all instruments with a single network. Nevertheless, several runs are characterized by many error values above the baselines, which means that the application of NAS should not applied only once and can potentially be stuck in a local optima.

Table 5. Bayesian search: Balanced relative errors on the test dataset of the best re-trained network architectures of each search run. The best value per class ist are highlighted darker. The best value for both tuners is marked **bold**. Values better than the baseline are highlighted brighter.

Run	Bass	Brass	Drums	Guitar	Organ	Piano	Pipe	Reed	Strings	Mean
#1	0.0167	0.0387	0.0154	0.0878	0.0035	0.0977	0.0655	0.0654	0.0323	0.0472
#2	0.0126	0.0266	0.0310	**0.0657**	**0.0010**	**0.0668**	0.0528	0.0446	0.0251	0.0363
#3	0.0178	0.0439	0.0514	0.1108	0.0031	0.0989	0.0912	0.1030	0.0334	0.0615
#4	0.0205	0.0680	0.0225	0.1533	0.0064	0.1772	0.1383	0.1682	0.0418	0.0885
#5	0.0159	0.0241	0.0127	0.0900	0.0020	0.1105	0.0439	0.0706	0.0257	0.0438
#6	0.0176	0.0753	0.0157	0.1183	0.0018	0.1647	0.0557	0.0604	0.0352	0.0605
#7	0.0174	0.0595	0.0242	0.1048	0.0035	0.1083	0.0632	0.0570	0.0368	0.0507
#8	0.0181	0.0396	0.0207	0.0907	0.0027	0.0999	0.0946	0.0597	0.0249	0.0501
#9	0.0166	0.0430	0.0220	0.0862	0.0017	0.0912	0.0578	0.0724	0.0237	0.0461
#10	0.0193	0.0268	**0.0113**	0.0816	0.0017	0.1190	0.0491	0.0512	0.0382	0.0442

Table 6. Hyperband search: Balanced relative error on the test dataset of the best re-trained network architectures of each search run. The best value per class ist are highlighted darker. The best value for both tuners is marked **bold**. Values better than the baseline are highlighted brighter.

Run	Bass	Brass	Drums	Guitar	Organ	Piano	Pipe	Reed	Strings	Mean
#1	0.0169	0.0383	0.0197	0.1680	0.0031	0.1136	0.0597	0.0844	0.0305	0.0594
#2	0.0140	0.0329	0.0271	0.0712	**0.0010**	0.0796	**0.0376**	**0.0364**	**0.0135**	0.0348
#3	0.0150	0.0299	0.0196	0.0810	0.0030	0.0791	0.0769	0.0472	0.0220	0.0415
#4	0.0180	0.0579	0.0157	0.1593	0.0047	0.1915	0.0807	0.0704	0.0708	0.0743
#5	0.0176	0.0234	0.0339	0.0879	0.0027	0.0957	0.0433	0.0503	0.0378	0.0436
#6	**0.0124**	0.0291	0.0152	0.0828	0.0016	0.0849	0.0426	0.0509	0.0283	0.0387
#7	0.0131	0.0287	0.0291	0.1040	0.0021	0.1110	0.1132	0.0713	0.0533	0.0584
#8	0.0143	0.0301	0.0271	0.0985	0.0014	0.1252	0.0685	0.0815	0.0277	0.0527
#9	0.0172	0.0325	0.0305	0.0880	0.0041	0.1037	0.0379	0.0406	0.0256	0.0422
#10	**0.0124**	**0.0201**	0.0149	0.0765	0.0012	0.0730	0.0442	0.0448	0.0198	**0.0341**

5 Results: Best Architectures

In the following, we present the best network architectures that were found by both search algorithms. The best networks by mean BRE for Bayesian and Hyperband tuners are shown in Tables 7 and 8, respectively.

When comparing the best architectures, we can observe that most of them are only using three convolution blocks. The networks prefer larger filter sizes like 128 and 256, especially in the deeper layers. The numbers of convolutional layers and their kernel sizes, however, show no distinct pattern. The best networks used maximum pooling layers between the convolution blocks at different locations, but unexpectedly not after each and every block.

Table 7. Architecture blocks for the best networks found by the Bayesian search runs, sorted by their mean BRE (**Mean**). The instruments (**Instr.**) that are classified the best by this architecture are listed. Keras layers: $C(s, f)$: Conv2D(kernel_size = (s,s), filters = f); p: MaxPooling2D; d(r): Dropout(r); G: GlobalMaxPooling2D; A: GlobalAveragePooling2D; B: BatchNormalization; f(n): Dense(relu, units = n); F: Dense($sigmoid$).

Run	#2	#5	#10	#9
Mean	0.0363	0.0438	0.0442	0.0461
Instr.	Bass, Guitar, Organ, Piano, Reed	Brass, Pipe	Drums	Strings
	$[C(3, 128) \times 2]$pd(0.25)	$C(5, 64)$d(0.5)	$[C(3, 128) \times 3]$p	$C(5, 256)$pd(0.5)
	$C(5, 256)$pd(0.5)	$[C(5, 128) \times 3]$p	$C(3, 256)$d(0.5)	$C(3, 256)$d(0.5)
	$C(3, 128)$p	$C(3, 256)$d(0.5)	$C(3, 256)$	$C(5, 256)$
	$C(5, 32)$	Af(1024)BF	Af(1024)d(0.5)BF	$C(3, 256)$pd(0.5)
	$[C(5, 16) \times 2]$d(0.5)			Af(1024)F
	Af(1024)F			

Table 8. Architecture blocks for the best networks found by the Hyperband search runs, sorted by their mean BRE (**Mean**). The instruments (**Instr.**) that are classified the best by this architecture are listed. Keras layers: $C(s, f)$: Conv2D(kernel_size = (s,s), filters = f); p: MaxPooling2D; d(r): Dropout(r); G: GlobalMaxPooling2D; A: GlobalAveragePooling2D; B: BatchNormalization; f(n): Dense(relu, units = n); F: Dense($sigmoid$).

Run	#10	#2	#6
Mean	0.0341	0.0348	0.0387
Instr.	Brass, Drums, Piano	Guitar, Organ, Pipe, Reed, Strings	Bass
	$[C(3, 32) \times 2]$d(0.25)	$[C(3, 64) \times 3]$d(0.25)	$[C(5, 128) \times 3]$p
	$[C(3, 256) \times 3]$pd(0.25)	$[C(5, 256) \times 2]$pd(0.25)	$C(3, 256)$d(0.25)
	$[C(5, 128) \times 2]$pd(0.25)	$[C(3, 256) \times 2]$	$C(5, 256)$pd(0.25)
	Af(1024)d(0.5)BF	Af(1024)d(0.5)F	Gf(1024)BF

In contrast to the baselines, almost all of the listed architectures use global average pooling instead of global maximum pooling. For the hidden dense layer, it is clear that 1024 neurons is the best value. For the dropout layers and the batch normalization layers, that are placed just in front of the final output layer, also no distinct pattern is observable.

6 Conclusions and Future Work

In this work, we have showed how neural architecture search can be successfully applied to the recognition of musical instruments in polyphonic audio signals using convolutional neural networks. For all nine tested instrument classes, it was possible to significantly reduce the classification errors using optimized architectures and hyperparameters after the application of either Bayesian or Hyperband tuner, compared to three fixed baseline architectures from previous related works. The strongest reduction of the balanced classification error to 34.34% of the best baseline value was achieved for the recognition of reed instruments (from 0.1060 to 0.0364), and the weakest for bass instruments (to 77.02%, from 0.0161 to 0.0124).

Although the neural architecture search itself has shown to be time intensive, it is required only once for model optimization, while the end-use application of instrument recognition can be performed fast and is necessary only once per music track. From both tuning methods, the tuner based on the Hyperband algorithm not only performed better than the Bayesian tuner on average with respect to smallest classification errors, but also required less epochs due to its sophisticated strategy to allocate resources.

Future work may, on the one hand, include the exploration of a larger hyperparameter search space, e.g., including layers of other types, in order to further optimize neural architectures for instrument recognition, and rigorous statistical comparison of different network models. On the other hand, regarding our results, it seems highly appropriate to attempt an application of neural architecture search to other music information retrieval tasks, like onset detection, chord recognition, tempo estimation, music segmentation, and similar. Furthermore, approaching multi-objective optimization scenarios like minimization of network size against maximization of classification quality could be promising.

References

1. Akhtar, N., Mian, A.: Threat of adversarial attacks on deep learning in computer vision: a survey. IEEE Access 6, 14410–14430 (2018)
2. Bogdanov, D., Won, M., Tovstogan, P., Porter, A., Serra, X.: The MTG-Jamendo dataset for automatic music tagging. In: Proceedings of Machine Learning for Music Discovery Workshop at the International Conference on Machine Learning (ICML) (2019)
3. Çakir, E., Parascandolo, G., Heittola, T., Huttunen, H., Virtanen, T.: Convolutional recurrent neural networks for polyphonic sound event detection. IEEE/ACM Trans. Audio Speech Language Process. 25(6), 1291–1303 (2017)

4. Choi, K., Fazekas, G., Sandler, M.B., Cho, K.: Convolutional recurrent neural networks for music classification. In: Proceedings of the IEEE International Conference on Acoustics, Speech and Signal Processing (ICASSP), pp. 2392–2396. IEEE (2017)
5. Damböck, M., Vogl, R., Knees, P.: On the impact and interplay of input representations and network architectures for automatic music tagging. In: Proceedings of the 23rd International Society for Music Information Retrieval Conference (ISMIR) (2022)
6. Dutta, A., Sil, D., Chandra, A., Palit, S.: CNN based musical instrument identification using time-frequency localized features. Internet Technol. Lett. **5**(1), e191 (2020)
7. Elsken, T., Metzen, J.H., Hutter, F.: Neural architecture search: a survey. J. Mach. Learn. Res. **20**, 55:1-55:21 (2019)
8. Eronen, A., Klapuri, A.: Musical instrument recognition using cepstral coefficients and temporal features. In: Proceedings of the IEEE International Conference on Acoustics, Speech, and Signal Processing (ICASSP), vol. 2, pp. 753–756. IEEE (2000)
9. Fuhrmann, F.: Automatic musical instrument recognition from polyphonic music audio signals. Ph.D. thesis, Universitat Pompeu Fabra, Department of Information and Communication Technologies (2012)
10. Garcia, H.F., Aguilar, A., Manilow, E., Pardo, B.: Leveraging hierarchical structures for few-shot musical instrument recognition. In: Proceedings of the 22nd International Society for Music Information Retrieval Conference (ISMIR), pp. 220–228 (2021)
11. Gururani, S., Summers, C., Lerch, A.: Instrument activity detection in polyphonic music using deep neural networks. In: Proceedings of the 19th International Society for Music Information Retrieval Conference (ISMIR), pp. 569–576 (2018)
12. Han, Y., Kim, J., Lee, K.: Deep convolutional neural networks for predominant instrument recognition in polyphonic music. IEEE/ACM Trans. Audio Speech Language Process. **25**(1), 208–221 (2017)
13. He, K., Zhang, X., Ren, S., Sun, J.: Deep residual learning for image recognition. In: Proceedings of the IEEE Conference on Computer Vision and Pattern Recognition (CVPR), pp. 770–778. IEEE Computer Society (2016)
14. Hinrichs, R., Gerkens, K., Ostermann, J.: Classification of guitar effects and extraction of their parameter settings from instrument mixes using convolutional neural networks. In: Martins, T., Rodríguez-Fernández, N., Rebelo, S.M. (eds.) EvoMUSART 2022. LNCS, vol. 13221, pp. 101–116. Springer, Cham (2022). https://doi.org/10.1007/978-3-031-03789-4_7
15. Jamieson, K., Talwalkar, A.: Non-stochastic best arm identification and hyperparameter optimization. In: Proceedings of the 19th International Conference on Artificial Intelligence and Statistics (AISTATS), pp. 240–248. JMLR.org (2016)
16. Kim, T., Lee, J., Nam, J.: Sample-level CNN architectures for music auto-tagging using raw waveforms. In: Proceedings of the IEEE International Conference on Acoustics, Speech and Signal Processing (ICASSP), pp. 366–370. IEEE (2018)
17. Kingma, D., Ba, J.: Adam: a method for stochastic optimization. In: Proceedings of the 3rd International Conference on Learning Representations (ICLR) (2015)
18. Krizhevsky, A., Sutskever, I., Hinton, G.E.: ImageNet classification with deep convolutional neural networks. Commun. ACM **60**(6), 84–90 (2017)
19. Lemley, J., Bazrafkan, S., Corcoran, P.: Smart augmentation learning an optimal data augmentation strategy. IEEE Access **5**, 5858–5869 (2017)

20. Li, L., Jamieson, K., DeSalvo, G., Rostamizadeh, A., Talwalkar, A.: Hyperband: a novel bandit-based approach to hyperparameter optimization. J. Mach. Learn. Res. **18**(185), 1–52 (2018)
21. Li, X., Wang, K., Soraghan, J., Ren, J.: Fusion of Hilbert-Huang transform and deep convolutional neural network for predominant musical instruments recognition. In: Romero, J., Ekárt, A., Martins, T., Correia, J. (eds.) EvoMUSART 2020. LNCS, vol. 12103, pp. 80–89. Springer, Cham (2020). https://doi.org/10.1007/978-3-030-43859-3_6
22. Marques, J., Moreno, P.J.: A study of musical instrument classification using Gaussian mixture models and support vector machines. Cambridge Res. Lab. Techn. Rep. Ser. CRL **4**, 143 (1999)
23. McFee, B., et al.: Librosa: audio and music signal analysis in python. In: Proceedings of the Python Science Conference, pp. 18–25 (2015)
24. Instruments, N.: Komplete 11 Ultimate. Native Instruments North America Inc., Los Angeles (2016)
25. O'Malley, T., et al.: Kerastuner (2019). https://github.com/keras-team/keras-tuner. Accessed 7 Feb 2023
26. Ostermann, F., Vatolkin, I.: AAM: artificial audio multitracks dataset (2022). https://doi.org/10.5281/zenodo.5794629. Accessed 7 Feb 2023
27. Pons, J., Nieto, O., Prockup, M., Schmidt, E.M., Ehmann, A.F., Serra, X.: end-to-end learning for music audio tagging at scale. In: Proceedings of the 19th International Society for Music Information Retrieval Conference (ISMIR), pp. 637–644 (2018)
28. Shahriari, B., Swersky, K., Wang, Z., Adams, R., de Freitas, N.: Taking the human out of the loop: a review of Bayesian optimization. Proc. IEEE **104**(1), 148–175 (2016)
29. Simonyan, K., Zisserman, A.: Very deep convolutional networks for large-scale image recognition. In: Proceedings of the 3rd International Conference on Learning Representations (ICLR) (2015)
30. Solanki, A., Pandey, S.: Music instrument recognition using deep convolutional neural networks. Int. J. Inform. Technol. 1–10 (2019)
31. Vatolkin, I., Preuß, M., Rudolph, G., Eichhoff, M., Weihs, C.: Multi-objective evolutionary feature selection for instrument recognition in polyphonic audio mixtures. Soft. Comput. **16**(12), 2027–2047 (2012)
32. Vatolkin, I., Adrian, B., Kuzmic, J.: A fusion of deep and shallow learning to predict genres based on instrument and timbre features. In: Romero, J., Martins, T., Rodríguez-Fernández, N. (eds.) EvoMUSART 2021. LNCS, vol. 12693, pp. 313–326. Springer, Cham (2021). https://doi.org/10.1007/978-3-030-72914-1_21
33. Wang, Y., Salamon, J., Cartwright, M., Bryan, N.J., Bello, J.P.: Few-shot drum transcription in polyphonic music. In: Proceedings of the 21th International Society for Music Information Retrieval Conference (ISMIR), pp. 117–124 (2020)
34. Zhou, X., Qin, A.K., Gong, M., Tan, K.C.: A survey on evolutionary construction of deep neural networks. IEEE Trans. Evol. Comput. **25**(5), 894–912 (2021)

AI-rmonies of the Spheres

Adrián García Riber[1]([✉]) [iD] and Francisco Serradilla[2] [iD]

[1] Universidad Politécnica y Real Conservatorio Superior de Música de Madrid, Madrid, Spain
adrian.griber@alumnos.upm.es
[2] Departamento de Sistemas Informáticos., Universidad Politécnica de Madrid, 28031 Madrid, Spain
francisco.serradilla@upm.es
http://pdmusica.etsist.upm.es/

Abstract. Thanks to the efforts and cooperation of the international community, nowadays it is possible to analyze astronomical data captured by the observatories and telescopes of major space agencies around the world from a personal computer. The development of virtual observatory technology (VO), and the standardization of the formats it uses, allow professional and amateur astronomers to access astronomical data and images through internet with relative ease. Immersed in this environment of global accessibility, this article presents an astronomical data-driven unsupervised music composition system based on *Deep Learning*, aimed at offering an automatic and objective review on the classical topic of the *Harmonies of the Spheres*. The system explores the MILES stellar library from the Spanish Virtual Observatory (SVO) using a variational autoencoder architecture to cross-match its stellar spectra via *Pitch-Class Set Theory* with a music score generated by a LSTM with attention neural network in the style of late-renaissance music.

Keywords: Deep Learning · LSTM · Sonification · Music · Astronomy

1 Introduction

The Universe, understood not only as a source of inspiration but also as a source of musical harmony, has occupied the mind of mathematicians, musicians and astronomers from the times of ancient Greece. Updating this concept to the current available technology, the connection of different astronomical data streams with the generation and control of sound variables, opens a wide window of possibilities for Sound Design and Music Composition. This work explores the potential in the use of *Deep Learning* techniques to provide an unsupervised perspective of the classical concept of the *Music of the spheres*, focused around the figure of one of its biggest names, Johannes Kepler, and what was understood as music during the writing process of his treatment *Harmonices Mundi*, published in 1619, and containing his third law for planetary motion.

ⓒ The Author(s), under exclusive license to Springer Nature Switzerland AG 2023
C. Johnson et al. (Eds.): EvoMUSART 2023, LNCS 13988, pp. 132–147, 2023.
https://doi.org/10.1007/978-3-031-29956-8_9

In summary, this work converts the almost 1000 stellar spectra of MILES stellar library from the Spanish Virtual Observatory (SVO) into a data base of "stellar chords", using a variational autoencoder architecture. These chords are cross-matched with the musical chords generated by a LSTM with attention neural network trained with over 1000 scores of music from the Italian composer Giovanni Pierluigi da Palestrina (1525-1594), considered one of the leading composers in Europe of late 16th-century. Finally, the spectral music composition driven by the style of Palestrina is generated with the "matched chords", that is, the musical chords present in the stellar chords library, or in a similar way, the stellar spectra auditory representations that fit the sounds of those times.

2   Concepts of Interest

Exploring the intersections between Music and Astronomy using Artificial Intelligence requires a quick overview of the historical framework as well as some useful definitions before delving into the technical details of the research.

2.1   Brief History of Astronomical Harmonies

It seems agreed to mention Pythagoras (VI century BC) and his *Music of the spheres*, as the first work of practical and theoretical reference in the field of Music and Astronomy. According to the ethnomusicologist Mark Ballora, the demonstration carried out by the Pythagorean school on the mono-chord of the relationships between intervals of perfect fifths and the distances to the earth of the bodies that at that time were believed to orbit around it, can also be seen as the first evidence on astronomical data sonification [1]. This idea of relating musical intervals to the distances and orbital velocities of the planets transcends from antiquity to Renaissance from the hand of Plato and Aristotle. Regarding Plato, through two main sources: *The myth of Er* and *The passage of Timaeus*, both belonging to his ten-volume work *The Republic* [2]. Regarding Aristotle, through his clear description of the theory of the harmony of the spheres in *De caelo* 290 b12 and in 291a 8, in which he affirms that this concept is Pythagorean [3]. However, and despite the fact that little is known with certainty about the Pythagorean doctrines, authors specialized in Greek theories of the harmony of the spheres such as Von Jan (Jan, V. 1893, cited in [2], p.23), point out that the astronomical references of the Pythagorean school should be interpreted as simple analogies and not as an astronomical theory due to their limited knowledge on the subject

In the second century A.D., Ptolemy also outlined the concepts of harmonies of celestial bodies in his book *Harmonics*, describing them as mere rational connections obeying the general laws of motion. For Ptolemy, ordered movement, both in the stars and in music, follows certain patterns so that the study of these patterns in one field can help in the understanding of other fields [2].

The writings of Boethius (480-524 AD) on Aristotelian logic and the *Quadrivium*, had a very influential role in the dissemination of Pythagorean musical theories during the Middle Ages. Since the beginning of the sixth century, Arithmetic, Geometry, Astronomy and Music, understood as the science of numbers that describe sound, represented the four main fields of quantitative science. The term *Quadrivium* was used to group these disciplines, considered as the four branches of Mathematics, capable of describing the knowledge of the natural world. Boethius established a tripartite classification of music -Mundane, Human and Instrumentalis-, which implied the acceptance of planetary relations with musical intervals from an apparently continuist position. *Mundane music*, referring to the harmony of the spheres of heaven, *Human music* dealing with the influence of music on the human soul and *Instrumentalis music*, what we currently know as music [2].

During the fifteenth century, humanists such as Coluccio Salutati or musicians such as Johannes Tinctoris rejected the idea of the existence of worldly or Mundane music, while music theorists such as Franchino Gaffurio claimed that it could only be heard by true virtuous people.

At the end of the sixteenth century, in an attempt to describe music in its entirety, Gioseffo Zarlino published the treatise *Istitutioni harmoniche* (1558), which extended the Pythagorean harmonic theory, and was the frame of reference for the musical theory of that time.

Far from the mysticism of the ancient Greeks, Ptolemy's approaches together with the discoveries of the empirical bases of the musical consonances of Vizenzo Galilei -father of Galileo- in his *Dialogo della musica antica, et della moderna* (1581), would inspire the work *Harmonices Mundi*, in which Johannes Kepler exposes how the planets move describing an elliptical orbit around the sun. The book was published in 1619, and it is considered the masterpiece of the interdisciplinary musical-astronomical thought. In *Book V* of this treatise, Kepler translated the parameters of motion and distance of the planets of the solar system into musical intervals, something that apparently led him to formulate the equation that allowed him to lay the foundations of astronomy. Analyzing his work as a model of planetary sonification, Kepler used the distances of aphelion and perihelion to obtain the relationships of musical intervals, matching angular velocities with frequencies and anticipating the identification of the concepts of frequency and height of a sound, made by Mersenne in his law of *L'harmonie universelle* [4]. In Kepler's own words, '*Astronomy and Music are various nationalities of the common homeland, Geometry*' (Kepler, 1619, quoted in [5]).

2.2   Sonification and Data-Driven Music

As defined by Herman [6], 'a technique that uses data as input, and generates sound signals (eventually response to optional additional excitation or triggering), may be called sonification, if and only if (C1) The sound reflects objective properties or relationships in the input data. (C2) The transformation is systematic. This means that there is a precise definition provided of how the data (and optional interactions) cause the sound to change. (C3) The sonification is reproducible: given the same data and identical interactions (or triggers) the resulting sound has to be structurally identical. (C4) The system can intentionally be used with different data, and also be used in repetition with the same data'.

Trying to maintain the accuracy of this definition in the use of terminology, we should also remark the main differences between Sonification and Data-Driven Music. Understanding Musification as the musical representation of data and according to Scaletti [7], 'perhaps the most important distinction between sonification and music is the difference in intent. The goal and purpose of data sonification is to aid in understanding, exploring, interpreting, communicating, and reasoning about a phenomenon, an experiment, or a model, whereas in sound art, the goal is to make an audience think by creating a flow of experience for them'. In this sense, the methods here described for unsupervised generation of music and scores should be considered under the Data-Driven Music paradigm, although the auditory exploration of stellar catalogs based on *Deep Learning* is also being used by the authors in scientific oriented approaches.

2.3   Stellar Spectra

Commonly used in star classification, a stellar spectrum is a two-dimensional graphical representation of the flux variations of the brightness of a star as a function of wavelength. It contains information used in the characterization of stars as, for instance, their effective temperature, luminosity and chemical composition. The *MK* system and the *OBAFGKM* temperature sequence of spectral types of stars are based in the detailed analysis of the absorption and emission lines revealed in these curves [8].

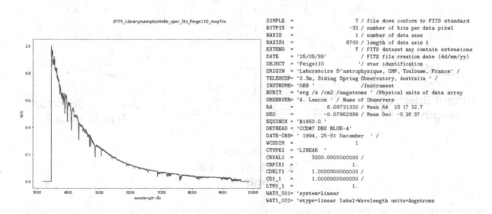

Fig. 1. Stellar spectra (left) and *FITS* file header fragment (right) of *Feige 110* star. STELIB library, Spanish Virtual Observatory (SVO) [9].

Stellar spectra are common data products publicly available in the standard *Flexible Image Transport System (FITS)* [10] files, that also provide identification metadata -as shown in Fig. 1-, including the name of the object, its position marked by right ascension and declination, the physical units, length and resolution of the axis, the date, and the instrument or mission of the observation.

All the information about this kind of data can be found in documents like the *Kepler Data Characteristics Handbook* [11] and the *MAST Kepler Archive Manual* [12,13].

3   Unsupervised Auditory Exploration of Stellar Catalogs

Providing an additional auditory dimension -understood as a complementary way of exploration- to the current graphical display systems for the virtual observation and analysis of astronomical data, opens up endless opportunities to be used in both, research and creative processes. The development of user-oriented tools focused on this dual role also represent a field of undoubted application for improving the accessibility of stellar catalogs, spectra and light curve databases for blind and visual impaired (BVI) users.

3.1   Autoencoders

Deep Learning is a subset of *Machine Learning* in which multilayered neural networks learn from a representative set of population data. Inside this category, *autoencoders* represent one of the unsupervised learning algorithms used to identify relationships within data.

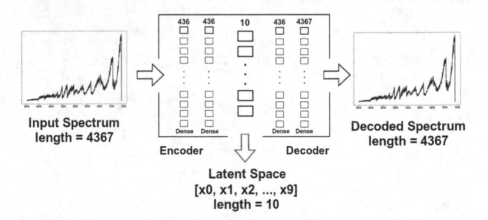

Fig. 2. Illustration of an autoencoder structure applied to stellar spectra input data. For the reduction of the 4367 values of each spectrum into a 10 axes latent space tensor, the encoder uses two hidden layers with 2,099,350 parameters. For the reconstruction of the decoded spectrum, the decoder uses two hidden layers with 1,913,175 parameters.

As defined by Goodfellow et al. [14] *'an autoencoder is a neural network that is trained to attempt to copy its input to its output'*. It is composed of two modules, an *encoder* and a *decoder*. Both are feed-forward neural networks with a variable number of hidden layers. Figure 2, presents an example of this architecture in which the *encoder* takes the input and compresses it to a lower-dimensional representation called the *latent space*. From this compressed representation, the *decoder* attempts to reconstruct the original input. The model is designed to be unable to learn to copy perfectly, being forced to prioritize which aspects of the input should be copied. This restriction makes the architecture to learn useful properties of the input data. Figure 3 shows the deviation between original and decoded spectra measured by the R^2 coefficient, which provides a value between 0 and 1 that represents the variance of the decoded output related to the variance of the original spectrum.

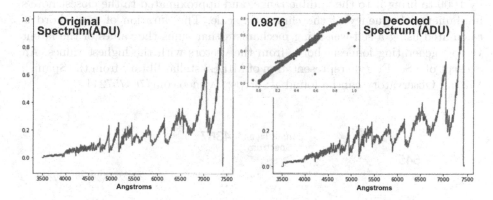

Fig. 3. Example using a 4D variational autoencoder. Original spectrum (left), deviation with $R^2 = 0.9876$ (center), and decoded output (right) for HD 017491 star with coordinates RA:02:47:55.90 and DEC:-12:27:38.16. MILES spectral library (SVO).

On the other hand, '*variational autoencoders have the added constraint that the encoded representation, the latent variables, follow some prior probability distribution. Usually, a Gaussian distribution is chosen for its generality*' [15]. In this approach, the encoder develops a conditional mean and standard deviation re-parameterized by an epsilon term -distributed normally-, to build the distribution of latent variables. In addition to the reconstruction loss function, *variational autoencoders* incorporate a *KL* loss function to maintain the shape of the latent distribution close to the normal. This feature is known as *manifold learning* or *representation learning* and provides a simple way of generating new realistic inputs by sampling the normal distribution of the latent space and decoding those values to obtain a synthetic output.

3.2 Sequential *Chordification* of MILES Stellar Library

One of the most interesting possibilities offered by the application of autoencoders to the sonification and musification of astronomical data, is the improvement of the accessibility of stellar spectra catalogs and databases for blind and visual impaired (BVI) scientists. This section describes a method for the conversion of stellar spectra into music chords to provide an auditory sequential representation of the curves from the MILES stellar spectra library [16], developed by the Spanish Virtual Observatory.

In this approach, a four-dimensional variational autoencoder is used. Figure 4 shows the structure of its encoder. For each spectrum, the four-values tensor generated by the encoder is translated into a four-notes musical chord. This musical structure was chosen in reference to the value of the tetra-chord as the unit of the harmonic system in ancient Greece [17], with the intention of generating complex but not-overwhelming sounds. Each latent value is multiplied by 1000 to bring it to the audible range, and approximated to the closest note's fundamental frequency of the chromatic scale. The duration of each chord is calculated using a self-weighting mechanism that sums the values of the latent vector, generating longest chords from the vectors with the highest values. An excerpt of a Sci-Fi style representation of MILES stellar library from the Spanish Virtual Observatory can be found at: https://vimeo.com/764757244

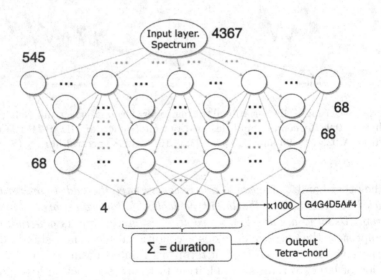

Fig. 4. Simplified architecture of the variational encoder used to extract latent features from each stellar spectrum of the MILES library (SVO). The numbers indicate the dimension of each *Dense* layer used, and the final four-dimensional latent space. The sample chord G4 G4 D5 A#4 is generated from the spectrum of the higher proper motion star BD+15 1305, with coordinates(J2000) RA 06:44:42.98 DEC 14:54:35.97.

3.3   Pipeline and Technical Details for the Musification of 4D Stellar Spectra Latent Space

Developed in *Jupyter notebook* [18], the proposed pipeline for the generation of stellar spectra latent space uses *astropy* [19], *numpy* [20] and *matplotlib* [21] libraries to allow the analysis of each stellar spectrum and to reduce it to a

4 dimensional latent space representation by a variational autoencoder. The encoder that generates each latent vector trying to mimic each stellar spectrum uses 4 *dense* layers and trains 2,427,624 parameters during 100 epochs. The neural network has been implemented using *tensorflow 2* [22] and, as shown in Fig. 5, its results are promising despite the simplicity of the network and the small number of curves used for the training (899). R^2 analysis values for the test and train sets are respectively 0.8472 and 0.8468 (variance weighted). The spectra set has been balanced through the repetition of the only three *O* type spectra included in the MILES library, increasing the total number of curves to 1124 spectra. The R^2 analysis values for the imbalanced set, using directly the 985 spectra of the library, were 0.8369 and 0.7978 (variance weighted). Finally, the Python music analysis library *music 21* [23] is used to generate a score with the sequence of chords that is rendered using the open software *MuseScore3* [24].

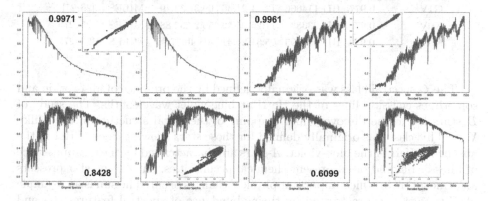

Fig. 5. Four examples generated by the 4D variational autoencoder over the MILES stellar spectra library (SVO). Original spectrum, deviation with R^2 results, and decoded output for HD 057061 (up-left), HD 095578 (up-right), HD 026965 (down-left), and M71 K169 (down-right).

3.4 OBAFGKM Evaluation

Spectral classification has traditionally been done by comparing unknown stellar spectra with those of known standard stars. Most current classification methods, including automation and artificial neural network pattern recognition, are based on the *Morgan-Keenan (MK)* spectral classification system and make use of the seven classical *OBAFGKM* types of stars, with *O* representing the hottest and *M* representing the cooler ones.

With the intention of evaluating the application of the four-dimensional variational autoencoder on stellar spectra feature extraction, a test with the samples of each type of star listed in *Table* 1 has been done. Analyzing the results presented in Fig. 6, despite the fact that the autoencoder generates an *O* type spectrum for the *O7Ia* input, which suggests that it has learned to differentiate this type of star with the only three replicated samples, this type of star presents the poorest response. The resulting OBAFGKM chord sequence can be heard at: https://vimeo.com/770510584.

Table 1. Stellar spectra samples used for the *OBAFGKM* evaluation. Summarizing the Morgan-Keenan (MK) luminosity classes, Ia and Ib correspond to luminous supergiants, II to bright giants, III are normal giants, IV are subgiants, V corresponds to main sequence dwarf stars, VI to subdwarfs and D to white dwarfs. The number allow scaling each type from 0 (the hottest) to 9 (the coolest).

Type of star	R^2	Name	RA(J2000)	DEC(J2000)	Library	Reference
O7Ia	0.988	HD 057060	07:18:40.38	−24:33:31.32	MILES	[25–27]
B5V	0.879	HD 003369	00:36:52.80	33:43:09.48	MILES	[25–27]
A0V	0.979	HD 031295	04:54:53.69	10:09:02.88	MILES	[26–28]
F1V	0.948	HD 222451	23:40:40.56	36:43:14.88	MILES	[25–27]
G1V	0.977	HD 114606	13:11:21.36	09:37:33.49	MILES	[26,27]
K0V	0.972	HD 233832	11:26:05.52	50:22:32.88	MILES	[25–27]
M1V	0.949	HD 036395	05:31:27.41	-03:40:37.99	MILES	[25–27]

The *O* spectral class was originally defined by the presence of absorption and sometimes emission lines of the He II at blue-violet wavelengths. The study of *B* stars led to the discovery and mapping of the spiral structure of the Milky Way. This class was originally defined for those stars showing lines of HeI in the absence of HeII in the blue-violet. *A*-stars are characterized by the disappearance of lines of HeI and often present chemical peculiarities. An *F*-type star presents an atmosphere in which important physical changes occur. The sun is an example of *G*-type star, characterized by their abundance of spectral features. *G*- and *K*-types are the most likely stars to have habitable planets around them while *M*-type stars are the most numerous of all classes [8].

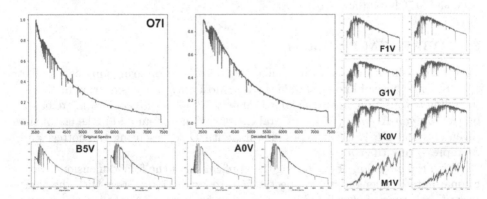

Fig. 6. Results for seven samples of the *OBAFGKM* types of stars. Original spectrum (left) and 4D variational autoencoder decoded output (right) for HD 057060, HD 003369, HD 031295, HD 222451, HD 114606, HD 233832, and HD036395.

3.5 Synthetic Stellar Spectra from Musical Chords

To close the circle of this review on the interdisciplinary possibilities involving stellar catalog exploration, variational autoencoders and music generation, some synthetic spectra can be created by sampling the latent space of the model with user defined chord inputs. Figure 7 shows an example of this approach that could be used in artistic contexts to generate synthetic stellar spectra with a piano keyboard.

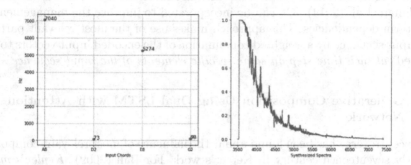

Fig. 7. Generation of synthetic stellar spectra from user defined chords. Input: *A8, D2, F8, G2.* Worth mentioning the appearance of several synthetic absorption lines.

4 Unsupervised Music Composition System Based on *Deep Learning*

With the double intention of exploring the possibilities of *deep music* generation from astronomical data, and providing an unsupervised and objective approach to Kepler's ideas of the music of the spheres, this paper offers some experiments realized with VAE-LSTM with attention neural networks trained with music corpus by Josquin Des Prez, Orlando di Lasso and Giovanni Pierluigi da Palestrina, as representative renaissance composers that could had inspired Kepler's musical thoughts.

4.1 RNN, LSTM Networks and *Attention* Mechanism

'A recurrent neural network (RNN) is a neural network specialized for processing a sequence of values' that *'can also process sequences of variable length'* through parameter sharing across the model. First described by Rumelhart et al.(1986a) [29], and specially useful in *sequence to sequence* models, it *'shares the same weights across several time steps'* so that *'each member of the output is produced using the same update rule applied to the previous outputs. This recurrent formulation results in the sharing of parameters through a very deep computational graph.'* [14]

Long short-term memory (LSTM) architectures, first presented by Graves et al. (2013) [30], are built on gated RNN to resolve the problem of vanishing and exploding gradients, appearing when RNNs try to learn long-term dependencies [31], through the introduction of context-dependent weighted self-loops. They propagate information through long sequences and allow previous outputs to be used as inputs along the hidden layers. LSTM networks improve the capacity to learn possible relationships between features over time and allow the maintenance in memory of relevant features from input data.

An additional *attention* mechanism, introduced on machine translation by Bahdanau et al. (2014) [32], can be incorporated to improve the management of long-term dependencies. This approach makes use of the most relevant parts of the input sequence by a weighted combination of the encoded input vectors to get focused *'at each time step on some specific elements of the input sequence'*. [15]

4.2   Generative Composition Using Dual LSTM with Attention Networks

Some controversy has been found about the influence of most relevant composers of early seventeenth century in Kepler's work. For Ball (2009) *'Kepler's musical world embraced the polyphonic opulence of Palestrina and Monteverdi'* [33] while Pesic (2005) affirms that *'he (Kepler) accepts Zarlino's system and refers only to Lasso and Artusi, never to Monteverdi'* [34]. Anyway, seems clear that although he didn't include any explicit mention in his treatment, Kepler could had been interested in what we know today as the Franco-Flemish school, especially focused on the music of Orlando di Lasso, and more specifically on his *In me transierunt* motet *'as approaching the ideal, unwritten celestial motet'* fitting his thoughts [34].

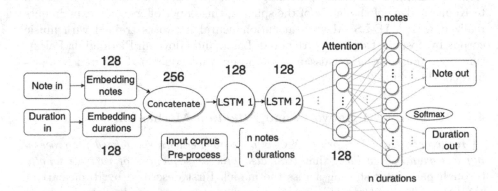

Fig. 8. Network diagram of the dual LSTM with attention architecture used to generate the unsupervised music compositions described in sects. 4.2 and 4.3, showing type and dimension for each layer.

Inspired by this references, several generative composition experiments have been done using music examples of the sixteenth century to train and test a LSTM with attention network based on one of the implementations described in Babcock and Bali(2021) [35]. The network, represented in Fig. 8, learns pitch and duration from each chord of the data set thanks to a dual-input dual-output architecture. Attending some technical details, its implementation uses *Tensorflow 2* and *Keras*, and includes a temperature-based sampling strategy after feeding the notes into the prediction function.

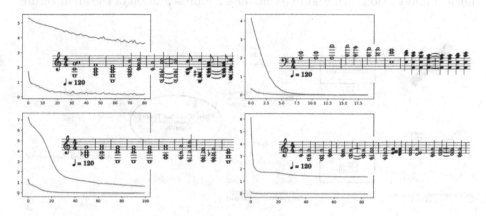

Fig. 9. Resulting scores and *Loss* function for pitch (blue) and durations (red) of four experiments with different training sets: one single Lasso's motet (up-left), the same piece repeated 100 times (up-right), 50 music pieces from Des Prez and Lasso (down-left) and 1318 pieces from Palestrina (down-right). Note how the Pitch Loss function reflects an unexpected fast learning rate for Palestrina's corpus (the biggest), probably motivated by the composer's intrinsic musical characteristics.

Figure 9 summarizes the experiments carried out. The up-left score is the result of testing the behavior of the network when trained with a single MIDI file. The 203 chords and 7 different note durations of Lasso's *In me transierunt* motet were used to train a 382,803 parameters network but, as expected, the results are only useful as starting point, providing a final pitch loss error of 3.6543 after 81 epochs. The up-right score was generated with an augmented corpus, repeating the motet 100 times. Worth mentioning how the network is able to mimic the piece after less than 10 epochs. The final pitch loss is 0.0118 after 20 epochs. The down-left experiment used 50 musical pieces, 25 by Josquin Des Prez and 25 by Orlando di Lassso. 3,245 "Chords" and 46 different durations that clearly increased the learning ability of the system, in this case, with 1,174,620 trainable parameters. After 100 epochs, the final pitch loss was 0.6431. The down-right score results from training the 3,643,876 parameters of the network with 12,884 "chords" and 15 different durations extracted from 1,318 pieces by Giovanni Pierluigi da Palestrina. Final pitch loss of 0.8879 after 41 epochs.

4.3 *Pitch Class Set Theory* Cross-Match

The last step in this approach to generate music from starlight is to cross-match the musical chords obtained with the neural network described in the previous section with those "stellar chords" obtained in Sects. 3.2 and 3.3 by an autoencoder exploring the MILES library, and generate a final score with the positive matched chords, which durations are function of their own summed latent space values. This cross-matching process is hold by the *Pitch Class Set Theory*, which provides one of the most used methods for reducing and labeling musical information [37]. Figure 10 provides a representation of the architecture.

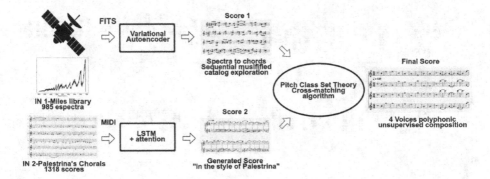

Fig. 10. Four voices astronomical data-driven unsupervised composition system based on *Deep Learning*. Block diagram of the complete VAE-LSTM with attention and *Pitch Class Set Theory* cross-match architecture.

According to the establishments of the theory, a *pitch class A*, is a group of all registers corresponding to that note with octave equivalence, *A0, A2, A4*, etc., and without distinguishing enharmonic equivalences like *A3 sharp* and *B3 flat*. Translating those classes into numbers, with *0* corresponding to *C*, and *11* corresponding to *B*, each generated chord is reduced to a single code that allows their comparison for finding positive matches between corpus. In this way, the chord *G4-E5-C5-G5* corresponds to the code *047*, as an example of a chord which is present in both generated stellar and musical chord scores, and corresponds to HD 049933 star with coordinates RA: 06:50:49.8309, DEC: -00:32:27.1675. This comparison is implemented using the *Chord.orderedPitchClassesString* method from *Music 21* library.

The resulting durations are forked and reduced to only 4 different figures -*whole, half, quarter* and *eight*- to maintain a slow cadence in the music that allows the synchronization with the graphical representation of the source star, in a multimodal exploration inspired mood. At the end of the process, the score is also generated using the *Music 21* library. Final musical results are rendered with commercial DAW software for best audio quality.

A synchronization of the generated music and score, with the images of each source star, and its spectrum for the Lasso-Des Prez corpus in an electroacoustic style render, is available at: https://vimeo.com/770493178.

A similar representation showing the results with the Palestrina corpus for flute, violin and piano can be found at: https://vimeo.com/746620075.

Finally, in order to obtain more general results, an experiment to train the network with the MAESTRO data set [36] has been conducted. This corpus offers about 200 h of virtuous piano performances by several composers from the 17th to the 20th century. A total of 82,231 unique notes and 118 different durations have been used to train the 21,492,526 parameters of the network during 525 epochs, four and a half hours using high performance GPU processing. The next video shows the resulting composition obtained when feeding the model with the notes and durations of Lasso's *In me transierunt* motet. https://vimeo.com/794718061.

4.4   Conclusion and Prospective

Assuming that *'for a sonification to represent information meaningfully, the information must be part of the experience of the representation'* Worrall [38], the unsupervised generation of scores from astronomical data has been proved as an effective way of composition that could be applied in the creation of original soundtracks for films and audiovisual content and, at the same time, having the potential of representing stellar information through sound. The method here described generates completely original pieces that could also be interpreted as an empirical and automated review of the classical *Harmonies of the Spheres* theories, framed in the context of Renaissance Music and Johannes Kepler's musical interests. This approach should be understood as a tribute to one of the most important pioneers of the concept that traces a new line of development for multimodal analysis and comparison of astronomical data through sound.

Thanks to the highly deterministic characteristics of renaissance music, it is possible to analyze the behavior of *Machine Learning* algorithms trained with renaissance music corpus through sound, which could also make this approach useful for neural network auditory monitoring, a branch of Sonification that could be more deeply explored and expanded. The creation of a big MIDI corpus of the Franco-Flemish school inspired by the MAESTRO concept is being considered, to generalize the model and potential results.

In the short term, additional efforts will be addressed to the search of interesting scientific case studies, and to the design of new experiments using selected corpus in both fields, Music and Astronomy. The collaboration and feedback between experts, as well as the evaluation of the results by specialized and non-specialized users, constitutes one of the main axis of future development in this field of research.

The highly interdisciplinary space in which this work is framed, and the polyhedral nature of its results, provide a wide field of resources aimed at using the Music-Astronomy binomial on science communication, outreach, and engagement, while improving the accessibility of astronomical catalogs and databases for blind and visual impaired users.

Acknowledgments. This research includes data from the STELIB and MILES library service developed by the Spanish Virtual Observatory in the framework of the IAU Commission G5 Working Group : Spectral Stellar Libraries.

All MIDI files used in the experiments have been downloaded from *music21* and *Musescore* open libraries.

References

1. Ballora, M.: Sonification, science and popular music: in search of the 'wow'. Organ. Sound **19**(1), 30–40 (2014). Cambridge University Press (2014)
2. Stephenson, B.: The music of the heavens: Kepler's harmonic astronomy. Princeton University Press (2014). https://doi.org/10.1515/9781400863822
3. Pabón, G. C.: Numerus-proportio en el De Música de San Agustín:(Libros I y VI): la tradición pitagórico-platónica. Universidad de Salamanca (2009)
4. Martín, R.G: La teoría de la armonía de las esferas en el libro quinto de Harmonices Mundi de Johannes Kepler, p. 71 (2009)
5. Smirnov, V. A.: Music theory and the harmony method in J. Kepler's work the harmony of the universe. Astron. Astrophys. Trans. **18**(3), 521–532 (1999)
6. Herman, T.: Taxonomy and definitions for sonification and auditory display. In: Proceedings of the 14th International Conference on Auditory Display, pp. 2, Paris, France (2008)
7. McLean, A., Dean, T. (ed.).: The OxfordHandbook of Algorithmic Music, Oxford University Press, pp. 377, New York, USA (2018)
8. Gray, R.O., Corbally, C.: Stellar spectral classification. Princeton University Press (2009)
9. Stelib stellar library. http://svocats.cab.inta-csic.es/stelib/. Accessed 16 Mar 2023
10. FITS standard. https://fits.gsfc.nasa.gov. Accessed 1 Feb 2023
11. Van Cleve, J. E., et al.: Kepler Data Characteristics Handbook (KSCI-19040005) (2016)
12. Thompson, S., Fraquelli, D., Van Cleve, J., Caldwell, D.: Kepler Science Document (KDMC-10008-006) (2016)
13. Mullally, S.: MAST Kepler archive manual (2020)
14. Goodfellow, I., Bengio, J., Courville, A.: Deep Learning. The MIT Press, Cambridge (2016)
15. Briot, J.P., Hadjeres, G., Pachet, F.D.: Deep learning techniques for music generation, pp.85. Springer, Switzerland (2020). https://doi.org/10.1007/978-3-319-70163-9
16. MILES stellar library. http://miles.iac.es/. Accessed 16 Mar 2023
17. Shirlaw, M.: The music and tone-systems of ancient Greece. JSTOR Music Lett. **XXXII**(2), 131–139 (1951)
18. Jupiter notebook. https://jupyter.org/. Accessed 1 Feb 2023
19. Astropy Collaboration: The astropy project: sustaining and growing a community-oriented open-source project and the latest major release (v5.0) of the core package. Astrophys. J. **935**(2), 167 (2022). https://doi.org/10.3847/1538-4357/ac7c74
20. Numpy library. https://numpy.org/. Accessed 1 Feb 2023
21. Matplotlib library. https://matplotlib.org/. Accessed 1 Feb 2023
22. Tensorflow library. https://www.tensorflow.org/guide. Accessed 1 Feb 2023
23. Cuthbert, M.S., Ariza, C.: Music21: a toolkit for computer-aided musicology and symbolic music data. ISMIR (2010)

24. Software Musescore. https://musescore.org/es. Accessed 1 Feb 2023
25. Prugniel, P., Soubiran, C.: New release of the ELODIE library (2004)
26. Sánchez-Blázquez, P., et al.: Medium-resolution Isaac Newton Telescope library of empirical spectra. Mon. Not. R. Astron. Soc. **371**, 703–718 (2006) https://doi.org/10.1111/j.1365-2966.2006.10699.x
27. Falcón-Barroso, J., et al.: An updated MILES stellar library and stellar population models (Research Note). Astron. Astrophys. **532**, A95 (2011). https://doi.org/10.1051/0004-6361/201116842
28. Valdes, F., Gupta, R., Rose, J.A., Singh, H.P., Bell, D.J.: The Indo-US library of Coudé feed stellar spectra. Astrophys. J. Suppl. Ser. **152**, 251–259 (2004)
29. Rumelhart, D.E., Hinton, G.E., Williams, R.J.: Learning representations by back-propagating errors. Nature **323**(6088), 533–536 (1986)
30. Graves, A., Mohamed, A. R., and Hinton, G.: Speech recognition with deep recurrent neural networks. In 2013 IEEE international conference on acoustics, speech and signal processing, pp. 6645–6649. IEEE (2013)
31. Torres, J.: Python deep learning: Introducción práctica con Keras y TensorFlow 2. Marcombo (2020)
32. Bahdanau, D., Cho, K., and Bengio, Y.: Neural machine translation by jointly learning to align and translate, arXiv preprint arXiv:1409.0473 (2014)
33. Ball, P.: The music instinct. How music works and why we can't do without it. Vintage Books (2009)
34. Pesic, P.: Earthly music and cosmic harmony: Johannes Kepler's interest in practical music, especially Orlando di Lasso. J. Seventeenth-Century Music **11**(1) (2005)
35. Babcock, J., Bali, R.: Generative AI with Python and TensorFlow 2: Create images, text, and music with VAEs, GANs, LSTMs, GPT models and more. Packt Publishing Ltd (2021)
36. Hawthorne, C., et al.: Enabling factorized piano music modeling and generation with the MAESTRO dataset. In: International Conference on Learning Representations (2019)
37. Forte, A.: The structure of atonal music (vol. 304). Yale University Press (1973)
38. Worrall, D.: Sonification Design: From Data to Intelligible Soundfields. HIS, Springer, Cham (2019). https://doi.org/10.1007/978-3-030-01497-1

SUNMASK: Mask Enhanced Control in Step Unrolled Denoising Autoencoders

Kyle Kastner[1,2(✉)], Tim Cooijmans[1,2], Yusong Wu[1,2], and Aaron Courville[1,2]

[1] University of Montreal, Montreal, QC H3T 1J4, Canada
kastnerkyle@gmail.com
[2] Mila - Quebec AI Institute, Quebec, Canada

Abstract. This paper introduces SUNMASK, an approach for generative sequence modeling based on masked unrolled denoising autoencoders. By explicitly incorporating a conditional masking variable, as well as using this mask information to modulate losses during training based on expected exemplar difficulty, SUNMASK models discrete sequences without direct ordering assumptions. The addition of masking terms allows for fine-grained control during generation, starting from random tokens and a mask over subset variables, then predicting tokens which are again combined with a subset mask for subsequent repetitions. This iterative process gradually improves token sequences toward a structured output, while guided by proposal masks. The broad framework for unrolled denoising autoencoders is largely independent of model type, and we utilize both transformer and convolution based architectures in this work. We demonstrate the efficacy of this approach both qualitatively and quantitatively, applying SUNMASK to generative modeling of symbolic polyphonic music, and language modeling for English text.

Keywords: Artificial neural networks · Non-autoregressive sequence modeling · Generative modeling

1 Introduction

Modern approaches to content generation frequently utilize probabilistic models, which can be parameterized and learned using artificial neural networks. Common types of neural probabilistic models used for generation can be broadly stratified to form two broad categories based on factorization: autoregressive models (AR), and non-autoregressive models (NAR). We introduce SUNMASK, a NAR generative model for structured sequences[1].

1.1 Autoregressive Models

AR modeling with deep neural networks has been a dominant approach for generative modeling and feature learning [36,54,62,65,66] which has many crucial

---

[1] https://github.com/SUNMASK-web/sunmask. Last accessed 8 February 2023.

A. Courville—CIFAR Fellow.

advantages in both training and inference. One key concern is the necessity of defining a "dependency chain" in the form of a (typically) directed acyclic graph (DAG). Sampling during inference can be accomplished in a straightforward manner using ancestral sampling - sampling from the first variable or variables in the DAG, using those to conditionally estimate a probability distribution for subsequent variables.

Many applications have straightforward orderings in which to define this chain of variables, based on domain knowledge. For example following the flow of time for timeseries modeling is often a logical choice, allowing models to make predictions into the future from the past. However in many other domains, for example images, language, or music, the process of defining a dependency chain over input variables (e.g. pixels, characters, words, or notes) is far from straightforward, as for any arbitrary ordering there can frequently be examples where this ordering *creates* long-term dependencies, or otherwise makes satisfaction of dependencies during training and evaluation more difficult than another alternative ordering.

This divide becomes further compounded in many creative applications to these domains, as creators typically iterate repeatedly: forming a concept, sketching out the concept, and seeing where the creative flow (based on the sketch) may lead to alterations in the original concept, thus "rewriting" sketched steps. Though the resulting output may be perceived in a time-ordered fashion (for example, reading a book or listening to a song), the initial creation was performed globally and holistically. This global view is often critical to creating elements such as foreshadowing and tension which make the resulting output interesting or enjoyable. This iterative process is directly at odds with a strict AR factorization, and requires well trained AR models to cope with a high degree of uncertainty and multi-modality for long range dependencies, which can lead to logical mistakes or other errors.

1.2   Non-Autoregressive Models

An alternative methodology for generative modeling is non-autoregression (NAR), broadly covering a large number of different modeling approaches which attempt to remove assumptions about variable ordering, instead either hand-defining per-exemplar orderings, or modeling variables jointly without resorting to chain rule factorization [23, 29]. One way to define an ordering over variables is via masking of inputs or intermediate network representations [20, 51, 63–65, 68], and indeed modern AR approaches such as transformers [67] use an autoregressive mask internally to define the chain of variables order. These masks can either be constant over all training (as in standard AR transformers and PixelCNN [65]) or dynamic per example (as in MADE [20]). When masks are dynamic per example, we begin to see the relationship between enforcing AR via masking and NAR methods, as although some ordering is assumed this ordering is no longer constant, and it becomes possible to use the same trained model to evaluate the probability of a particular output variable under *multiple* possible orderings.

Closely linked to masking methods are so called *diffusion models*, which relax the variable ordering problem through noise prediction [30, 59, 61]. Rather than predicting a new variable or variables given previous ones in an arbitrarily chosen DAG, diffusion models focus on predicting a less noisy version of many variables jointly, given a set of noisy input variables. Iteratively applying this learned denoising improvement operator should eventually result in predicting a fully clean output estimate, given either a noisy version of the target domain, or even starting from pure noise. Given this framing it is clear that diffusion models are closely linked to denoising methods in general, specifically denoising autoencoders, as well as modern density modeling approaches such as generative adversarial networks (GAN [21]), variational autoencoders (VAE [38]), flow-based models (NICE [13], RealNVP [14], Normalizing Flows [56], IAF [39], MAF [51]), iterative canvas sampling (DRAW [22]), and noise contrastive estimation (NCE [26]). Particular applications of this denoising philosophy such as BERT [12], WaveGrad [9], and GLIDE [50], have resulted in large quality improvements for feature learning and data generation for text, images, and audio [28,41,55].

1.3   Trade-Offs Between Autoregressive and Non-autoregressive Approaches

The choice between AR and NAR methods is not clear-cut. For many domains, high-quality models exist using both approaches but we can define some crucial parameters. Some NAR methods such as GAN or VAE are capable of generating output in only one inference step, however they are typically hard to train on certain data modalities (e.g. text data) comparing to AR counterparts. Other NAR methods such as diffusion models typically allow for choosing a diffusion length during inference, which is independent of that used at training. Choosing a low diffusion length can frequently lead to poor sample quality, and tuning this setting (among many others) is critical to high quality generation. However if the tuned diffusion length for a given sequence (of length T) is shorter than the length of those sequences, the NAR method has a computational advantage over the equivalent AR model (which would require T steps for a T length sequence). In addition, the ability to tune this diffusion length can be useful in interactive applications, or when a variety of output is desirable. This setting can also be a curse, as even well-trained models perform poorly with improper diffusion settings. Several branches of current research are focused on improving guarantees and convergence speed for diffusion models [35,37,40,60].

1.4   SUNMASK, A Non-autoregressive Sequence Model

We introduce SUNMASK, a NAR sequence model which uses masks over noised, discrete data to learn a self-improvement operator which transitions from categorical noise toward the data distribution in iterated steps. Given a target data representation, we train a model which can map from a noisy version of input data to a corrected form of the input. In this work, we use multinomial noise - namely entries are corrupted to 1 of P possible values (for a given set size P), with the number of noised entries in a sequence defining the relative noise level for the training example. This is similar to many diffusion approaches at a high level, and particularly shown to be an effective tool in SUNDAE [58] and Coconet [33]. In addition to the use of multinomial noise, we also form a mask representing *where* the data was noised, feeding this mask alongside the input data to form a conditional probability distribution.

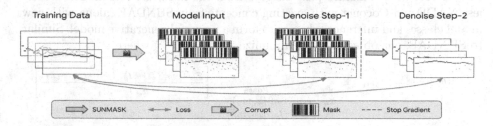

Fig. 1. Step-unrolled denoising training for SUNMASK on polyphonic music, unrolled step length 2. Training data (left) consists of four voices corrupted by sampling a random mask per voice and replacing the masked data (red) with random pitches (green). SUNMASK takes both mask and corrupted training data as input, predicting denoised original data as output. In the second step, the model takes a sampled version of the model step predictions and the same mask as input, outputting another prediction of the original data. (Color figure online)

2 Method

The relationship between discrete diffusion and denoising autoencoders has been explored in previous work [4,31,32,58]. We build upon this foundation, combined with many insights from prior orderless modeling approaches, crucially Orderless NADE [64], Coconet [33] (which is a more modern variant of Orderless NADE), and SUNDAE [58].

SUNMASK is built around a process $x_t \sim f_\theta(\cdot|x_{t-1}; m)$ on a space $X = \{1, \ldots, v\}^N$ of arrays with categorical variables. This parametric transition function f_θ takes an additional argument $m \in {0, 1}^N$. During training, m indicates variables that were not initially corrupted, and as a consequence we can use it during inference to tell f_θ which variables to trust.

Given a sequence of masks $m_0, \ldots, m_{T-1}$, the generating distribution of our model derives from a prior p_0 (typically uniform noise) and repeated application of f_θ:

$$p_T(x_T; m_0, \ldots, m_{T-1}) = \Big(\sum_{x_1, \ldots, x_{T-1} \in X} \prod_{t=1}^{T} f_\theta(x_t|x_{t-1}; m_{t-1}) \Big) p_0(x_0) \quad (1)$$

In practice, p_0 is typically elementwise iid uniform noise, and the masks $m_0, \ldots, m_{T-1}$ are drawn according to a schedule and may be held constant for several steps.

To train f_θ, we take a training example $x \sim p_{\text{data}}$ and draw a mask m. We apply the corruption procedure $x_0 \sim q(\cdot|x; m)$ to obtain x_0 which equals x

where the mask m is true and uniform random values elsewhere. Then we iterate $x_t \sim f_\theta(\cdot|x_{t-1}; m)$ with the aim of reconstructing x.

As in SUNDAE, the transition f_θ models the variables as conditionally independent of one another. However SUNDAE has no direct concept of masking. SUNMASK thus combines past insights from the masked NAR models Orderless NADE and Coconet with existing concepts from SUNDAE, along with new model classes and inference schemes to form a powerful generative model. Similar to SUNDAE, our objective is to minimize $\frac{1}{2}(L^{(1)} + L^{(2)})$ where

$$
L^{(t)}(\theta) = -\mathbb{E}_{m_0, \cdots, m_{t-1}} \mathbb{E}
\begin{array}{ll}
x & \sim & p_{\text{data}} \\
x_0 & \sim & q(\cdot|x, m_0) \\
x_1 & \sim & f_\theta(\cdot|x_0; m_0) \\
x_2 & \sim & f_\theta(\cdot|x_1; m_1) \\
& \cdots & \\
x_{t-1} & \sim & f_\theta(\cdot|x_{t-2}; m_{t-2})
\end{array}
\left[\frac{\sum_i (1 - m_{t-1}^{(i)}) \log f_\theta^{(i)}(x^{(i)}|x_{t-1}; m_{t-1})}{\sum_i 1 - m_{t-1}^{(i)}} \right]
$$

(2)

is the reconstruction loss for the elements of x that were corrupted. As in Orderless NADE [64] and Coconet [33], we weigh each term according to the size of the mask, to ensure that the overall weight on each conditional $f_\theta^{(i)}$ is uniform across i. Unlike previous methods, we target *only masked variables* in the loss. In practice we choose $m_0 = \cdots = m_{t-1}$ during training and $t = 2$. Since we only go to $t = 2$, keeping the mask constant is a close enough approximation to the masking schedule used in inference. The choice of $t = 2$ is driven by the ablation study in SUNDAE, where $t = 2$ was found to account for nearly all performance gains in translation experiments, with higher unrollings showing no additional benefit. In addition higher values of t unrolling generally increase memory usage, making the training of high order unrollings complicated.

SUNMASK allows for direct control at inference using both proposal masks and noising of variables, combining elements of both SUNDAE and Coconet. We show a high level example of the unrolled training scheme, mask proposals, and input data processing in Fig. 1.

The overall unrolled mask and iterative inference setting is largely independent of architecture choice, and as long as the internal architecture does not make any ordering assumption over the input data we can incorporate it into SUNMASK. We use two primary archetypes for the internal model in this paper: Attentional U-Net and Relative Transformer.

SUNMASK uses an unrolled training scheme, similar to that shown in SUNDAE, as well as a mask which is input to the model and defines manipulated variables as in Coconet. The loss is masked based on this manipulation mask, unlike Coconet or SUNDAE. The SUNMASK loss is further weighted by the total amount of masked variables. Comparisons of various high level modeling features between SUNMASK, Coconet, and SUNDAE are shown in Table 1.

2.1   Model Training

During training, the internal architecture is combined with a *step unrolled* training procedure, as highlighted by SUNDAE [58]. Rather than directly randomizing

Table 1. Comparing SUNMASK, Coconet, and SUNDAE

Model	SUNMASK	Coconet	SUNDAE
Mask input to model	✓	✓	X
Masked loss	✓	X	X
Re-weighted loss	✓	✓	X
Unrolled loss	✓	X	✓
Inference mask schedule	✓	✓	X
Sampling rejection step	✓	X	✓
Mask control preserves data	✓	X	X

positions, we re-write this as a masking scheme, first sampling a mask (with 0 randomize, 1 keep, which we denote as 0-active format) then performing randomization to one of P possibilities, for the masked subset of K variables. This random masking procedure is equivalent to the approach from SUNDAE, but using a mask allows us to further combine the mask information with the input data, in order to form a conditional probability estimate. In addition, this 0-active masking scheme makes direct comparison to masking schemes with absorbing states (such as OrderlessNADE [64], Coconet [33], VQ-Diffusion [24] and OA-ARDM [31]) simpler, as the mask can be directly multiplied with the data in a 0-active format.

Each training batch is randomly sampled from the training dataset, and a corresponding noise value drawn from $rand(N)$ for N examples in the minibatch. This per-example noise value is then used to derive a per-step mask over T timesteps, by comparing noise $rand(N) < rand(N, T)$. During training, this means some examples have a high per-example noise value (e.g. .99), and thus many values masked and noised in the training, while other examples may have a low noise value (e.g. .01) drawn instead. Combined with a training loss which learns to denoise the input and focuses on imputing information about masked corrupted inputs, the overall model will learn a chain to go from more noisy data to less noisy step-wise, resulting in a learned improvement operator [32, 58].

This improvement operator can be applied to noisy data or pure noise, and iterate toward a predictive sample from the training distribution. See Multinomial Diffusion [32] and SUNDAE [58] for more detail on this proof, as well as fundamental work on denoising autoencoders [2]. In SUNMASK, we combine the mask used to noise the input with the input data itself, while modifying the loss to predict *only masked variables*. In addition, we downweight the loss by $\frac{1}{1+\sum 1 - m_t}$ for each batch element, meaning that losses for heavily masked entries are downweighted compared to losses on examples with little masking, in a form of curriculum weighting based on expected estimation difficulty.

While a one step denoising scheme can be sufficient for learning the data manifold [2, 4], *unrolling* this denoising scheme into a multi-step process can have performance benefits. SUNMASK directly uses the unrolled loop scheme described in [58], using a step value of 2. For a detailed description of the step unrolled training scheme, see the overview description from SUNDAE [58]. The masked and unrolled training can be seen as a container for any internal model which does not make ordering assumptions, and we utilize both convolutional U-Net (a variant of the GLIDE [50] U-Net) and Relative Transformer [11,34,53] models for various experiments, shown in Sect. 4.

2.2   Convolutional SUNMASK

SUNMASK is most closely related to Coconet [33] and SUNDAE [58]. Coconet (as an instance of OrderlessNADE using convolutional networks), trains by sampling a random mask per training example, using this mask to set part of the input (in one hot format) to zero. The mask is further concatenated to the zeroed data along the channel axis, and this combined batch is passed through a deep convolutional network with small 3×3 kernels. Convolutional SUNMASK uses a downweighted loss over only variables masked in the input. However, SUNMASK additionally uses the unrolled training scheme, as well as a different inference procedure due to preserving the values of masked out variables during training and sampling.

Our best performing convolutional SUNMASK architecture takes hints from recent image transformer and vector quantized generators, exchanging the small kernels used in Coconet for extremely large kernels of shape $4 \times P$ over the time and feature dimensions, somewhat analogous to input patches, removing the model's translation invariance over the feature axis by setting kernel dimension equal to the total feature size. However this makes the number of parameters per convolutional layer extremely large. Convolutional SUNMASK adopts an attentional U-Net structure which reduces only across the time axis, modified from GLIDE [50], rather than the deep residual convolution network used by Coconet. Combined with the addition of step unrolled training, we are only able to train convolutional SUNMASK with a batch size of 1 (expanded to effective batch size 2 due to step unrolling) on commodity GPU hardware with 16GB VRAM.

Due to the design choice of extremely large kernel sizes which depend on the size of the domain, we only use convolutional SUNMASK for polyphonic music experiments, see Sect. 4 for more details.

Attention is applied on the innermost U-Net block size as well as the middle block, with 1 attention head [50]. Convolutions are used in all resampling, and all resampling happens only on the time axis, making the Attentional U-Net effectively a 1-D architecture. However, rather than learning both instrument and pitch relations across channels, we isolate pitch relations and instrument relations into separate axes of the overall processing, the "width" and "channels" axes, respectively assuming $(N, C, H, W) == (N, I, T, P)$ axes. As is standard in many U-Net designs, we double the number of hidden values for layers every time the resolution is halved, with the reverse process being used when upsampling. Though the parameter count here is large, it is similar in spirit to other approaches to small datasets on text [1].

2.3   Transformer SUNMASK

Transformer SUNMASK relates closely to the transformer used in SUNDAE. The architecture uses a relative multi-head attention [11,34] and has no autoregressive masking. SUNMASK transformer also uses larger batch sizes, typically 20 or larger, though this is far smaller than the batch sizes seen in the experiments of SUNDAE. Sequence length and data iterator strategy were both a critical aspect for training transformer SUNMASK. We found short sequences (from 32 to 128) worked best, along with iteration strategies that were example based.

Transformer SUNMASK was trained on every dataset used in this paper, and we show performance in Sect. 4, as well as comparisons to convolutional SUNMASK on symbolic polyphonic music modeling. Both convolutional and transformer based SUNMASK use the Adam optimizer, with gradient clipping by value at 3. Inference hyperparameter types and general sampling strategies used are the same with both models, though specific hyperparameter values may change between datasets.

There is a large discrepancy in model parameter count between our best performing convolutional models for JSB, and our best transformers. Training larger transformers can work well for generation [1], but our large parameter transformers (on the order of 400M parameters) had poor generative performance on JSB.

Pitch size / vocabulary size, sequence length, and batch size changed for the transformers used in the text experiments, but the global architecture remained in the style of "decoder only" transformers [54], similar to SUNDAE. Notably, we use vocabulary size 5.7k, sequence length 52, batch size 48 for EMNLP2017 News and vocabulary size 27, sequence length 64, batch size 20, and a slightly extended training step length of 150000 for text8.

2.4   Inference Specific Settings

Well-trained SUNMASK models should be applicable to full content generation, as well as a variety of partially conditional generative tasks such as infilling and human-in-the-loop creation. Basic sampling involves creating a set of variables, with all variables randomly set to 1 of P values in the domain (or partial randomization in the case of infilling) along with an accompanying mask, which is initially all 0 for full generation, or mixed 1s and 0s for partial generation tasks. Given this data and mask as input, the trained model then predicts a probability distribution over all possible P values, for all variables. Despite the use of masked losses in training, we sample these prediction distributions for *all* variables. These predictions are then accepted or rejected from the original set, resulting in a new variable set. We then sample a new mask (based on a predefined schedule) and combine it with the initial mask, then iterate this overall process, updating at least some of the variables at each step.

During inference we use several key techniques to improve generative quality. We use typicality sampling [47] on the output probability distribution and a variable number of diffusion steps, on the order of 100 to 2000. Masks are randomly sampled using the schedule defined in [33] which linearly decreases the number of masked variables over time according to $\alpha_n = \max(\alpha_{\min}, \alpha_{\max} - \frac{n}{\eta N}(\alpha_{\max} - \alpha_{\min}))$ with $\alpha_{\min} = .001$, $\alpha_{\max} = .999$, and $\eta = 3/4$, along with an optional triangular linear ramp-up and ramp-down schedule for the probability of accepting predictions from the model into the current variable set at each step, as shown in [58].

Tuning hyperparameters for inference is critical to success, as improper settings can drastically lower the performance of SUNMASK, see Sect. 4 for variance over various inference settings in different tasks. For human-in-the-loop applications, the existence of these controls can allow a number of fine-grained workflows to emerge, driven by expert users to create and curate interesting output [10,16], demonstrated in Fig. 3.

3   Related Work

We state here some key related approaches, as well as how our method differentiates from these previous settings. A number of recent publications on diffusion models and feature learning have incorporated masks as part of their overall training scheme [28,31], however these papers use masks for blanking, rather than as indicators over stochastic variables. Many infilling models [12,15], and masked image models [28] feature conditional modeling with a mask (blank) token, predicting the variables masked from the input for feature learning or generative modeling. XLNet [68] combines the infilling and autoregressive paradigms, learning arbitrary permuted orders over masked out variables, using blank-out masking and randomly generated autoregressive ordering similar to OrderlessNADE and Coconet. Conditional diffusion generators [48] and GAN generators [18] have the combination of mask indicators as well as preserving stochasticity of the masked variables. However these methods do not use an unrolled training scheme, and generally target image related tasks, with the notable exception of maskGAN. Many models use a concept of a working canvas, and do repeated inference steps for generation or correction of data [5,19,22], SUNMASK differs from these models due to architecture choices, training scheme, and loss weighting, as well as application domain [49,50,52,57].

4   Experiments

We demonstrate the use of SUNMASK for polyphonic symbolic music modeling on the JSB dataset [3,6]. The JSB dataset consists of 382 four-part chorales, originally written by Johann Sebastian Bach. These chorales are quantized at the 16th note interval, cut into non-overlapping chunks of length 128, skipping chunks which would cross the end of a piece. This processing results in a training dataset of 4956 examples, with each example being size $(4, 128)$. We train convolutional and transformer versions of both SUNMASK and SUNDAE for comparison, as well as the pretrained Coconet [33]. For polyphonic music, the quantized data was rasterized in soprano, alto, tenor, bass (SATB) order, as in Music Transformer [34] and BachBot [42], then chunked into non-overlapping training examples. Results are shown in Table 2. These results are evaluated on Bach ground truth data (Bach GT), BachMock Transformer (BachMock [17, 44]) (closely related to the decoder from VQ-CPC [27]), Coconet, SUNDAE (SD), and SUNMASK convolutional (SMc) and SUNMASK transformer (SMt). Model sampling variants are indicated as Typical Sampling (T).

4.1 Musical Evaluation

The grading function used for evaluation, referred to as BachMock here, is designed specifically to correlate with expert analysis on Bach. In particular using this metric to choose correct examples in a paired comparison test, outperforms novice, intermediate, and expert listeners by varying margins [17]. This indicates that scoring well on the aggregate metric should correlate to high sample quality. The metric has many sub-parts, ranking various musical attributes crucial to codifying the style of J.S. Bach. AugGen [44] incorporated this metric into an iterative training and sampling scheme which improved final generative capability for a fixed model, showing the effectiveness of BachMock in practice for ranking machine generated samples. For each grading function in the Bach Mock grading evaluation, we show the median value and $\pm$ standard deviation (showing the average of each interval SATB performance for brevity), as well as the overall grade. Lower values for all metrics are better, and we see the strongest results for convolutional SUNMASK with typicality sampling. Combined with final top-N ($N = 20$) selection out of a candidate set of 200 samples, the overall sample quality outperforms strong baselines. This high quality subset (SMc-T BEST20) rivals both the "BachMock" transformer and the dataset itself on this metric. We find SUNMASK generations are qualitatively good and listenable overall, even though some SUNMASK samples do fare poorly by the grading metrics.

Table 2. Quantitative results from the Bach Mock grading function [17]. Top rows compare to existing literature, bottom rows show ablation study of SUNMASK style models. Lower values represent better chorales.

Model	Note	Rhythm	Parallel Errors	Harmonic Quality	Interval	Repeated Sequence	Overall
Bach Data	0.24 ± 0.15	0.23 ± 0.14	0.0 ± 0.69	0.41 ± 0.2	0.55 ± 0.4	1.29 ± 0.88	*4.91 $\pm$ 1.63*
BachMock	0.37 ± 0.22	0.26 ± 0.14	2.16 ± 3.22	0.54 ± 0.31	0.71 ± 0.68	1.86 ± 2.81	8.94 ± 4.64
SMc-T BEST20	0.39 ± 0.16	0.53 ± 0.26	0.0 ± 0.81	0.68 ± 0.27	0.75 ± 0.42	1.44 ± 0.52	**7.16 ± 0.97**
AugGen	–	–	–	–	–	–	8.02 ± 2.92
Coconet	**0.44 ± 0.23**	1.85 ± 0.39	2.61 ± 6.56	1.38 ± 0.39	**0.86 ± 0.73**	6.07 ± 1.76	17.00 ± 6.58
SD	0.59 ± 1.82	0.93 ± 0.84	6.42 ± 4.11	0.98 ± 0.67	1.99 ± 5.68	2.45 ± 2.39	23.25 ± 21.45
SD-T	0.63 ± 2.40	0.60 ± 0.96	3.82 ± 4.98	0.96 ± 0.64	2.50 ± 5.03	**1.52 ± 3.43**	20.09 ± 23.88
SMc	0.87 ± 2.05	0.63 ± 0.77	1.38 ± 6.00	1.02 ± 0.49	2.07 ± 5.72	2.32 ± 2.31	22.47 ± 20.80
SMc-T	0.57 ± 1.79	0.69 ± 0.35	1.28 ± 3.73	**0.93 ± 0.49**	1.10 ± 4.68	1.81 ± 0.83	**13.43 ± 19.27**
SMt	3.00 ± 1.85	0.74 ± 0.90	**0.00 ± 1.95**	1.64 ± 0.70	7.90 ± 5.58	3.10 ± 2.97	42.87
SMt-T	3.74 ± 2.16	**0.58 ± 0.56**	0.00 ± 2.56	1.73 ± 0.73	7.74 ± 4.73	2.35 ± 1.79	46.21 ± 17.30

4.2   Text Datasets

The EMNLP 2017 News dataset is a common benchmark for word-level language modeling [7], containing a large number of news article sentences [45]. Preprocessing steps collapse to sentences containing the most common 5700 words, resulting in a training set of 200k sentences with a test set of 10k. The overall maximum sentence length is 51. Common processing for this dataset includes padding all sentences up to this maximum length, different than the standard long sequence chunking commonly used in other language modeling tasks.

We show the results of several SUNMASK models for generating sentences similar to EMNLP2017News, comparing to benchmarks using the standard Negative BLEU/Self-BLEU evaluation [7,70] over generated corpora of 1000 sentences in Fig. 2. This set of scores, varied across temperature, is compared against baseline scores [8,25,43,46,67,69], similar to the evaluation shown in SUNDAE [58]. These reference benchmarks used 10000 sentences to form performance estimates.

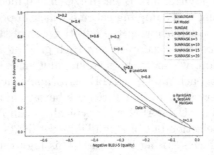

Fig. 2. Negative BLEU/Self-BLEU scores on EMNLP2017 News. Left (x-axis) is better, lower (y-axis) is better. Quality/variation is controlled by changing the temperature (t), and varying diffusion schedule (s). For SUNMASK, *typical* sampling results [47] are shown.

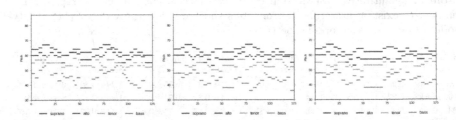

Fig. 3. SUNMASK harmonization (bass, tenor, alto) of an existing melody (left), with a mask which highlights the left half (0 to 64) soprano voice (middle), or a left half mask but replacing right half melody as well (right)

4.3   Music Control

Given the flexibility of masking at inference, we perform a number of qualitative queries to inspect how the model adapts based on noise and mask value. Figure 3 demonstrates the use of SUNMASK for musical inpainting, holding the top voice (soprano) either fully or partially fixed to the well-known melody "Ode to Joy", by Ludwig van Beethoven.

4.4   Text Control

Masking can also used to variably increase or decrease the weight on various pre-specified terms, held fixed throughout inference. The combination of these words, and their mask status can be seen to influence the overall tone of the selected text passages which showed the strongest effect in a particular inference batch. The following qualitative samples using masks for word influence are drawn from SUNMASK Transformer on EMNLP2017News dataset. Though the generation quality is flawed, we clearly see a relationship between the masked word and the emergent surrounding context, for example highlighting **disaster** draws forth injured, displaced, and pressure, while **success** instead references happy, nice, good, and playing.

Success unmasked, *disaster* masked

- I think I want to leave **success** at the end of the *disaster*, but because that 's a nice to say it' s not good to be the challenge and this is a very good thing <eos>
- That was the job I was **success** to have to pay my *disaster* but hopefully I have been able to pull playing in the first couple of the season, I ' ve been happy to go through this team, he said <eos>

Success masked, **disaster** unmasked

- Although more than 80, 000 *success* have been displaced in the **disaster** since the last year , more than 700 , 000 lives have been injured in the country, and 70 of them were killed, according to the UN media <eos>
- I haven ' t had a *success* at the league, the **disaster** and picked running with the door ago we have Champions, and I was a couple of pressure . . . and it was a lot of times <eos>

5   Conclusion

We introduce SUNMASK, a method for masked unrolled denoising modeling of structured data. SUNMASK separates the role of masking and correction by conditioning predictions on the mask, allowing for fine-grained control at inference. When applied to text as well as symbolic polyphonic music, SUNMASK is competitive with strong baselines, outperforming reference baselines on music modeling. Leveraging the separation of mask and noise allows for subtle control at inference, paving the way for a variety of domain specific applications and generative pipelines for human-in-the-loop creation.

References

1. Al-Rfou, R., Choe, D., Constant, N., Guo, M., Jones, L.: Character-level language modeling with deeper self-attention. In: Proceedings of the AAAI Conference on Artificial Intelligence, vol. 33, pp. 3159–3166 (2019)
2. Alain, G., Bengio, Y.: What regularized auto-encoders learn from the data-generating distribution. J. Mach. Learn. Res. **15**(1), 3563–3593 (2014)
3. Allan, M., Williams, C.: Harmonising chorales by probabilistic inference. In: Advances in Neural Information Processing Systems 17 (2004)
4. Austin, J., Johnson, D.D., Ho, J., Tarlow, D., van den Berg, R.: Structured denoising diffusion models in discrete state-spaces. Adv. Neural. Inf. Process. Syst. **34**, 17981–17993 (2021)
5. Bachman, P., Precup, D.: Data generation as sequential decision making. In: Advances in Neural Information Processing Systems 28 (2015)
6. Boulanger-Lewandowski, N., Bengio, Y., Vincent, P.: Modeling temporal dependencies in high-dimensional sequences: application to polyphonic music generation and transcription. In: Proceedings of the 29th International Conference on International Conference on Machine Learning, pp. 1881–1888 (2012)
7. Caccia, M., Caccia, L., Fedus, W., Larochelle, H., Pineau, J., Charlin, L.: Language GANs falling short. In: International Conference on Learning Representations (2020)
8. Che, T., et al.: Maximum-likelihood augmented discrete generative adversarial networks. arXiv preprint arXiv:1702.07983 (2017)
9. Chen, N., Zhang, Y., Zen, H., Weiss, R.J., Norouzi, M., Chan, W.: WaveGrad: estimating gradients for waveform generation. In: International Conference on Learning Representations (2020)
10. Crowson, K., et al.: VQGAN-CLIP: Open domain image generation and editing with natural language guidance. In: Avidan, S., Brostow, G., Farinella, G.M., Hassner, T. (eds.) Computer Vision – ECCV 2022. ECCV 2022. LNCS, vol. 13697, pp. 88–105. Springer, Cham (2022). https://doi.org/10.1007/978-3-031-19836-6_6
11. Dai, Z., Yang, Z., Yang, Y., Carbonell, J.G., Le, Q., Salakhutdinov, R.: Transformer-xl: Attentive language models beyond a fixed-length context. In: Proceedings of the 57th Annual Meeting of the Association for Computational Linguistics, pp. 2978–2988 (2019)
12. Devlin, J., Chang, M.W., Lee, K., Toutanova, K.: BERT: pre-training of deep bidirectional transformers for language understanding. In: Proceedings of NAACL-HLT, pp. 4171–4186 (2019)
13. Dinh, L., Krueger, D., Bengio, Y.: Nice: Non-linear independent components estimation. arXiv preprint arXiv:1410.8516 (2014)
14. Dinh, L., Sohl-Dickstein, J., Bengio, S.: Density estimation using real NVP. In: International Conference on Learning Representations (2017)
15. Donahue, C., Lee, M., Liang, P.: Enabling language models to fill in the blanks. In: Proceedings of the 58th Annual Meeting of the Association for Computational Linguistics, pp. 2492–2501 (2020)
16. Esser, P., Rombach, R., Ommer, B.: Taming transformers for high-resolution image synthesis. In: Proceedings of the IEEE/CVF Conference on Computer Vision and Pattern Recognition (CVPR), pp. 12873–12883 (2021)
17. Fang, A., Liu, A., Seetharaman, P., Pardo, B.: Bach or mock? a grading function for chorales in the style of J.S. Bach. arXiv preprint arXiv:2006.13329 (2020)

18. Fedus, W., Goodfellow, I., Dai, A.M.: MASKGAN: better text generation via filling in the _. In: International Conference on Learning Representations (2018)
19. Ganin, Y., Kulkarni, T., Babuschkin, I., Eslami, S.A., Vinyals, O.: Synthesizing programs for images using reinforced adversarial learning. In: International Conference on Machine Learning, pp. 1666–1675. PMLR (2018)
20. Germain, M., Gregor, K., Murray, I., Larochelle, H.: Made: Masked autoencoder for distribution estimation. In: International Conference on Machine Learning, pp. 881–889. PMLR (2015)
21. Goodfellow, I., et al.: Generative adversarial nets. In: Advances in Neural Information Processing Systems 27 (2014)
22. Gregor, K., Danihelka, I., Graves, A., Rezende, D., Wierstra, D.: Draw: a recurrent neural network for image generation. In: International Conference on Machine Learning, pp. 1462–1471. PMLR (2015)
23. Gu, J., Bradbury, J., Xiong, C., Li, V.O., Socher, R.: Non-autoregressive neural machine translation. In: International Conference on Learning Representations (2018)
24. Gu, S., et al.: Vector quantized diffusion model for text-to-image synthesis. In: Proceedings of the IEEE/CVF Conference on Computer Vision and Pattern Recognition, pp. 10696–10706 (2022)
25. Guo, J., Xu, L., Chen, E.: Jointly masked sequence-to-sequence model for non-autoregressive neural machine translation. In: Proceedings of the 58th Annual Meeting of the Association for Computational Linguistics, pp. 376–385 (2020)
26. Gutmann, M., Hyvärinen, A.: Noise-contrastive estimation: a new estimation principle for unnormalized statistical models. In: Proceedings of the Thirteenth International Conference on Artificial Intelligence and Statistics. pp. 297–304. JMLR Workshop and Conference Proceedings (2010)
27. Hadjeres, G., Crestel, L.: Vector quantized contrastive predictive coding for template-based music generation. arXiv preprint arXiv:2004.10120 (2020)
28. He, K., Chen, X., Xie, S., Li, Y., Dollár, P., Girshick, R.: Masked autoencoders are scalable vision learners. In: Proceedings of the IEEE/CVF Conference on Computer Vision and Pattern Recognition, pp. 16000–16009 (2022)
29. Hill, F., Cho, K., Korhonen, A.: Learning distributed representations of sentences from unlabelled data. In: Proceedings of the 2016 Conference of the North American Chapter of the Association for Computational Linguistics: Human Language Technologies, pp. 1367–1377 (2016)
30. Ho, J., Jain, A., Abbeel, P.: Denoising diffusion probabilistic models. Adv. Neural. Inf. Process. Syst. **33**, 6840–6851 (2020)
31. Hoogeboom, E., Gritsenko, A.A., Bastings, J., Poole, B., van den Berg, R., Salimans, T.: Autoregressive diffusion models. In: International Conference on Learning Representations (2022)
32. Hoogeboom, E., Nielsen, D., Jaini, P., Forré, P., Welling, M.: Argmax flows and multinomial diffusion: Learning categorical distributions. In: Advances in Neural Information Processing Systems 34 (2021)
33. Huang, C.Z.A., Cooijmans, T., Roberts, A., Courville, A.C., Eck, D.: Counterpoint by convolution. In: Proceedings of the 18th International Society for Music Information Retrieval Conference, pp. 211–218. ISMIR, Suzhou, China (2017)
34. Huang, C.Z.A., et al.: Music transformer: generating music with long-term structure. In: International Conference on Learning Representations (2018)
35. Huang, C.W., Lim, J.H., Courville, A.C.: A variational perspective on diffusion-based generative models and score matching. In: Advances in Neural Information Processing Systems 34 (2021)

36. Kalchbrenner, N., Danihelka, I., Graves, A.: Grid long short-term memory. arXiv preprint arXiv:1507.01526 (2015)
37. Kingma, D.P., Salimans, T., Poole, B., Ho, J.: Variational diffusion models. In: Advances in Neural Information Processing Systems (2021)
38. Kingma, D.P., Welling, M.: Auto-encoding variational bayes. In: International Conference on Learning Representations (2014)
39. Kingma, D.P., Salimans, T., Jozefowicz, R., Chen, X., Sutskever, I., Welling, M.: Improved variational inference with inverse autoregressive flow. In: Advances in Neural Information Processing Systems 29 (2016)
40. Lam, M.W., Wang, J., Su, D., Yu, D.: BDDM: bilateral denoising diffusion models for fast and high-quality speech synthesis. In: International Conference on Learning Representations (2022)
41. Lee, J., Mansimov, E., Cho, K.: Deterministic non-autoregressive neural sequence modeling by iterative refinement. In: EMNLP (2018)
42. Liang, F.: BachBot: automatic composition in the style of Bach Chorales. Univ. Cambridge 8, 19–48 (2016)
43. Lin, K., Li, D., He, X., Zhang, Z., Sun, M.T.: Adversarial ranking for language generation. In: Advances in Neural Information Processing Systems 30 (2017)
44. Liu, A., Fang, A., Hadjeres, G., Seetharaman, P., Pardo, B.: Incorporating music knowledge in continual dataset augmentation for music generation. arXiv preprint arXiv:2006.13331 (2020)
45. Lu, S., Zhu, Y., Zhang, W., Wang, J., Yu, Y.: Neural text generation: past, present and beyond. arXiv preprint arXiv:1803.07133 (2018)
46. de Masson d'Autume, C., Mohamed, S., Rosca, M., Rae, J.: Training language GANs from scratch. In: Advances in Neural Information Processing Systems 32 (2019)
47. Meister, C., Pimentel, T., Wiher, G., Cotterell, R.: Typical decoding for natural language generation. arXiv preprint arXiv:2202.00666 (2022)
48. Meng, C., et al.: SDEdit: guided image synthesis and editing with stochastic differential equations. In: International Conference on Learning Representations (2021)
49. Mittal, G., Engel, J., Hawthorne, C., Simon, I.: Symbolic music generation with diffusion models. arXiv preprint arXiv:2103.16091 (2021)
50. Nichol, A.Q., et al.: Glide: towards photorealistic image generation and editing with text-guided diffusion models. In: International Conference on Machine Learning, pp. 16784–16804. PMLR (2022)
51. Papamakarios, G., Pavlakou, T., Murray, I.: Masked autoregressive flow for density estimation. In: Advances in Neural Information Processing Systems 30 (2017)
52. Pati, A., Lerch, A., Hadjeres, G.: Learning to traverse latent spaces for musical score inpainting. In: Flexer, A., Peeters, G., Urbano, J., Volk, A. (eds.) Proceedings of the 20th International Society for Music Information Retrieval Conference, ISMIR 2019, Delft, The Netherlands, 4–8 November 2019, pp. 343–351 (2019)
53. Payne, C.: Musenet. https://openai.com/research/musenet (2019)
54. Radford, A., Wu, J., Child, R., Luan, D., Amodei, D., Sutskever, I.: Language models are unsupervised multitask learners. OpenAI Blog 1(8), 9 (2019)
55. Ramesh, A., Dhariwal, P., Nichol, A., Chu, C., Chen, M.: Hierarchical text-conditional image generation with clip latents (2022)
56. Rezende, D., Mohamed, S.: Variational inference with normalizing flows. In: International Conference on Machine Learning, pp. 1530–1538. PMLR (2015)
57. Rombach, R., Blattmann, A., Lorenz, D., Esser, P., Ommer, B.: High-resolution image synthesis with latent diffusion models (2021)

58. Savinov, N., Chung, J., Binkowski, M., Elsen, E., van den Oord, A.: Step-unrolled denoising autoencoders for text generation. In: International Conference on Learning Representations (2022)
59. Sohl-Dickstein, J., Weiss, E., Maheswaranathan, N., Ganguli, S.: Deep unsupervised learning using nonequilibrium thermodynamics. In: International Conference on Machine Learning, pp. 2256–2265. PMLR (2015)
60. Song, Y., Durkan, C., Murray, I., Ermon, S.: Maximum likelihood training of score-based diffusion models. In: Advances in Neural Information Processing Systems 34 (2021)
61. Song, Y., Ermon, S.: Generative modeling by estimating gradients of the data distribution. In: Advances in Neural Information Processing Systems 32 (2019)
62. Theis, L., Bethge, M.: Generative image modeling using spatial LSTMs. In: Advances in Neural Information Processing Systems 28 (2015)
63. Uria, B., Côté, M.A., Gregor, K., Murray, I., Larochelle, H.: Neural autoregressive distribution estimation. J. Mach. Learn. Res. **17**(1), 7184–7220 (2016)
64. Uria, B., Murray, I., Larochelle, H.: A deep and tractable density estimator. In: International Conference on Machine Learning, pp. 467–475. PMLR (2014)
65. Van Oord, A., Kalchbrenner, N., Kavukcuoglu, K.: Pixel recurrent neural networks. In: International Conference on Machine Learning, pp. 1747–1756. PMLR (2016)
66. Vasquez, S., Lewis, M.: MelNet: a generative model for audio in the frequency domain. arXiv preprint arXiv:1906.01083 (2019)
67. Vaswani, A., et al.: Attention is all you need. In: Advances in Neural Information Processing Systems 30 (2017)
68. Yang, Z., Dai, Z., Yang, Y., Carbonell, J., Salakhutdinov, R.R., Le, Q.V.: XLNet: generalized autoregressive pretraining for language understanding. In: Advances in Neural Information Processing Systems 32 (2019)
69. Yu, L., Zhang, W., Wang, J., Yu, Y.: SeqGAN: sequence generative adversarial nets with policy gradient. In: Proceedings of the AAAI Conference on Artificial Intelligence, vol. 31 (2017)
70. Zhu, Y., et al.: Texygen: a benchmarking platform for text generation models. In: The 41st International ACM SIGIR Conference on Research & Development in Information Retrieval, pp. 1097–1100 (2018)

SketchSynth: Cross-Modal Control of Sound Synthesis

Sebastian Löbbers[✉], Louise Thorpe, and György Fazekas

Centre for Digital Music, Queen Mary University of London, London, UK
{s.lobbers,l.thorpe,george.fazekas}@qmul.ac.uk

Abstract. This paper introduces a prototype of *SketchSynth*, a system that enables users to graphically control synthesis using sketches of cross-modal associations between sound and shape. The development is motivated by finding alternatives to technical synthesiser controls to enable a more intuitive realisation of sound ideas. There is strong evidence that humans share cross-modal associations between sound and shapes, and recent studies found similar patterns when humans represent sound graphically. Compared to similar cross-modal mapping architectures, this prototype uses a deep classifier that predicts the character of a sound rather than a specific sound. The prediction is then mapped onto a semantically annotated FM synthesiser dataset. This approach allows for a perceptual evaluation of the mapping model and gives the possibility to be combined with various sound datasets. Two models based on architectures commonly used for sketch recognition were compared, convolutional neural networks (CNNs) and recurrent neural networks (RNNs). In an evaluation study, 62 participants created sketches from prompts and rated the predicted audio output. Both models were able to infer sound characteristics on which they were trained with over 84% accuracy. Participant ratings were significantly higher than the baseline for some prompts, but revealed a potential weak point in the mapping between classifier output and FM synthesiser. The prototype provides the basis for further development that, in the next step, aims to make *SketchSynth* available online to be explored outside of a study environment.

Keywords: Sound synthesis control · Sound sketching · Cross-modal mapping · Musical timbre perception · Deep learning · Sketch recognition · Human-computer interaction

1 Introduction

Digital technology is now ubiquitous in music production. Eliminating the need for expensive analogue equipment and studio space enables a larger number of people to produce music. As a result, the sound of contemporary music is fundamentally shaped by digital synthesisers, audio effects and sample libraries. However, these tools are typically organised in reference to technical concepts,

C. Johnson et al. (Eds.): EvoMUSART 2023, LNCS 13988, pp. 164–179, 2023.
https://doi.org/10.1007/978-3-031-29956-8_11

which can make it difficult to realise sound ideas in a straightforward way. Recent developments seek to close this gap by centering their designs around human perception, often with the help of machine learning. The aim of this research is to develop *SketchSynth*, a system that allows for the exploration of a synthesiser space by sketching one's visual association with sound. The design is informed by multiple studies that asked participants to sketch their sound associations. The results show that similarities in sketched representations exist between participants, but individual human factors introduce significant noise to the data, a common challenge in cross-modal research. When asking participants to sketch a sound it cannot be determined with certainty which characteristic primarily influenced their representation. For example, a sound that could be described semantically as *noisy* and *thin* might be represented with focus on only one of these descriptors. This raises the question whether sound-sketches show greater similarity among participants if they are produced while imagining a sound from a semantic description, rather than listening to sounds directly. The work presented in this paper follows two objectives: first to implement and evaluate a proof-of-concept prototype of *SketchSynth* and second to collect a dataset of sketches that were produced to semantic prompts describing a sound. The results provide the basis for future work that will develop *SketchSynth* to be tested in a music practice context.

The paper is structured as follows: Sect. 2 introduces related work in music production tools and outlines relevant research about sound-shape associations and sketch recognition; Sect. 3 describes the design of *SketchSynth* and evaluation methods followed by the results, discussion and conclusion in Sects. 4, 5 and 6.

2 Related Work

Many recent developments that aim to simplify music production with the help of artificial intelligence can be summarised under the umbrella term Intelligent Music Production (IMP) [31]. IMP research is increasingly implemented into commercial software; for example *XO* by XLN audio [37] is based on perception-informed re-organisation of sample libraries for easier sound exploration and retrieval [4,12], or iZotope's mixing and mastering plugins [21] which build on research into automatic and assisted mixing [7]. Other works explore different modes of interaction, for example, retrieving synthesiser or audio effect parameters from sounds or vocal mimicry [11,30,34], and synthesis control through gestures [39] or visual sound metaphors [13]. By implementing a functioning prototype for the first time, this research extends the proposal for a sketch-based sound retrieval tool by Knees and Andersen [22]. Through interviews with music producers, they found that mental concepts of sound are often rooted in the visual domain.

2.1 Sound-Shape Associations

People frequently reference visual concepts like colour, brightness, shapes and contour when they think of sound [29]. Sound-shape were first described by the *bouba/kiki* effect [23,32] that was shown to be robust across cultures and different demographics [5], the visually impaired [2] as well as between shapes and musical instruments [1] or abstract sounds [15]. Recent studies asked participants to sketch their personal associations with sound rather than using existing visual stimuli [9,24,25]. The results can be categorised into figurative and abstract representations, with the latter showing correlations between visual and sound features that align with prior sound-shape research. An evaluation study showed that participants can successfully match these abstract representations and their corresponding sounds [26]. Engeln et al. trained an end-to-end autoencoder that can be used for sketch-based sound query [10]. A similar approach is deployed in this work, however here a real-time sketch input is mapped to sounds using a model that is trained to predict sound characteristics.

2.2 Sketch Recognition

Sketch recognition is typically used in the context of image retrieval but can also find application for cross-modal mapping tasks like *SketchSynth*. The established approach using convolutional neural networks (CNNs) significantly outperforms conventional machine learning methods at the benchmark task of classifying handwritten digits with the MNIST dataset [8]. Increasingly popular are recurrent neural networks (RNNs) that can take advantage of the sequential vector format in which digital sketches are typically saved. Seminal work by Ha and Eck [16] introduced the *Quick, Draw!* dataset and the Sketch-RNN architecture for sketch classification and generation. *Quick, Draw!* [14] is a large open-source dataset with over 50 million sketches that enables researchers to experiment and pre-train models for specific tasks. While a RNN classifier can already outperform a CNN on complex sketch classes by learning temporal relationships, CNNs might be more suited for learning abstract visual structures [38]. While *SketchRNN* produces impressive results for sketch classification and generation, algorithmic approaches might be more suitable for describing the shape of a sketch. Wolin et al.'s *ShortStraw* algorithm [35] provides a simple, effective tool to extract corner points. Xiong et al. [36] extended the algorithm to also recognise curve points. Szegin et al. [33] further show that information can not only be extracted from a sketch's shape but also from the drawing speed.

3 Methods and Material

This section describes the design and evaluation of the *SketchSynth* prototype. A deep learning approach is used to predict sound characteristics from a sketch input, which is then mapped onto an annotated FM synthesiser dataset. Two architectures, a CNN and a RNN, were compared and a study was conducted to obtain participant ratings of the predicted sounds. The study was designed as a between-subject multivariate test where two different model architectures were tested against a random baseline. The following hypotheses were postulated:

- Sketches produced from a semantic prompt describing a sound can achieve higher accuracy than sketches produced from a sound stimulus.
- A binary classifier can be trained to distinguish between sketches of two sound classes above the random baseline of 50% accuracy.
- A mapping model using this binary classifier will receive significantly higher ratings from human participants compared to the random baseline.
- The CNN and RNN architectures will perform similarly for classification accuracy and participant ratings.

3.1   Sketch Dataset

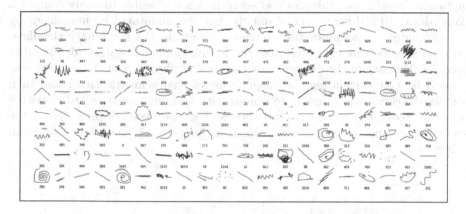

Fig. 1. A representative subset of sketches from the *Sketching Sounds* dataset [27].

Both classifiers were trained on the *Sketching Sounds* dataset, illustrated in Fig. 1, that contains 1760 sound-sketches collected in a study with 88 participants that followed the design of an earlier study by the author [25]. Participants sketched their association with synthesiser sounds described in Sect. 3.2 with a digital interface similar to the one shown in Fig. 5. In addition, each participant created two sketches from a prompt (*Draw a calm/noisy sound*) in a pre-task test without audio. While larger sketch datasets exist as discussed in Sect. 2.2, *Sketching Sounds* is the only dataset of sketched synthesiser sound representations to date.

3.2   Sound Dataset

The *Sketching Sounds* dataset was created on a subset of 20 sounds sourced from a dataset by Hayes et al. [18] that includes 364 synthesiser sounds. Each sound was saved as a set of parameters that belong to a browser-based FM synthesiser. [17] This dataset was chosen because it includes annotations from participants that enable perceptual analysis of the sound-sketches. Thirty music producers created the sounds from prompts (*bright, rough, thick*), which were semantically rated by a subset of 24 English speaking participants along a scale of 30 sound descriptors [19]. Factor analysis of these annotations found five semantic factors (*sharpness, mass, clarity, percussiveness, rawness*) that explained 74% of data variance.

3.3   Sketch-to-sound Mapping Using Deep Learning

As shown in Fig. 2, the sketch-to-sound mapping consists of two parts: (1) a binary classifier predicting the sound category from a sketch input and (2) the selection of a suitable sound from the FM synthesis dataset described in Sect. 3.2. This simple architecture was chosen to allow for a transparent, perceptual interpretation. In addition, this modular setup makes it possible to easily connect a sketch-input to a different set of sounds that was annotated by humans or by an automated music-information retrieval approach. For the binary classification, six subsets were extracted from the *Sketching Sounds* dataset described in Sect. 3.1. For each of the five semantic factors described in Sect. 3.2 a subset was created by first calculating the mean rating of that semantic factor and then sorting sketches corresponding to sounds with a rating one standard deviation below or above the mean into two classes. The test sketches produced to the semantic prompts *calm* and *noisy* described in Sect. 3.1 formed an additional subset.

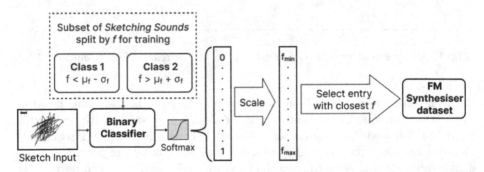

Fig. 2. Mapping logic between sketch input and sound output. A binary classifier predicts the sound-category from two opposing classes (e.g. *noisy* and *calm*) which is then used to pick a suitable sound from an annotated FM synthesiser dataset. Multiple binary classifiers can be trained to simultaneously predict multiple sound categories. A demonstration of the setup is available at https://youtu.be/ca1LYn8Yy-g.

Two deep classification architectures commonly used for sketch recognition tasks, as discussed in Sect. 2.2, were compared: a CNN and a RNN. Both models were implemented in Keras with cross-entropy as the loss function and accuracy as the evaluation metric. To address the relatively small sample size, sketches were augmented through rotation, scaling, dropouts and Perlin noise applied to sketch points [6, 16]. In addition, the RNN architecture was pre-trained on the geometric categories *squiggle, zigzag, square, triangle, circle* and *line* from the *Quick, Draw!* dataset that represent similar abstract structures as the *Sketching Sounds* dataset. Figure 3 shows the RNN architecture that uses the encoder part of the Sketch-RNN variational autoencoder (VAE) proposed by Ha and Eck [16] for feature extraction and three dense layers for classification. After pre-training, the feature-extractor layers were frozen to accelerate the training time for the sound-sketches. Figure 4 shows the CNN architecture that uses a structure commonly used for image classification like MNIST handwritten digit classification [8]. This architecture was chosen because *Sketching Sounds* contains simple, monochromatic representations similar to MNIST. In addition, a simple model might provide faster predictions when used in a real-time, client-side setup for future work, as discussed in Sect. 5. The classifiers' softmax output was scaled to the range of the semantic factor ratings in the FM dataset to retrieve the sound with the closest annotated value. For the *calm* and *noisy* subset, the output was scaled to the ratings of the *noisy* descriptor, where a negative rating represents a *calm* sound. For the random baseline, synthesiser sounds were picked randomly from the FM dataset after a sketch input was received.

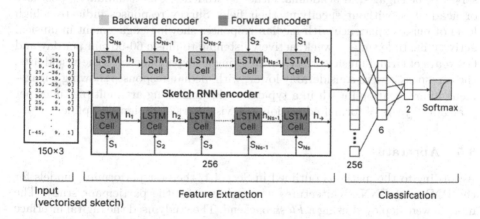

Fig. 3. Classification model using the encoder architecture from Sketch-RNN [16]. The feature extraction part of the network was pre-trained on a subset of the *Quick, Draw!* dataset as explained in Sect. 3.3.

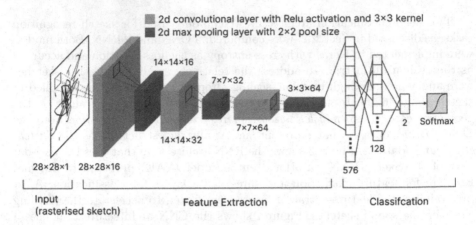

Fig. 4. CNN Classification model.

3.4   Participants

Sixty-two participants were recruited internally at the author's institution and externally through the ISMIR mailing list [20]. Twenty-two identified as female, 37 as male, 2 as other and one participant preferred not to disclose this information. Ages ranged from to 22 to 58 ($\mu = 30.16$, $\sigma = 6.66$). The majority of participants (46) work in engineering, computer science or psychology, some with a focus on music. Only three were outside of these fields. Thirteen described themselves to be engaged in academia, either as students, PhD candidates, postdocs or academics, without specifying their field. Survey responses indicate a high level of music experience with median responses showing engagement in musical activity multiple times a week, actively listening to music 60-90 min per day and 6-9 years of formal music education. Relating to experience with the visual arts, these responses were considerably lower, with median responses showing engagement in art creation for 0 h in a typical week, consuming art multiple times per year and 0 years of formal education in a visual art or design discipline.

3.5   Apparatus

According to the methods outlined in Sect. 4.1, the best performing models for the RNN and CNN architectures were selected for the participant study. The models were deployed using a *Flask* backend. The study used the digital interface shown in Fig. 5. Each round started with a prompt displayed in the middle of the canvas that faded out after starting a sketch. Finished sketches were sent to the backend to be saved in a database and to predict a suitable sound. The sound parameters were then returned to the participant who rated the synthesised sound on a scale from 0 (no match) to 100 (perfect match).

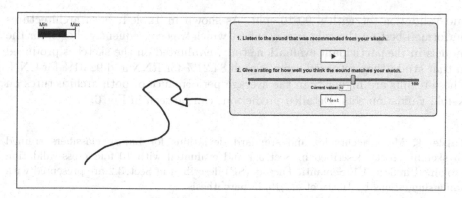

Fig. 5. Screenshot of the online study interface that was. A similar interface was used to collect the *Sketching Sounds* dataset described in Sect. 3.1. To encourage simple, abstract representations, the sketch length was limited to a range visualised by a meter in the top left corner of the canvas.

3.6   Procedure

Participants were first presented with a set of information ensuring that they use a laptop or desktop and were able to listen to sound either through headphones or loudspeakers. This was followed by a short introduction of the study with guidelines on representing sounds in an abstract rather than figurative way, a short explanation of the sketching interface and a guide to adjusting playback volume to a comfortable level. The main task consisted of six rounds in randomised order in which participants sketched according to a prompt and rated the resulting sound as explained in Sect. 3.5. The prompts were structured as *Draw a [descriptor] sound* using the descriptors: *calm, noisy, clean, rough, thin, thick*. The descriptors were derived from the FM synthesis study by Hayes et al. [18] introduced in Sect. 3.2. *Bright/dark* were replaced with *noisy/calm* in this study, because, as described in Sect. 4.1, the models trained on the *calm/noisy* subset were chosen for participant evaluation. The study concluded with a survey collecting information about participants' demographic data, experience with music and art, hardware that was used to complete the study and feedback about their overall experience.

4   Analysis and Results

This section describes the evaluation of the classifiers and the analyses and results of participant ratings and evaluation.

4.1   Model Evaluation

K-fold cross-validation with 10 folds was used to evaluate the deep classifiers. For each run, one fold was used as a test set and the remaining nine formed the train and validation set with a 90-10 split. As shown in Table 1, both architectures performed best on the *calm/noisy* subset which was consequently chosen for the models in the participant evaluation study. Evaluated on the sketches produced in that study, they returned accuracies of 84.21% for RNN and 92.31% for CNN. These results are higher than the average performance of both architectures on K-fold validation sets. Detailed predictions can be seen in Fig. 6.

Table 1. Mean accuracies and standard deviations for binary classifiers trained on sketch subsets described in Sect. 3.3 and evaluated with 10 fold cross-validation explained in Sect. 4.1. Semantic Factors (SF) described in Sect. 3.2 are presented with name suggestions by Hayes et al. [19] in parenthesis.

	CNN		RNN	
Subset	Mean Acc. [%]	Std. Dev. [%]	Mean Acc. [%]	Std. Dev. [%]
calm/noisy	**71.82**	**9.80**	**73.41**	**10.86**
SF1 (Sharpness)	52.66	3.81	51.74	4.30
SF2 (Mass)	60.23	3.50	64.06	7.48
SF3 (Clarity)	53.81	3.53	54.48	2.70
SF4 (Percussiveness)	60.72	3.05	64.18	5.88
SF5 (Rawness)	55.35	6.54	62.38	2.68

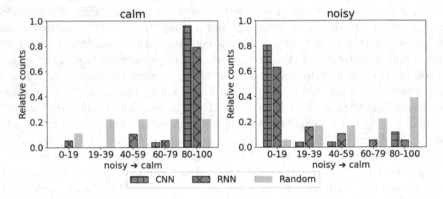

Fig. 6. Prediction histograms for the *calm* and *noisy* prompts. The x-axis of each subplot corresponds to the softmax output of the *calm/noisy* classification model with 0 referring to a completely *noisy* and 100 to a completely *calm* prediction.

4.2 Sound Ratings

A main objective of this study was to find out whether satisfactory sound predictions could be made with a simple, generalised mapping model. To test for significant differences, the Mann-Whitney U-test was used to compare between ratings for the random baseline model and the CNN and RNN models that were trained on the *calm/noisy* subset. Figure 7 shows the ratings participants gave for the *calm* and *noisy* prompt with annotated significance levels in comparison to the random baseline. CNN and RNN both received significantly higher ratings for the *calm* prompt (p<.01 for CNN and p<.05 for RNN); however, no significant difference could be found for the *noisy* prompt. Interestingly, CNN predictions were rated significantly higher for *clean* despite not being trained on this semantic class. This could be explained with correlations between the sound descriptors *calm* and *clean* that were also found by Hayes et al. [19].

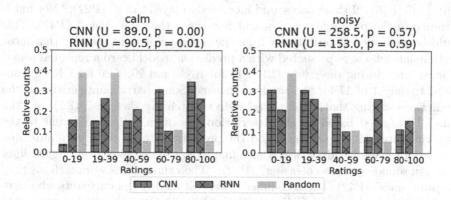

Fig. 7. Rating histograms for classification models and random baseline. Statistics and p-values are reported for Mann-Whitney U test between distributions of the respective model and the random baseline.

4.3 Survey Responses

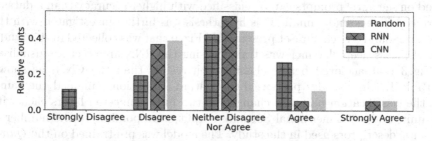

Fig. 8. Answer to system evaluation question: *I thought that this system produced suitable sounds from my sketches.*

Participants' overall experience with the system was collected on a five point Likert scale and presented separately for each mapping scheme as shown in Fig. 8. Distributions look similar for CNN, RNN and the random baseline with most participants giving neutral ratings for the system. Responses skewed slightly negative for random (82.35% said *neither agree/disagree* or *disagree*) and RNN

(89.47% said *neither agree/disagree* or *disagree*) and slightly positive for CNN (69.23% said *neither agree/disagree* or *agree*). Significant differences to the random baseline were not expected for the overall experience as the classification model was only trained to recognise *calm* and *noisy* sketches. General feedback was submitted as free-form text and summarised with thematic analysis [3] to identify common remarks. Multiple participants found that sound predictions were either too similar or that the same sound was played multiple times: "[...] there was not much change between them. I encountered 3 different sounds basically [...]" (P16); "I think one sound incorrectly played twice" (P22); "Not much variance in the proposed sound - had few times the same ones." (P42). This could be confirmed quantitatively from the sound predictions showing that most participants who were presented with a prediction model heard a repeated sound at least once during the study (13 of 19 for RNN and 20 of 26 for CNN), compared to only 1 of 17 for the random baseline. Some participants criticised the predictions stating that, "the sound didn't match my sketches" (P1), or "the sound I imagined based on the description was often entirely different to the sound I heard" (P33). Others found the predictions to be accurate: "pure sine tone from the circle I drew was a nice mapping, the mapping from jagged lines to rough sounds was also pleasing" (P12); "The calm sounds work well, as they are pure tones" (P42). These comments were all made by participants who were presented with either the CNN or RNN. Overall, multiple positive comments about the study were left that suggest a wider interest in the system: "This was really interesting. I enjoyed using the system a lot." (P61); "Anyways I loved the experience!" (P31).

4.4   Evaluation of Semantic Sound-Sketches

The model evaluation in Sect. 4.1 suggests that using a sound-sketch dataset based on semantic prompts can be classified with higher accuracy than a dataset produced from sound stimuli. This hypothesis was further investigated with the semantic sound-sketch dataset presented in Fig. 9 that was collected in this study. A multi-class deep classifier was trained using the RNN architecture visualised in Fig. 3 that achieved better classification results than the CNN, as shown in Table 1. The encoder part used for feature extraction remained the same, but the classification part was changed to two fully connected layers each with 128 units and a 6-dimensional Softmax output corresponding to the number of semantic descriptors used in the study. The model was pre-trained on the *Quick, Draw!* subset described in Sect. 3.3. The encoder was frozen for training with the semantic sound-sketch dataset with a 84-16 training-test split. The results of the evaluation with 72 test sketches shown in Fig. 10 show significantly higher accuracy than the random baseline of 16.6% (2 in 12 correct predictions) for all prompts except for *thick*. The results further suggest similarities between *noisy* and *rough* sketches with both classes being most often misclassified as the respective other. Surprisingly, *calm* and *clean* sketches were not confused with one another as suggested in Sect. 4.2 but were most often misclassified as *thin*. Predictions for *thick* are spread across multiple classes, however qualitatively

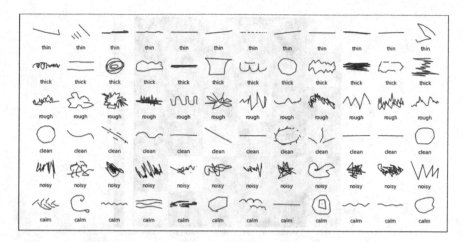

Fig. 9. A representative selection from the *SketchSynth* dataset [28] that was collected in this study. Each row shows sketches that were produced to a different semantic prompt.

assessing *thick* and *thin* sketches in Fig. 9 suggests that a binary classifier might be able to distinguish between these opposing classes.

5   Discussion

The results of the classification model presented in Sect. 4.1 indicate that sound-sketches can be distinguished more easily with a deep learning model when they are produced to semantic prompts like *Draw a calm/noisy sound*. It is not always clear which sound characteristics participants represented in their sketches when listening to a sound, which can lead to a larger variance in representation compared to using prompts. However in Sect. 4.4, the evaluation of the semantic sound-sketch dataset collected in this study suggests participants do not represent all perceptual sound dimensions with distinctly different sketch approaches. Similar approaches can be found between classes; in Fig. 9, for example, circular shapes are seen for *thick, clean* and *calm*. However, these shapes do not appear in the opposing classes *thin, rough* and *noisy*. While this hypothesis needs to be investigated systematically, it does hint that a model using multiple binary classifiers between opposing semantic classes (*noisy/calm, clean/rough, thick/thin*) might be able to achieve higher accuracies and would enable the simultaneous prediction of multiple perceptual classes. Section 4.1 suggests that the Sketch-RNN architecture does not provide a significant advantage over the simpler CNN architecture for this specific use. This might be due to the abstract structure of the sketches which are less complex than many of the *Quick, Draw!* categories and might, therefore, not require a complex architecture. Contrary to expectation, both models achieved higher classification accuracies with the sketches collected in this study compared to the *Sketching Sounds* dataset. This might

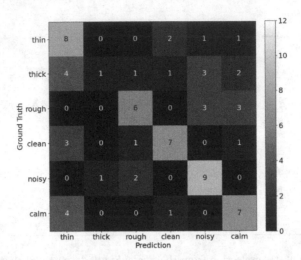

Fig. 10. Confusion matrix for the multi-class classifier that was trained on sketches created to semantic prompts. The test set consisted of 12 sketches for each prompt.

be resulting from the different study design or participant pool that drew primarily from a population with high music expertise. Comparing Figs. 6 and 7 show that, while predictions for *noisy* and *calm* prompts were highly accurate, participants' ratings of the suggested sounds were less positive, implying a bottleneck in the mapping between deep learning model output and synthesiser parameters. This is reflected in some of the qualitative feedback presented in Sect. 4.3 with participants noticing similar or same sounds being produced multiple times. This can be explained with the behaviour of classification models that push their output to 0 or 1 which, following the mapping model shown in Fig. 2, leads to sounds with the maximal or minimal annotated value for *noisy* being selected more often. These results show that a single, generalised mapping architecture can predict suitable sounds to some level; however, the performance can be improved through a number of approaches: (1) parameters of the current classification models could be fine-tuned for this specific task or the architecture could be improved, for example by combining RNNs and CNNs; (2) the current prototype only returns a sound after a sketch is submitted. A design that returns sounds while sketching would provide immediate feedback to a user allowing them to adjust their sketches accordingly and continuously explore the synthesiser sound space; (3) the softmax output of the classifier could be interpreted as a relative change (e.g. increasing the value for *noisy* to select a sound), which would prevent the over-representation of a small number of sounds; (4) the modular mapping architecture could be replaced with an end-to-end model that directly predicts synthesis parameters; however, this would make a perceptual interpretation difficult; (5) predictions could adjust to personal preferences, for example through reinforcement learning. The *SketchSynth* prototype presented in this work fulfilled its function as a proof-of-concept, but future work will need to move away from a perceptual research environment to evaluate the concept in a music practice context. A next step could invite music producers to explore the system and reflect on how they might integrate it into their practice.

6   Conclusion

A first prototype of *SketchSynth* was implemented successfully and evaluated in a participant study. The results show that suitable sounds can already be predicted with a simple, generalised mapping architecture. The prompt-based sound-sketch dataset collected from the study provides a basis for extending the prediction model to additional perceptual categories. Participant feedback and quantitative analyses of the architectures will inform the future development of this system. The next step seeks to release a readily-available online version of *SketchSynth* that will allow for interaction outside of a study environment.

Acknowledgements. EPSRC and AHRC Centre for Doctoral Training in Media and Arts Technology (EP/L01632X/1).

References

1. Adeli, M., Rouat, J., Molotchnikoff, S.: Audiovisual correspondence between musical timbre and visual shapes. Front. Hum. Neurosci. **8**, 352 (2014). https://doi.org/10.3389/fnhum.2014.00352
2. Bottini, R., Barilari, M., Collignon, O.: Sound symbolism in sighted and blind. the role of vision and orthography in sound-shape correspondences. Cognition **185**, 62–70 (2019). https://doi.org/10.1016/j.cognition.2019.01.006
3. Braun, V., Clarke, V.: Using thematic analysis in psychology. Qual. Res. Psychol. **3**(2), 77–101 (2006). https://doi.org/10.1191/1478088706qp063oa
4. Bruford, F., Barthet, M., McDonald, S., Sandler, M.B.: Groove explorer: An intelligent visual interface for drum loop library navigation. In: Proceedings of the ACM IUI Workshops. CEUR-WS.org, Los Angeles, USA (2019)
5. Ćwiek, A., et al.: The Bouba/Kiki effect is robust across cultures and writing systems. Philosop. Trans. Royal Soc. B: Biol. Sci. **377**(1841), 20200390 (2022). https://doi.org/10.1098/rstb.2020.0390
6. Das, A., Yang, Y., Hospedales, T., Xiang, T., Song, Y.Z.: SketchODE: learning neural sketch representation in continuous time. In: Proceedings of International Conference on Learning Representations. OpenReview.net, virtual (2022)
7. De Man, B., Reiss, J., Stables, R.: Ten years of automatic mixing. In: Proceedings of the Workshop on Intelligent Music Production. Salford, U.K. (2017)
8. Deng, L.: The MNIST database of handwritten digit images for machine learning research. IEEE Signal Process. Mag. **29**(6), 141–142 (2012). https://doi.org/10.1109/MSP.2012.2211477
9. Engeln, L., Groh, R.: CoHEARence of audible shapes—a qualitative user study for coherent visual audio design with resynthesized shapes. Pers. Ubiquit. Comput. **25**(4), 651–661 (2020). https://doi.org/10.1007/s00779-020-01392-5

10. Engeln, L., Le, N.L., McGinity, M., Groh, R.: Similarity analysis of visual sketch-based search for sounds. In: Proceedings of Audio Mostly 2021, pp. 101–108. Association for Computing Machinery, Trento, Italy (2021). https://doi.org/10.1145/3478384.3478423
11. Esling, P., Masuda, N., Chemla-Romeu-Santos, A.: FlowSynth: simplifying complex audio generation through explorable latent spaces with normalizing flows. In: Proceedings of International Joint Conference on Artificial Intelligence, pp. 5273–5275 (2020). https://doi.org/10.24963/ijcai.2020/767
12. Garber, L., y Ciencia, M.A., Ciccola, T., Amusategui, J.C.: AudioStellar, an open source corpus-based musical instrument for latent sound structure discovery and sonic experimentation. In: Proceedings of International Computer Music Conference, pp. 86–91. Santiago, Chile (2021). https://hdl.handle.net/2027/fulcrum.t435gg568
13. Giannakis, K.: Sound mosaics: a graphical user interface for sound synthesis based on audio-visual associations, Ph. D. thesis, Middlesex University (2001)
14. Google: Quick, Draw! (2017). https://quickdraw.withgoogle.com/. Accessed 8 Feb 2023
15. Grill, T., Flexer, A.: Visualization of Perceptual Qualities in Textural Sounds. In: Proceedings of International Computer Music Conference, pp. 589–596. Michigan Publishing Services, Ljubljana, Slovenia (2012). http://hdl.handle.net/2027/spo.bbp2372.2012.110
16. Ha, D., Eck, D.: A neural representation of sketch drawings. arXiv preprint arXiv:1704.03477 (2017)
17. Hayes, B.: FM synth study (2020). https://github.com/ben-hayes/fm-synth-study. Accessed 8 Feb 2023
18. Hayes, B., Saitis, C.: There's more to timbre than musical instruments: semantic dimensions of FM sounds. In: Proceedings of International Conference on Timbre. Timbre 2020, Thessaloniki, Greece (2020)
19. Hayes, B., Saitis, C., Fazekas, G.: Disembodied timbres: a study on semantically prompted fm synthesis. J. Audio Eng. Soc. **70**(5), 373–391 (2022). https://doi.org/10.17743/jaes.2022.0006
20. ISMIR: Homepage (2022). https://ismir.net/. Accessed 8 Feb 2023
21. iZotope: Mix & Master Bundle Advanced (2023). https://www.izotope.com/en/shop/mix-master-bundle-advanced.html. Accessed 8 Feb 2023
22. Knees, P., Andersen, K.: Searching for audio by sketching mental images of sound: a brave new idea for audio retrieval in creative music production. In: Proceedings of International Conference on Multimedia Retrieval, pp. 95–102. Association for Computing Machinery, New York, USA (2016). https://doi.org/10.1145/2911996.2912021
23. Köhler, W.: Gestalt psychology. Liveright (1929)
24. Küssner, M.B., Tidhar, D., Prior, H.M., Leech-Wilkinson, D.: Musicians are more consistent: gestural cross-modal mappings of pitch, loudness and tempo in real-time. Front. Psychol. **5**, 00789 (2014). https://doi.org/10.3389/fpsyg.2014.00789
25. Löbbers, S., Barthet, M., Fazekas, G.: Sketching sounds: an exploratory study on sound-shape associations. In: Proceedings of International Computer Music Conference, pp. 299–304. Michigan Publishing Services, Santiago, Chile (2021). https://hdl.handle.net/2027/fulcrum.t435gg568
26. Löbbers, S., Fazekas, G.: Seeing sounds, hearing shapes: a gamified study to evaluate sound-sketches. In: Proceedings International Computer Music Conference, pp. 174–179. Michigan Publishing Services, Limerick, Ireland (2022). https://hdl.handle.net/2027/fulcrum.nk322g689

27. Löbbers, S., Fazekas, G.: Sketching Sounds Dataset (1.0) [Data set] (2023). https://doi.org/10.5281/zenodo.7590916
28. Löbbers, S., Fazekas, G.: SketchSynth Dataset (1.0) [Data set] (2023). https://doi.org/10.5281/zenodo.7591067
29. Martino, G., Marks, L.E.: Synesthesia: strong and weak. Curr. Dir. Psychol. Sci. **10**(2), 61–65 (2001)
30. Mehrabi, A., Dixon, S., Sandler, M.B.: Vocal imitation of synthesised sounds varying in pitch, loudness and spectral centroid. J. Acoust. Soc. Am. **141**(2), 783–796 (2017)
31. Moffat, D., Sandler, M.B.: Approaches in intelligent music production. Arts **8**(4), 125 (2019). https://doi.org/10.3390/arts8040125
32. Ramachandran, V.S., Hubbard, E.M.: Synaesthesia-a window into perception, thought and language. J. Conscious. Stud. **8**(12), 3–34 (2001)
33. Sezgin, T.M.: Feature point detection and curve approximation for early processing of free-hand sketches, Ph. D. thesis, Massachusetts Institute of Technology (2001)
34. Singh, S., Bromham, G., Sheng, D., Fazekas, G.: Intelligent control method for the dynamic range compressor: a user study. J. Audio Eng. Soc. **69**(7/8), 576–585 (2021). https://doi.org/10.17743/jaes.2021.0028
35. Wolin, A., Eoff, B., Hammond, T.: ShortStraw: a simple and effective corner finder for polylines. In: Proceedings of Eurographics Workshop on Sketch-Based Interfaces and Modeling. The Eurographics Association, Annecy, France (2008). https://doi.org/10.2312/SBM/SBM08/033-040
36. Xiong, Y., LaViola, J.J.: Revisiting shortStraw: improving corner finding in sketch based interfaces. In: Proceedings of Eurographics Symposium on Sketch-Based Interfaces and Modeling, pp. 101–108. Association for Computing Machinery, New Orleans USA (2009). https://doi.org/10.2312/SBM/SBM09/101-108
37. XLN Audio: XO product page (2023). https://www.xlnaudio.com/products/xo. Accessed 8 Feb 2023
38. Xu, P., et al.: SketchMate: deep hashing for million-scale human sketch retrieval. In: Proceedings of Conference on Computer Vision and Pattern Recognition, pp. 8090–8098. IEEE Computer Society, Salt Lake City, USA (2018)
39. Zbyszyński, M., Di Donato, B., Visi, F.G., Tanaka, A.: Gesture-timbre space: multi-dimensional feature mapping using machine learning and concatenative synthesis. In: Kronland-Martinet, R., Ystad, S., Aramaki, M. (eds.) CMMR 2019. LNCS, vol. 12631, pp. 600–622. Springer, Cham (2021). https://doi.org/10.1007/978-3-030-70210-6_39

Towards the Evolution of Prompts
with MetaPrompter

Tiago Martins⬡, João M. Cunha⁽✉⁾⬡, João Correia⬡,
and Penousal Machado⬡

CISUC, Department of Informatics Engineering, University of Coimbra,
Coimbra, Portugal
{tiagofm,jmacunha,jncor,machado}@dei.uc.pt

Abstract. The dissemination of open-source text-to-image generative
models and the increasing quality of their output has led to a growth in
interest in the field. The quality of the images greatly depends on the
prompt used, *i.e.* a phrase that includes descriptive terms to be used as
input on text-to-image model. However, choosing the right prompt is a
complex task, often relying on a trial-and-error approach. In this paper,
we introduce an evolutionary approach to prompt generation where users
begin by creating a blueprint for what might be a candidate prompt
and then initiate an evolutionary process to interactively explore the
space of prompts encoded by the initial blueprint and according to their
preferences. Our work is a step towards a more dynamic and interactive
way to generate prompts that lead to a wide variety of visual outputs,
with which users can easily obtain prompts that match their goals.

Keywords: Image Generation · Text-to-Image · Stable Diffusion ·
Interactive Evolutionary Computation

1 Introduction

In the past two years, we have witnessed a growing interest in text-to-image Artificial Intelligence (AI) systems caused by an increase in their performance and
output quality. This wave of development can be linked to the appearance of multimodal models, such as Contrastive Language-Image Pre-Training (CLIP) [14].
CLIP is a contrastive language-visual model, trained on a dataset of 400 million
text-image pairs collected from the internet. It results in the compression of two
models at once (language and visual), establishing a connection between the
two and allowing the estimation of semantic similarity between an image and a
given text. While prior image generators were greatly limited to the classes of
the datasets used in their training process (*e.g.* MS-COCO [7]), the introduction
of these language-visual models enabled a larger scale text-to-image generation
with few restrictions on what can be produced.

One of the first text-to-image generation models to use contrastive models is
DALL-E [15]. However, the lack of public access triggered the interest in open-
source alternative approaches, leading to the development of multiple openly

© The Author(s), under exclusive license to Springer Nature Switzerland AG 2023
C. Johnson et al. (Eds.): EvoMUSART 2023, LNCS 13988, pp. 180–195, 2023.
https://doi.org/10.1007/978-3-031-29956-8_12

Fig. 1. Snapshot of a population of prompts evolved using the presented system. The images in each column are generated from different seeds using the same prompt, which is shown at the bottom.

available systems that also use CLIP for guiding image generation (*e.g.* Vector Quantized Generative Adversarial Network (VQGAN)-CLIP) [1], Stable Diffusion [17], *etc.*). These text-conditioned generative systems produce images from text inputs that are referred to as prompts. The choice of the words used in the prompt is key for producing images that match individual preferences and goals – adding specific keywords to the prompt can greatly improve results as reported by independent users (*e.g.* "unreal engine" improves images generated with VQGAN and CLIP[1]) and prior research studies [9,13]. However, the task of constructing effective prompts, known in the field of Natural Language Processing (NLP) as *prompt engineering* [16], has an open-ended nature and often consists of a highly experimental and iterative process [9]. This trial-and-error process often involves the use of prompt parts without a clear understanding of their impact on the results, leaving users with a sense of randomness [19].

The importance of prompts for producing high-quality images, both in terms of aesthetics and alignment with user goals, is reflected in the number of resources related to prompt engineering under development. One example that shows the value of prompt engineering expertise is the platform *PromptBase*,[2] which centres its business model on the monetisation of prompts, allowing users to buy and sell prompts for different generative models.

On the other hand, the growing accessibility of text-to-image generation models led to the emergence of communities, whose members (from artists to researchers) are devoted to the open-source development of these systems and to the creation and sharing of resources, such as guides (*e.g. A Traveler's Guide to the Latent Space* [18]). Researchers have focused on the development of computational approaches for the generation of prompts [13] and addressed the identification of design guidelines to write effective prompts [9,12].

In this paper, we aim to contribute to the set of resources that facilitate the creation of prompts. We propose a method where users *(i)* design a *meta prompt*

---

[1] https://twitter.com/arankomatsuzaki/status/1399471244760649729 accessed 2023.
[2] https://promptbase.com/ accessed 2023.

that encodes a space of different alternative prompts and then *(ii)* interactively evolve a population of prompts (see Fig. 1) that explores multiple points of this space according to their preferences. Our goal with the proposed method is to offer a more dynamic and interactive way to find and tune prompts with which users can obtain images that match their preferences.

Overall, the experimental results indicate the ability of the proposed method to evolve prompts that result in images that meet preferences expressed interactively by the user. Furthermore, we can generate a wide variety of visual outputs and also converge to prompts whose images share visual characteristics, namely their content and style.

The remainder of this paper is organised as follows: Sect. 2 summarises related work on prompt engineering, tools for prompt construction and prompt generation; Sect. 3 explains the proposed system; Sect. 4 describes the experimentation conducted to explore and analyse the possibilities created with the presented system and presents the obtained results; finally, Sect. 5 presents conclusions and directions for future work.

2   Related Work

Our work aims to facilitate the production of prompts with which users can obtain images that match their preferences. We present related work divided into three sections: prompt engineering, tools for prompt construction and generation.

2.1   Prompt Engineering

Prompt Engineering can be understood as the task of composing a textual instruction or description with the goal of obtaining a specific result from a language-based model [8]. Although the most recent interest in prompt engineering is related to text-to-image systems, the term has originated in relation to text-to-text systems [9] and researchers have investigated ways of formulating prompts for specific tasks (*e.g.* summarisation [8]).[3]

Regarding text-to-image systems, a prompt usually consists of a description of the image(s) that the user wishes to produce. Several authors have identified and studied different ways of structuring prompts (*prompt templates*), such as [Subject] in the style of [Style] [9], [Medium] [Subject] [Artist(s)] [Details] [Image repository support] [12] or [Subject], by [Artist] (and [Artist]), [Modifier(s)],... [18]. As observed in these templates, prompts can be divided into parts that can be ontologically categorised, from which the most important is considered to be the subject [9]. Other prompt parts are often referred to as *modifiers* and consist of words or phrases that are added to the prompt to alter the style or improve the quality of the results [11].

The discovery of new modifiers usually occurs in trial-and-error approaches and good results may lead to the widespread of modifiers in the community (*e.g.* "greg rutkowski" [6]). The identification of these modifiers has been a key

---

[3] See examples in https://beta.openai.com/examples accessed 2023.

task in the field, resulting in the development of multiple prompt guides by community members (*e.g.* the *Stable Diffusion Prompt Book*[4]) and modifier lists.[5]

Researchers have also devoted their attention to studying aspects of prompt engineering. Oppenlaender [12] describes an ethnographic study with online communities that resulted in the proposal of a prompt modifier taxonomy with six different types: subject terms, image prompts, style modifiers, quality boosters, repetitions, and magic terms. Liu and Chilton [9] conducted a study using nine configurations of the [Subject][Style] template and propose a set of guidelines for prompt engineering – *e.g.* focus on subject and style instead of connecting words and generate between 3 to 9 seeds to get a representative set.

Some aspects of prompt engineering, *e.g.* weight assignment and repetition, despite being mentioned by multiple authors, have not yet been thoroughly studied in terms of their impact on results.

2.2   Tools for Prompt Construction

Given the difficulty of producing good-quality prompts, a number of tools have been implemented to aid in prompt engineering. An example is *Prompt Builder*,[6] which is described as a user-friendly tool that provides aid in constructing prompts and generating images. The user starts by selecting a model (*e.g.* Stable Diffusion) and then is presented with an interface for prompt building with multiple sections: *(i)* add an image prompt, *(ii)* add prompt parts, *(iii)* select a base image, *(iv)* add details and *(v)* mimic an art style or artist. Regarding prompt parts, the user is presented with an input field where a subject can be introduced but they also have the option to add more fields for extra prompt parts. Each field can be assigned a weight. The details section allows the user to select from different categories (*e.g.* "Art Medium") and sub-categories (*e.g.* "Drawing").

A common method used for producing prompts involves getting inspiration from prompts of already generated images, to identify keywords that may improve the prompts being created. There are multiple online platforms that allow users to explore generated images, for example, *Krea.ai*, *Lexica.art* and *OpenArt.ai*. In addition to these platforms, there are freely accessible datasets of prompts and generated images that can be used to investigate prompt engineering. An example is *Open Prompts* (used to build *krea.ai*), whose authors invite further research with it as an alternative to retrain models for quality improvement.[7] Another example is the work developed by Wang et al. [19], who present an open-source large-scale text-to-image prompt dataset (DIFFUSIONDB), containing 2 million images generated by Stable Diffusion. The authors suggest multiple applications for their dataset, including *prompt autocomplete* (*i.e.* keyword suggestion) and *prompt auto-replace* (*i.e.* exchanging prompt keywords for more effective ones). These functions can be considered as part of what we address in the following section: *prompt generation*.

---

[4] https://openart.ai/promptbook accessed 2023.
[5] https://proximacentaurib.notion.site/2b07d3195d5948c6a7e5836f9d535592 ac. 2023.
[6] https://promptomania.com/stable-diffusion-prompt-builder/ accessed 2023.
[7] https://github.com/krea-ai/open-prompts accessed 2023.

2.3   Prompt Generation

Several approaches have been explored for prompt generation. Although a great part of the work has been done within the context of text generation [8] and does not have the visual domain as a target [9], text generators can be useful for purposes of text-to-image generation. On the one hand, text generators can be used as sources of inspiration, aiding users in the production of prompts. This still holds even if they are not specifically developed for prompt generation (*e.g. drawingprompt.com*). On the other hand, text generation can be used specifically for the task of producing prompts for text-to-image generation, both directly and indirectly. An example of the latter is the work by Ge and Parikh [5]. The authors [5] implement a pipeline to generate prompts by using a Bidirectional Encoder Representations from Transformers (BERT) language model to predict masked words in templates (*e.g.* the moon is like a [MASK]), which are then used to establish an analogical relation between different concepts and produce a prompt. The prompt is used to produce visual conceptual blends [3] with BigSleep and DeepDaze. Regarding direct prompt production, an example is the *Stable Diffusion Prompt Generator*,[8] which is a GPT-2 model (Generative Pre-trained Transformer) that generates prompts from text input, trained with data retrieved from *Lexica.art* Stable Diffusion image repository.

Text generation can also be integrated as part of a bigger system. An example is the work by Liu et al. [10], who propose a system that brings together DALL-E, GPT-3 and CLIP within a computer-aided design software. The system uses 3D keywords sampled from a set of high-frequency words, styles and design parts generated with GPT-3 and keywords given by the user to produce text prompts, which are then used in DALL-E to obtain 3D designs.

Other approaches focus on prompt optimisation, for example by using nature-inspired algorithms. Pavlichenko and Ustalov [13] follow a human-in-the-loop approach and use a Genetic Algorithm (GA) to find the keyword set that produces the most aesthetically appealing images with Stable Diffusion, from a list of 100 keywords. They use 60 different image descriptions to produce prompts with the following template: [keyword,...] [description] [keyword,...]. They report that their keyword sets produce better results than the most popular keywords used by the community. A different approach is used in the system *EvoGen*[9], which combines an evolutionary algorithm, Stable Diffusion and an aesthetics model. An initial prompt population is randomly produced and evolved using the highest-rated prompts based on the aesthetic quality of the images generated with them. To produce prompts, they sample from different lists, *e.g.* artists and genres keywords, an English dictionary, among others.

Although our work aligns with some of the described work, *e.g.* by using nature-inspired computation, our approach differs from the ones described. Our goal is to facilitate the task of finding appropriate prompts for a given user. Instead of using metrics of aesthetics, we explore an interactive approach in which users can evolve prompts according to their preferences.

---

[8] https://huggingface.co/Gustavosta/MagicPrompt-Stable-Diffusion accessed 2023.

[9] https://github.com/MagnusPetersen/EvoGen-Prompt-Evolution accessed 2023.

3   Approach

The generation of images using a text-to-image generative model usually begins with the writing of a prompt that is then passed to the model to create images. Although there are several resources with numerous examples of tested prompts that anyone can modify and use, we believe that the process of finding and tuning prompts can be more dynamic and interactive. First, instead of dealing with prompts individually, which requires the selection and sequencing of individual terms and then the testing of each possibility by passing it to the model to generate images, we suggest a method where the user designs a blueprint for what might be a candidate prompt. As a result, rather than writing a specific prompt, the user writes a *meta prompt* which encodes a space of prompts. This allows variation using a set of terms given directly as input as well as terms automatically obtained through a method of conceptual extension [2] (*i.e.* from an initial term, *e.g.* "animal", obtaining others, *e.g.* "dog"). Second, instead of having the user manually tune the prompt, we propose an interactive approach in which an evolutionary system promotes solutions based on user feedback.

The presented method integrates two core modules. The first module takes as input a special type of prompt, which we call meta prompt, and translates it into a set of individual prompts. The second module provides an interactive way to explore and test the prompts produced by the first module. In the following subsections, we describe these two modules in more detail.

3.1   Creating Meta Prompts to Represent Spaces of Prompts

Our approach is based on a strategy in which the user does not input a specific prompt but a blueprint that can be used to produce multiple prompts. The creation of meta prompts is achieved using a syntax made especially for this task. To inform the design of a functional and flexible syntax, we analysed prompt examples retrieved from different sources and studied prompt taxonomies by other authors. Differently from other taxonomies, *e.g.* [12], we distinguish between *components* (*e.g. subject*) and *functions* (*e.g. repetition*). Moreover, we refer to different options of a given component as *terms* instead of the commonly used expression "modifiers", which we believe is not suitable for all components (*e.g.* different *subject* options are not exactly "modifiers"). The proposed syntax is based on four main functions that are described in the following paragraphs.

Define Meta Prompt Components — Like in any prompt or sentence, the creation of a meta prompt requires the definition of a structure or pattern consisting of a sequence of components, *e.g.* subject, verb and then object. Each component can contain one or more words. In our meta prompt syntax, a dynamic component can be identified by enclosing it inside a less-than sign (<) and a greater-than sign (>). For example, the meta prompt `<person> eating <fruit>` explicitly identifies two dynamic components and a static one (`eating`), which is directly printed to the final prompts. This process of identifying the components of the meta prompt is essential for the next function.

Enumerate Possible Terms for Each Meta Prompt Component — For each dynamic component defined in the meta prompt, we can enumerate options (terms) that can be selected and used to create prompt variations from the initial meta prompt. For example, the meta prompt `<farmer|policeman> eating <banana|kiwi|orange>` encodes two possible terms for the first dynamic component and three terms for the second one. By recombining these options, we can obtain six different prompts. This example illustrates the specification of different terms made directly in the meta prompt using vertical bars (|) to separate them. In addition to this method, we can link the dynamic component to an external list of terms. This method, which is illustrated in the example presented at the end of this subsection as lists B, C and D, not only facilitates the input of larger sets of terms but more importantly enables the dynamic creation and modification of such sets. For example, we can make use of existing sources or tools to retrieve related terms and use them as options for a dynamic component (*i.e.* conceptual extension [2]).

Combine Terms in a Prompt Component — In addition to specifying the set of terms that can be used in a given component, we can indicate that multiple terms can be combined. Specifically, we can set the number or an interval (minimum and maximum) of terms that should be selected and combined (the default number is 1). It is also possible to set the text that is used to join multiple terms (the default join text is a space). For example, the meta prompt `god eating <banana|kiwi|orange:1-2: and >` presents a dynamic component that has four possible terms and specifies the minimum and the maximum number of terms that can be used (1 and 2, respectively) as well as the text that should be used to join the selected terms (`and`). This information is indicated inside the component and is separated by colons (:). There is an extra option related to the possible repetition of the terms. By default, the system will try not to repeat the selected terms. However, we can make the system skip this check by inserting an asterisk (*) after the interval or number of terms.

Repeat Prompt Components — We can specify in the meta prompt the number of times that a given dynamic component should be repeated in the resulting prompts. This function will repeat the group of one or more terms selected for that dynamic component. For example, the meta prompt `astronaut riding a <<yellow|green>4>` horse may result in two possible prompts where the selected colour will be repeated 4 times. This function can be useful to give more weight to a given prompt component. Repetition and weight assignment are functions that we identified when analysing existing prompts, although there is scarce information on how they exactly work and impact the resulting images. Despite this, we considered that they should be included in our syntax.

With the proposed meta prompt syntax, we are not limited to any type of predefined grammar or sentence constructions. On the contrary, it is flexible by allowing the encoding of varied types of prompts, which in turn can spawn a vast set of alternative prompts.

To illustrate the functions explained above, we present a simplified example of a meta prompt below. The first text line is the meta prompt and the second shows the lists of terms that can be used to fill (replace) specific components of the meta prompt.

```
<A1|A2> of <B> with <C:1-2: and >, <D:2-5*:, >
B=[B1,B2]    C=[C1,C2,C3,C4]    D=[D1,D2,D3,D4,D5,D6]
```

From the meta prompt above, we can create several different prompts (over half a million prompts). Some examples of these prompts are presented below.

```
A1 of B1 with C3 and C2, D3, D4, D2, D3, D2
A2 of B2 with C4 and C2, D4, D2, D2, D2
A1 of B1 with C4 and C3, D3, D5
A2 of B2 with C3, D5, D4
A2 of B2 with C3 and C1, D3, D6, D2
A1 of B2 with C2, D1, D3
```

3.2 Exploring Spaces of Prompts in an Interactive Fashion

The range of individual prompts that can be generated from a meta prompt can easily grow and might result in a wide range of imagery. However, this also poses the challenge of finding the prompts that result in images that please the user. To facilitate this search process, we use an Interactive Evolutionary Computation (IEC) method, in particular an Interactive Genetic Algorithm (IGA), to evolve a population of prompts and this way interactively explore the space of prompts created from an input meta prompt.

Each individual in the population represents a prompt. Each prompt is encoded as a list of lists, where each inner list relates to a component specified in the meta prompt and stores integers representing indexes of selected terms for that component. Using these indexes, and applying the functions that may be specified in the meta prompt (see Sect. 3.1), we create each individual prompt.

The recombination of evolving prompts is achieved with a crossover operator which exchanges inner lists corresponding to the same component of the meta prompt. In the presented version of the approach, we use a two-point crossover operator. Regarding the mutation of prompts, we use an operator that is capable of deleting, replacing or/and inserting indexes in the inner lists. Each one of these procedures can occur independently with preset rates and according to the meta prompt configuration (*e.g.* minimum and maximum number of terms; or the possibility of repeating terms).

The initial population is seeded with random prompts. For each prompt in the population, a preset number of images is generated and displayed. In this process, we use fixed random seeds so that the first image of each prompt is created from a given seed, the second image of each prompt is created from another seed, and so on. This way, we can re-create any image produced during the evolutionary process. Furthermore, it facilitates the comparison of images generated with different evolved prompts.

The user takes a key role in the evolutionary process by looking at the images generated with the evolved prompts and selecting the preferred sets (prompts) to create the next generation, *i.e.* the fitness of the evolved prompts is determined by the user selection. The idea is that users can regard the population of prompts as a dynamic repository of images, and their prompts, and interactively generate variations of the preferred ones.

3.3   Implementation

One of our initial goals was to make the approach easy to use by anyone, ideally without the need for complicated technical configurations and installation of necessary dependencies on the computer for the approach to work. This way, we created a Google Colaboratory notebook that enables anyone to run the presented approach, which is implemented in Python, using a web browser. The source code of the project is publicly available.[10]

To get our approach running on a Colab notebook, we had to come up with a way to allow users to visualise the population of prompts being evolved, select the preferred ones, and ask the system to evolve the next generation of prompts, all this in a notebook. The result is a graphical interface implemented as a web page which is dynamically created and embedded in an output cell of the Colab notebook (see Fig. 1). Once each new generation of prompts is generated, the population is displayed to the user to select the preferred prompts and continue to the next generation. For each prompt in the population, we present the prompt string and a preset number of images generated using that prompt.

4   Experimentation

To assess the validity of the developed approach and its generative potential, we tested it under different conditions, divided into two experimental scenarios. For the tests hereon the system is deployed as an IEC system, *i.e.*, users guide evolution by selecting the individuals they like the most. As explained in the previous section, our system is designed to allow the input of external lists of terms to be used as options for the meta prompt components. In our experiments, we used two data sources. First, we produced a dataset of terms by collecting a total of 1,725 "modifiers" from different sources (*e.g.* lists of stable diffusion modifiers). Then, we removed the terms for which we had no class information and manually selected one class for those that had more than one, resulting in a dataset with a total of 1,237 terms, belonging to 25 different classes (*e.g. medium, material, style, etc.*). When we define the prompt component <MEDIUM>, we access the list of terms from the *medium* class (if it is given as input). Second, we implemented a method of conceptual extension that automatically retrieves related terms from the platform *relatedwords.org* using an initial input.

---

[10] The source code of the presented approach can be found at: https://cdv.dei.uc.pt/metaprompter.

4.1   Scenario 1: Study with Users

For a first experimental scenario, we designed and conducted a user study with the goal of assessing the potential of the approach. We used the platform *Drawing Prompt Generator* to generate random prompt-like phrases (*e.g.* "Naughty dog stealing a piece of pizza off the table"). From these phrases, we produced three different meta prompts (A, B and C):

a <MEDIUM> of a <COLOR> <ANIMAL>, exploring a pirate shipwreck
a <MEDIUM> of a <ANIMAL> stealing a piece of <VEGETABLE> off the table
a <ANIMAL> in their <HOUSE> in the style of <STYLE>

For these meta prompts, <MEDIUM>, <COLOR> and <STYLE> use terms from the produced dataset, while <ANIMAL>, <VEGETABLE> and <HOUSE> use terms obtained through conceptual extension.

We asked participants with background in graphic design to use the system to evolve prompts. In each run of the system, a set of three random seeds is produced – the seeds stay fixed and are used to generate the phenotype of each individual (composed of three images generated with the same prompt). As such, the individuals were always based on the same seeds, allowing image comparison. In each generation, the participant would identify the individual that they considered most aesthetically pleasing, select it and produce a new generation. We established a limit of ten generations. In the end, they would save the best individual and conduct the following tasks: (T1) identify a style shared by the three images; (T2) rate the ability of the system to evolve according to their taste from 1 (very bad) to 5 (very good); (T3) rate the aesthetic quality of the selected set of images from 1 (very bad) to 5 (very good); (T4) rate how well the selected set of images represent the corresponding prompt from 1 (very bad) to 5 (very good). The participants were also asked for comments.

For the IGA setup, we used the setting of Table 1, with the exception of population size (set to 6) and tournament size (set to 2).

Results

In total, ten users participated in the study, three with meta prompt A, three with B and four with C. All participants reached the tenth generation, except for two (one did an extra generation and one only reached the eighth). Regarding the tasks, one of the participants did not provide answers.

To assess if the system was able to converge using the interactive feedback of the users, we calculated the number of different words ($\neq$W) in the prompts of the individuals in each generation and the standard deviation (σ) of this value among generations. For all runs, $\sigma <= 1$ considering all generations and $\sigma <= 1.24$ considering only the first and the last (note that $\sigma < 0.86$ in all runs except one). As such, this shows that prompts had a constant word length and, consequently, a reduction of $\neq$W throughout the run would indicate a convergence. We observe a reduction in all runs – in 5 out of the 10 runs $\neq$W reduced to values between 52–54% of the value calculated for the initial population; in the other 5 runs, this percentage was higher but below 67%. Interestingly, the minimum of $\neq$W was achieved in one of the last two generations only in six of the runs; in the other four, the minimum $\neq$W was reached around generation 4–7, which suggests that the system converged but then the user preferences changed. This result is aligned with the comment provided by a participant, who reported having changed their style goal during the run.

Regarding T1, all participants indicated that they identify a shared style; some of the participants even identified a specific style, *e.g.* "crayon-like". For T2–T4, we calculated mode (mo), median ($\tilde{x}$), mean ($\overline{x}$) and standard deviation (σ), obtaining the following results: T2 (evolution) $mo = 2$, $\tilde{x} = 3$, $\overline{x} = 3$, $\sigma = 1$; T3 (aesthetics) $mo = 4$, $\tilde{x} = 4$, $\overline{x} = 3.88$, $\sigma = 0.78$; T4 (representation) $mo = 3$, $\tilde{x} = 3$, $\overline{x} = 2.88$, $\sigma = 1.166$. Regarding T2, the results indicate that the users do not fully perceive the system adaptation to their preferences. We believe this result may be related to the setup of the IGA for this experience, specifically a low tournament size (2), which does not foster selection for recombination and mutation of the individual(s) selected by the user as the best. Regarding the images produced, these were considered of good aesthetic quality (T3) but of only medium representation quality (T4), meaning that the images were not considered good representations of the prompt. This latter value may be related to the low number of inferences steps (10), which was chosen to reduce the time of the image generation but consequently has an impact on the connection between text and image.

4.2   Scenario 2: Variety and Convergence

In a second experimental scenario, the system is used to converge into a visual style. Thus, we perform a fixed number of generations and aim for convergence, obtaining a prompt or a set of prompts that produce images in a style that matches our preferences. We took into consideration the conclusions and user comments from Scenario 1 in the following experimentation with the system. The setup used to conduct this experiment can be viewed in Table 1.

Table 1. IGA setup parameters.

Parameter	Setting	Parameter	Setting
population size	10	insert term	0.1
elite size	1	generations	10
tournament size	5	image size	512×512
crossover rate	0.7	inference steps	20
delete term	0.1	images per individual	3
replace term	0.25		

For this experiment, we were inspired by a widely known *Openai*'s DALL-E prompt: "a photo of an astronaut riding a horse". From this prompt, we defined the following meta prompt and lists:

```
<M_of|SM_of|STYLE:0-1> astronaut riding a <COLOR:0-1> <horse|ANIMAL>
<of_S:0-1>
STYLE = ['3D',...]    MEDIUM = ['cartoon',...]    COLOR = ['cinna-
mon',...]
ANIMAL = get_related_terms('animal')
M_of = ['a {} of an'.format(m) for m in MEDIUM]
of_S = ['in the style of {}'.format(s) for s in STYLE]
SM_of = ['a {} {} of an'.format(s, m) for m in MEDIUM for s in STYLE]
```

The lists with terms used in the dynamic components of the meta prompt are composed as follows: MEDIUM contains 121 types of medium (*e.g.* acrylic painting, cartoon), ANIMAL contains 6 terms related with animals (*e.g.* mammal, vertebrate, fish), COLOR contains 309 different colours described with text (*e.g.* CMYK, cinnamon), and STYLE contains 228 different visual styles (*e.g.* fractal, acrylic artwork, pixel art). In total, the search space is composed of over 5 million possible prompts, among which is the initial *Openai*'s prompt.

Results

Figure 2 shows the initial population generated with the experimental settings shown in Table 1. It is possible to observe the diversity of prompts and images that are generated. As shown in Fig. 2, the dynamic nature of the meta prompt enables the generation of multiple prompts – all prompts are distinct. We can also observe that we get different results even with the same prompt when rendered with different random seeds.

The process is carried out with the interactive evaluation of the generated prompts and corresponding images. In this experiment, we aimed at convergence while picking individuals we considered fit. Note that we can select more than

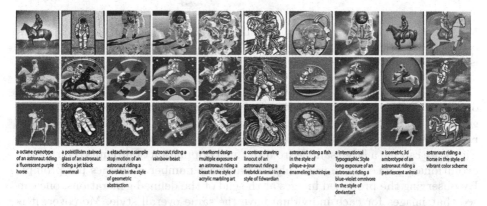

Fig. 2. Prompts created in the initial generation. The images in each column are generated from different seeds using the same prompt, which is shown at the bottom.

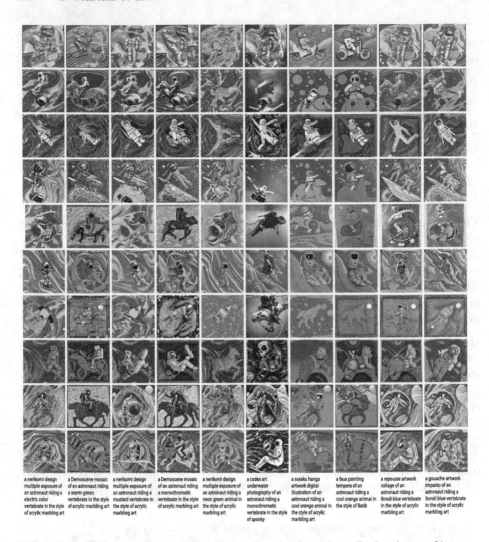

a nerikomi design multiple exposure of an astronaut riding a electric color vertebrate in the style of acrylic marbling art

a Demoscene mosaic of an astronaut riding a warm green vertebrate in the style of acrylic marbling art

a nerikomi design multiple exposure of an astronaut riding a mustard vertebrate in the style of acrylic marbling art

a Demoscene mosaic of an astronaut riding a monochromatic vertebrate in the style of acrylic marbling art

a nerikomi design multiple exposure of an astronaut riding a neon green animal in the style of acrylic marbling art

a codex art underwater photography of an astronaut riding a monochromatic vertebrate in the style of spooky

a sosaku hanga artwork digital illustration of an astronaut riding a cool orange animal in the style of acrylic marbling art

a faux painting tempera of an astronaut riding a cool orange animal in the style of Batik

a repousse artwork collage of an astronaut riding a Bondi blue vertebrate in the style of acrylic marbling art

a gouache artwork impasto of an astronaut riding a Bondi blue vertebrate in the style of acrylic marbling art

Fig. 3. Images generated from a population of evolved prompts. Each column of images is obtained from the same prompt, which is shown at the bottom. Each row of images is generated from the same random seed.

one individual per generation. After the evolutionary process, we can pick up and generate more images based on the evolved prompts via the modification of the random seed of the generator prior to the stable diffusion rendering process. In Fig. 3, we can see a large sample from the final population of prompts. During the evolutionary process, the user only sees a preset number of images per prompt. By observing the produced images at the end of the defined generations, one can see that images for each individual have the same overall style. Moreover, it is possible to see that the individuals of the population share visual characteristics, suggesting convergence. To further assess convergence, we analysed the prompts using the metric based on the number of different words ($\neq$W) used in Scenario 1. One difference that we observe is that a direct comparison of $\neq$W value does not suffice (59 in the first generation and 50 in the last), as there is a higher variation of prompt length – standard deviation of 2.76 considering all runs and 3.38 considering first and last. As such, we also calculated the percentage of words that are unique in the population prompts, obtaining 49.5% in the first generation and 31.4% in the last (more than 2/3 were repeated words), showing that we ended up with a converged population where the prompts have similar terms. Moreover, we note that there are still terms of the prompt present in the final population that were in the initial population. The results further the idea of guidance and convergence as well as support the validation and utility of the process of evolving prompts with multiple terms.

In summary, we were able to guide evolution towards a point of convergence in terms of style despite having different textual outputs. Visuals are distinct in some cases, but we can see the influence of the <COLOR> and <STYLE>, which conveys the idea of convergence in this experiment. From the set of images from Fig. 3 we can notice the impact of the random seeds, leaving a trail of common artefacts and objects across all prompts from the last generation (namely, in the 1st, 2nd, 3rd, 6th, 8th and 9th rows). If we ignore the colour, we can see similar styles and objects between individuals, even though the prompts differ. Overall, this showcases the ability of the approach to generate several visual outputs with prompts that can be diverse but convey the same visual traits.

5   Conclusions and Future Work

Recent development and increase in the output quality of text-to-image computational approaches have led to a growing interest in the field. Multiple communities have emerged and are dedicated to the development of open-source resources for text-to-image generative models, *e.g. Stable Diffusion.* One key aspect is the relation between the input prompt and the quality of the output. For this reason, there is a general venture to identify better prompts as well as better ways of conducting this search.

We presented an approach in which users can define a prompt blueprint (meta prompt), which is used to produce prompts, and interact with the system in order to produce solutions that match their preferences. We have tested our approach with two experimental scenarios: *(i)* a user study using three meta prompts, and *(ii)* a study using a meta prompt inspired by a widely known prompt example. The results show that our approach allows users to converge to specific styles, obtaining prompts that can be further used to produce images in the same style. It is important to mention that although there are differences between generative models and prompt syntax, obtaining results of different quality with the same prompt [19], our meta prompt approach is flexible and can be easily adapted to work with different models.

Our experimentation also allowed us to identify future research directions. First, we have mostly used terms that have been previously experimented with, being retrieved from other authors' works. However, there is potential to be explored in the identification of new terms that may lead to "hotspots" in the latent space. For example, Daras and Dimakis [4] investigate the existence of a "hidden vocabulary" in DALL-E-2 – apparently nonsensical prompts result in a given type of visual output (*e.g.* "apoploe vesrreaitais" is reported to produce birds). Second, there is still work to be done in identifying the best terms to use based on the prompt subject, *e.g.* 3DALL-E [10] uses terms specifically related to 3D modelling. Third, the system could be coupled with a prompt validator based on NLP approaches, aimed to analyse parts of speech and improve the quality of the prompts before generating the images. Additionally, the approach could also be further developed to also allow the evolution of meta prompts with the goal of finding the optimal configuration for each user.

The current approach has similarities with Grammatical Evolution (GE) approaches despite being a conventional IGA system. Therefore, the evolutionary engine can be enhanced by GE mechanisms, such as variation operators or initialisation methods and genotype-to-phenotype mapping approaches.

Another avenue of research is the automation of evaluation to explore a different dimension of the presented approach. Methods to automatically evaluate the generated images and the prompts could be used to build enhanced and automatic fitness function schemes, *e.g.*, use aesthetic evaluation models to evolve visually appealing images. Mechanisms to measure and improve the distinctness of generated outputs are also a hypothesis.

Our main goal is to facilitate the process of finding the best prompts, which is aligned with attempts of strengthening the relationship between the user and AI, fostering a collaborative interaction. In this sense, future developments can be made to the interface to improve this interaction and increase usability.

Acknowledgements. This research was partially funded by the FCT - Foundation for Science and Technology, I.P./MCTES through national funds (PIDDAC), within the scope of CISUC R&D Unit - UIDB/00326/2020 or project code UIDP/00326/2020.

References

1. Crowson, K., et al.: VQGAN-CLIP: open domain image generation and editing with natural language guidance. In: Avidan, S., Brostow, G., Farinella, G.M., Hassner, T. (eds.) Computer Vision – ECCV 2022. ECCV 2022. LNCS, vol. 13697, pp. 88–105. Springer, Cham (2022). https://doi.org/10.1007/978-3-031-19836-6_6
2. Cunha, J.M.: Generation of concept representative symbols: towards visual conceptual blending, Ph. D. thesis, University of Coimbra (2022)
3. Cunha, J.M., Martins, P., Machado, P.: Let's figure this out: a roadmap for visual conceptual blending. In: Proceedings of the Eleventh International Conference on Computational Creativity (2020)
4. Daras, G., Dimakis, A.G.: Discovering the hidden vocabulary of DALLE-2. CoRR abs/2206.00169 (2022)
5. Ge, S., Parikh, D.: Visual conceptual blending with large-scale language and vision models. In: Proceedings of the 12th International Conference on Computational Creativity (2021)

6. Heikkila, M.: This artist is dominating AI-generated art. And he's not happy about it. https://www.technologyreview.com/2022/09/16/1059598/ (2022). Accessed Jan 2023
7. Lin, T.-Y., et al.: Microsoft COCO: common objects in context. In: Fleet, D., Pajdla, T., Schiele, B., Tuytelaars, T. (eds.) ECCV 2014. LNCS, vol. 8693, pp. 740–755. Springer, Cham (2014). https://doi.org/10.1007/978-3-319-10602-1_48
8. Liu, P., Yuan, W., Fu, J., Jiang, Z., Hayashi, H., Neubig, G.: Pre-train, prompt, and predict: a systematic survey of prompting methods in natural language processing. CoRR abs/2107.13586 (2021). https://arxiv.org/abs/2107.13586
9. Liu, V., Chilton, L.B.: Design guidelines for prompt engineering text-to-image generative models. In: Barbosa, S.D.J., Lampe, C., Appert, C., Shamma, D.A., Drucker, S.M., Williamson, J.R., Yatani, K. (eds.) CHI '22: CHI Conference on Human Factors in Computing Systems, New Orleans, LA, USA, 29 April 2022–5 May 2022, pp. 1–23. ACM (2022)
10. Liu, V., Vermeulen, J., Fitzmaurice, G., Matejka, J.: 3DALL-E: Integrating text-to-image AI in 3D design workflows. arXiv preprint arXiv:2210.11603 (2022)
11. Oppenlaender, J.: The creativity of text-to-image generation. In: 25th International Academic Mindtrek conference, Academic Mindtrek 2022, Tampere, Finland, 16–18 November 2022, pp. 192–202. ACM (2022)
12. Oppenlaender, J.: A taxonomy of prompt modifiers for text-to-image generation. arXiv preprint arXiv:2204.13988 (2022)
13. Pavlichenko, N., Ustalov, D.: Best prompts for text-to-image models and how to find them. CoRR abs/2209.11711 (2022)
14. Radford, A., et al.: Learning transferable visual models from natural language supervision. In: Meila, M., Zhang, T. (eds.) Proceedings of the 38th International Conference on Machine Learning, ICML 2021, 18–24 July 2021, Virtual Event. Proceedings of Machine Learning Research, vol. 139, pp. 8748–8763. PMLR (2021)
15. Ramesh, A., et al.: Zero-shot text-to-image generation. In: Meila, M., Zhang, T. (eds.) Proceedings of the 38th International Conference on Machine Learning, ICML 2021, 18–24 July 2021, Virtual Event. Proceedings of Machine Learning Research, vol. 139, pp. 8821–8831. PMLR (2021)
16. Reynolds, L., McDonell, K.: Prompt programming for large language models: beyond the few-shot paradigm. In: Kitamura, Y., Quigley, A., Isbister, K., Igarashi, T. (eds.) ACM CHI Conference on Human Factors in Computing Systems. ACM (2021)
17. Rombach, R., Blattmann, A., Lorenz, D., Esser, P., Ommer, B.: High-resolution image synthesis with latent diffusion models. In: IEEE/CVF Conference on Computer Vision and Pattern Recognition, CVPR 2022, New Orleans, LA, USA, 18–24 June 2022, pp. 10674–10685. IEEE (2022)
18. Smith, E.: A traveler's guide to the latent space. https://sweet-hall-e72.notion.site/A-Traveler-s-Guide-to-the-Latent-Space-85efba7e5e6a40e5bd3cae980f30235f (2022). Accessed Jan 2023
19. Wang, Z.J., Montoya, E., Munechika, D., Yang, H., Hoover, B., Chau, D.H.: DiffusionDB: a large-scale prompt gallery dataset for text-to-image generative models. CoRR abs/2210.14896 (2022)

Is Writing Prompts Really Making Art?

Jon McCormack(✉)⬤, Camilo Cruz Gambardella⬤, Nina Rajcic⬤,
Stephen James Krol⬤, Maria Teresa Llano⬤, and Meng Yang⬤

SensiLab, Monash University, Caulfield East, Victoria 3145, Australia
{Jon.McCormack,Camilo.Cruzgambardella,Nina.Rajcic,Stephen.Krol,
Teresa.Llano,Meng.Yang}@monash.edu

Abstract. In recent years Generative Machine Learning systems have advanced significantly. A current wave of generative systems use text prompts to create complex imagery, video, even 3D datasets. The creators of these systems claim a revolution in bringing creativity and art to anyone who can type a prompt. In this position paper, we question the basis for these claims, dividing our analysis into three areas: the limitations of linguistic descriptions, implications of the dataset, and lastly, matters of materiality and embodiment. We conclude with an analysis of the creative possibilities enabled by prompt-based systems, asking if they can be considered a new artistic medium.

Keywords: Artificial Intelligence · Diffusion Models · Art · Neural Networks

1 Introduction

We live in an age defined by technological innovation, while our world floods and burns with increasing veracity and severity. Over the last decade a seemingly endless wave of innovations in generative machine learning (ML) models have allowed the generation of photo-realistic images of non-existent people [14], coherent paragraphs of text [29], conversion of text directly to runable computer code and, most recently from text descriptions to images [24], video [28], and even 3D models.

Systems such as DALL-E 2, MidJourney and Stable Diffusion allow the generation of detailed and complex imagery from short text descriptions. These Text-to-Image (TTI) systems allow anyone to write a brief English description and have the system respond with a series of images that depict the scene described in the text, typically within 5–30 seconds. An example is shown in Fig. 1. Despite being quite recent and still an on-going development, these systems have become highly popular and an abundance of machine synthesised examples can be regularly found on social media, NFT sites, and other online platforms[1].

---

[1] MidJourney's interface is through the popular on-line messaging application *Discord*, allowing anyone with permission to the appropriate channel access to both the prompts and generated images.

© The Author(s), under exclusive license to Springer Nature Switzerland AG 2023
C. Johnson et al. (Eds.): EvoMUSART 2023, LNCS 13988, pp. 196–211, 2023.
https://doi.org/10.1007/978-3-031-29956-8_13

The obvious source of these systems' popularity is that they offer something entirely new: being able to generate an image just by describing it, without having to go to the trouble of learning a skill – such as illustration, painting or photography – to actually make it. And importantly, the quality and complexity of the images generated is often comparable to what an experienced human illustrator or designer could produce. Moreover, TTI systems demonstrate a semantic interpretation of the input text and can convert those semantics so that (in some cases) they are more-or-less coherently represented in the generated images. This new-founded capability has inspired many useful image generation and manipulations possibilities, such as "outpainting" where a pre-existing image can have its edges extended with coherent and plausible content, or as an "ideation generator" where new versions of a set of input images are generated.

Fig. 1. Image generated by stable diffusion from the text prompt: *"still life with human skulls of different sizes, a rose, the most beautiful image ever seen, trending on art station, hyperrealistic, 8k, studio lighting, shallow focus, unreal engine"*.

The makers of such tools pitch them as artistic, in a press release from LMU Munich (the University responsible for training Stable Diffusion), Björn Ommer, group leader of the Machine Vision & Learning Group at LMU claimed that, "the model removes a barrier to ordinary people expressing their creativity"[2], and the first release of *Stable Diffusion* claimed that it "will empower billions of people to create stunning art within seconds"[3].

As with many previous new technologies in the Arts, some artists have expressed resistance to embracing them as legitimate artistic tools. This is unsurprising, initial efforts following the invention of the photographic camera resulted in much technical innovation and significant creative implications, but there was an initial reluctance to accepting photography as a legitimate art form. It took many decades for photography to gain acceptance within the Art Academy, and finally gain wide inclusion as a valid artistic medium. It is also important to note that using a camera does not necessarily make you an artist. Millions of people happily snap photographs with their smartphones every day, but the images captured are not typically presented as works of art. However, one important difference between TTI systems and other technologies adopted for artistic purposes is that TTI systems are specially pitched as being artistic tools, i.e. their primary intended application is to create "art".

In this paper we look at some of the implications and concerns expressed in this idea. We analyse the creative and artistic value of these new tools beyond the bold claims of press releases of those with a large commercial stake in the success of such models, looking at the implications of Text-to-Image creativity and the underlying ethical considerations of such systems. We unpack the drawbacks and the opportunities of TTI systems employed as artistic medium, offering an analysis of their potential for cultural contribution.

We divide our analysis of TTI systems into three key topics: the limitations of linguistic descriptions, the implications of the dataset, and lastly, matters of materiality and embodiment in TTI systems. Following this analysis we conclude with a discussion around the artistic and creative opportunities enabled by treating TTI systems as a new artistic medium.

2   The Limitations of Linguistic Description

If you could say it in words, there would be no reason to paint

— Attributed to Edward Hopper

An obvious limitation of any system that attempts to interpret text into another medium (such as an image) is that one must be able to express the desired output linguistically (i.e. *in words*). We only need to look at art history to see that this is often very difficult or impossible, particularly – but not exclusively – for non-figurative images. Visual art has often claimed to be able to express the inexpressible [10] – to show through visual media what cannot be directly

---

[2] https://www.lmu.de/en/newsroom/news-overview/news/revolutionizing-image-generation-by-ai-turning-text-into-images.html.

[3] https://stability.ai/blog/stable-diffusion-announcement.

expressed in words. One only has to think of the works of Mark Rothko, Willem de Kooning, Clyfford Still or Bridget Riley as examples of this proposition. We note the important difference between an image expressing something that is beyond words and using words to describe a desired image (that may not yet exist).

In discussing what can be expressed linguistically, we need to differentiate that which cannot be expressed *a priori* and which might be expressed *a posteriori*. For example, phenomenological experiences cannot be fully expressed linguistically. Individual mental lexicons vary across human populations[4]. Furthermore, one may experiment with language in deriving appropriate text in a text to image system, learning from prior examples or growing their lexicon in response to working with such systems.

It could be argued that one may be able to create an image "a-la-Rothko" by providing an excruciatingly detailed set of physical instructions, an "iconic" form of representation that could be compared to the lists of commands used in *turtle graphics*, where the meaning of the instructions has no relationship to the perceived meaning of what they produce. The appeal of TTI systems lies on their capability of translating symbolic representations – textual descriptions – into iconic ones – images (see [12]). However, the way this translation is implemented can be perceived as a limitation in two key ways:

The "understanding" that TTI systems have of images is literal, as they are built upon millions of formal descriptions of visual material, where words or descriptive phrases (noun+adjective or verb+adverb pairings) capture the visual identity and characteristics of elements present in an image, their spatial relationships (e.g. "a red flower in a vase", "a human skull on a table"), as well as some more general stylistic features (e.g. "still life", "unreal engine"). Consequently the images produced by these systems can only reflect the content of a prompt literally. Any further meaning, intention or encoded information assigned to a specific textual construct, be it metaphorical or culturally charged, will be lost in translation. One example of this is how notoriously bad TTI systems are at producing diagrammatic or abstract images, where lines, annotations, dimensions, shading colour and texture are loaded with information beyond their mere graphic expression (Fig. 2).

Secondly, and especially under the paradigm of creative practice as a reflective process [26], in which designers and artists "discover" novel alternatives by following cues that emerge from in the work through their own creative actions. These come in the form of non-textual constructs – diagrams, sketches, motifs – that illuminate pathways to what things could become, by displaying new physical and/or perceptual order [1]. Explicitly verbalising something that has yet to be formalised – i.e. ex-ante description – is akin to describing a colour that does not exist [16].

Such limitations may explain why almost all the images produced by TTI systems are figurative and literal (e.g. Fig. 1). One might say that these systems

---

[4] For example a 20-year old native English speaker's vocabulary may range between approximately 27,000 to 51,000 lemmas.

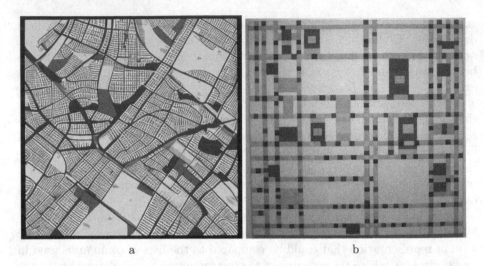

a   b

Fig. 2. Two maps: a) Broadway generated by Stable Diffusion from the text prompt: *"An abstract map of Broadway in yellow, blue and red"..* b) Broadway Boogie Woogie by Piet Mondrian (image credit: Wally Gobetz) (Color figure online)

are "object-oriented" in that they are trained predominantly on literal descriptions of images or using machine-trained image description technologies, such as CLIP, which identify objects in an image. Such systems have no understanding of metaphor, analogy or visual poetry, hence they cannot understand what is "in" an image in the way that a human artist or designer implicitly would[5].

2.1   Levels of Control

Another issue with such high-level text descriptions is the level of control and malleability one has over image making via language only. Painting, drawing, photography – all human image-making techniques – employ the body centrally in their production. This raises issues of embodiment, materiality, and material agency that traditionally come into play when making a visual artwork, something elaborated on in Sect. 4. In many aspects of art-making, modes of cognition such as verbalisation are not the primary means of art production, for example the paintings of Jackson Pollock reflect the interplay between body movement, paint and physics, which are all critically important.

Even for those aspects of image production that TTI systems are able to control, their conceptual, relational and semantic understanding is not like that of a human [27]. For example, prompting a TTI system for "an astronaut riding a horse" gives a literal representation of that description, but prompting for "a horse riding an astronaut" gives much the same imagery as seen in Fig. 3.

---

[5] These limitations have been extensively documented in previous deep learning systems, a classic example being the painting by Margritte, *La Trahison des Images*.

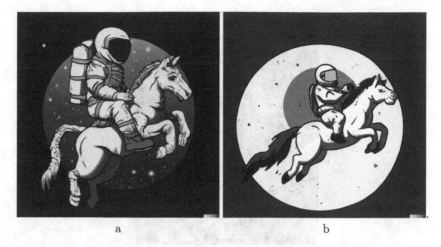

Fig. 3. Two images generated using DALL-E 2: a) Text prompt: *"an astronaut riding a horse"* b) Text prompt: *"a horse riding an astronaut"*

Issues of control are even more vexed when we consider the underlying training data and how systems establish statistical associations between words and images. Similar to Internet search engines, linguistic terms return the most common (or most populous in the dataset) interpretations of language-image associations. So if one asks a TTI system for a picture of a "beautiful man" the system returns an Internet stereotypical image of a man: white, heterosexual, youthful, professional, athletic. When used as part of prompts, basic artistic concepts such as "beauty" follow the statistical patterns expressed in the dataset, precluding cultural differences, homogenising representations and reinforcing biases.

2.2 Authorship

A fundamental question in the debate about AI art has been the level of autonomy that computer systems have in making decisions that are essential for the creative process. In June 2022, Cosmopolitan magazine published an issue with what they claimed was "The First Artificially Intelligent Magazine Cover" (using DALL-E 2) which they further claimed "only took 20 s to make" (see Fig. 4). However, the article that describes the experiment in the magazine[6] reveals that a digital artist and other members of the magazine's team had heavily intervened in the design and production of the cover, which actually took around 24 h of work.

This current hype around AI, particularly in the last couple of years, has shown an inclination to overstate the role of AI systems and understate the role of the human artist, exploiting the framing of "autonomous AI artists" (who are neither autonomous or artists) for marketing and publicity reasons [6,8,18], and creating ethical issues around authorship.

---

[6] https://www.cosmopolitan.com/lifestyle/a40314356/dall-e-2-artificial-intelligence-cover/.

Fig. 4. Cosmopolitan cover created by digital artist Karen X. Cheng using DALL-E 2 and the text prompt *"wide-angle shot from below of a female astronaut with an athletic feminine body walking with swagger toward camera on Mars in an infinite universe, synthwave digital art"*

3   Data Implications

Parasite *(n)* an organism that lives in or on an organism of another species (its host) and benefits by deriving nutrients at the other's expense

3.1   AI as Parasite

We argue that deep learning models trained on a large dataset of human created art – such as TTI systems – are *parasitic* to human art and creativity, in that they derive and sustain themselves from human art, ultimately at the expense of current human art. In Biology, for a relationship between species to be parasitic, the parasite must benefit at the expense of the host. This is different from other relationships, such as symbiosis or mutualism, where both may benefit. The direct dependency of TTI models trained on human art datasets needs no further clarification, so to justify our claim of parasitism we need to demonstrate a negative effect on human art. Hence, if human art is diminished by TTI and similar systems then the relationship can be justified as parasitic.

  As discussed in Sect. 2, TTI systems reinforce the statistical norms and biases of their training sets through popular interpretations of language and images associated with them. Through this reinforcement (a kind of cultural colonisation), biases become normalised, less popular interpretations are more difficult to specify linguistically, so become marginalised, and eventually may be forgotten or lost. Additionally, the training data is *backward-focused* in that it learns only from prior, existing imagery. Even more limiting is that this imagery must be accessible on the Internet. Of course, human art is not created *ex nihilo*. Artists are influenced and inspired by those who came before them [7]. For example, Van Gogh used direct imitation in some of his work [13] and research has shown that viewing and imitating unfamiliar work is a powerful tool for art students to learn about art [20]. However, unlike these TTI models, human artists draw from more than just previous art to inspire their own creations. A simple example is the depiction of nature in art. Van Gogh's work "A Wheatfield, with Cypresses" and Monet's "The Water-Lily Pond" not only draw from previous work, but are also inspired by the artist's experience in those settings. This experience is more than just an image of a wheat field or a pond of lilies. It is a combination of emotion, feeling, mindset and other attributes that are difficult to convey linguistically.

Additionally, while TTI systems can generate "new" images (in the sense that the particular image has not existed before), they are all statistical amalgamations of pre-existing images, precluding the possibility of stylistic innovation, let alone any conceptual innovation (discussed further in Sect. 4). In the terminology of Boden [3], they support only combinatorial creativity (the combination of existing elements), not transformational creativity (where elements are transformed into something new). In simple terms, TTI systems could not "invent" Cubism if they were only trained on data prior to 1908, yet this was possible for human artists.

If TTI systems become an increasingly significant part of cultural production[7] it is possible that human art and creativity will be diminished for the reasons given above. Moreover, as such systems replace human illustrators the pool of human talent able to competently illustrate may diminish as paying a human illustrator is less viable commercially than using an AI.

3.2 Parasitic Meaning

This idea that generative computer algorithms that mimic human art are "parasitic" isn't new. The philosopher Anthony O'Hear formed this view almost 30 years ago [19]. O'Hear considered computer programs that could generate art similar to human art, first emphasising the need to understand that appreciation of visual art was more than just the consumption of visual images. An important criteria of something being considered "art" meant that there was a direct connection between art as a product and its "history or mode of production" [19]. Hence, art was *more* than just pleasant visual appearances or surface aesthetics, it relied on human *intention*: the intention to engender a particular experience, to communicate an idea, or elicit a particular emotional response. O'Hear conceded that not *every* work of art necessarily achieved this; there may be ambiguity or some other failure where the intention wasn't realised for the work's audience. Nevertheless, in O'Hear's view, this intention and authenticity were necessary conditions for something to be considered as "art", as opposed to the more general class of visual images that saturate contemporary life.

---

[7] With major image creation platforms such as Shutterstock teaming with Stability AI it appears that AI synthesised images will increasingly become part of the cultural vernacular.

O'Hear specifically considers the possibility of computer-generated works of art that "drew on repertoires derived from existing works, but which were not copies of any complete works". He argues that any meaning(s) attributed to them come from those that originated through human thought. Hence they "would be parasitically meaningful, deriving their meanings from the techniques and conventions which human artists had developed in their works." [19].

We are not proposing that TTI systems are unable to communicate human intention, rather that much (or all) of that intention is derived from preexisting human art, hence any meaning inferred is parasitic, in O'Hear's sense of the term. The author of the prompt that generated an image may also have some intention (the intention to make a "beautiful" image, to combine or juxtapose unlikely elements, to create something that mimics an existing human style, even to convey a particular feeling or mood), but this intention can only ever be peripherally enacted in the image itself, since it is the TTI system, not the prompt author, that is actually making the image.

3.3 Data Laundering

One of the key factors that contributes to the capability of TTI models is their access to massive datasets used for training and validation. Achieving the visual quality and diversity that they are capable of reproducing requires a very large corpus of human-created imagery, which is typically scraped from the Internet. In a practice that has been dubbed "data laundering", scraped datasets – which include large amounts of copyrighted media – rely on special exemptions for "academic use" to avoid any legal barriers preventing their use, or for copyright owners to claim against [2]. For example, Stability AI (the "creators" of Stable Diffusion) funded the Machine Vision & Learning research group at the Ludwig Maximilian University of Munich to undertake the model training and a small nonprofit organisation, LAION, to create the training dataset[8] of approximately 5.85 billion images, many of which are copyrighted, and in general appropriated for this purpose without the image creator's direct permission.

Concern has been raised by artists about the ethical and moral implications of their work being used in such systems. These concerns include the appropriation of an individual artist's "style", mimicry, and even the replacement of a human artist or illustrator. Furthermore, there is no easy way to be excluded or removed from such datasets. Such is the concern raised by TTI systems that websites such as https://haveibeentrained.com/ have emerged, allowing anyone to search to see if their work has been used as part of the training data. A companion project allows people to opt-in or opt-out of their data being included. At the time of writing (November 2022) around 60% of respondents chose to opt-out of being included in any training data.

The use of copyrighted images in datasets creates another issue, raising the question that whether training models on copyrighted data should be considered plagiarism. Being able to easily generate an image in an artist's style without

---

[8] https://laion.ai/projects/.

paying for that artist to create it (or paying any royalties or licensing fees), allows users of such technology to bypass the traditional economic, legal and moral frameworks that have supported artists traditionally. Generating copyright-free images immediately for commercial use without the cost or time involved in securing copyright from a human artist may become an attractive proposition, raising the interesting legal question of who would be the defendant in any copyright infringement case brought about by this scenario.

4 Materiality and Embodiment

AI can't handle concepts: collapsing moments in time, memory, thoughts, emotions – all of that is a real human skill, that makes a piece of art rather than something that visually looks pretty.

— Anna Ridler

The vast majority of visual art currently made by humans is made *physically*, that is it involves embodied physicality, situated awareness, and the interplay of material agencies. As expanded upon in Sect. 2, TTI systems rely solely on language for the production of digital images, excluding any possibility of interaction that precedes representation. This limited interface excludes the phenomenological and embodied modalities that typically comprise creative practice, presenting a number of implications for TTI systems as artistic medium.

Central to human creativity is the cultivation of an artistic practice. The honing and refinement of an artistic craft contributes not only to the quality of the work, but also embeds tacit knowledge within the wider community, shaping the direction of future work created in a given medium. In the case of TTI systems, the artistic "practice" involves the iterative prompting and re-prompting of the TTI model, often touted by AI artists as veritable skill which has been generously dubbed "prompt-engineering"[9]. Some prompt engineers are unwilling to reveal their specific prompts for fear that others may easily generate the same results, thereby diminishing the "originalilty" of the imagery they have been able to find. Indeed, the refinement and elaboration of the text prompt does lead to "better" (a subjective measure) images, or at least grants one more control over the resulting image.

Prompt-engineering falls short as an artistic practice for a number of reasons. Firstly, although the skill is developed through countless hours spent engaging with the medium, it is not learned in any tacit, embodied, or individualised way. Prompts can be directly shared, copied, stolen, and in an instant the entire body of knowledge and invested time is transferred. Prompt-engineering, if shared, does indeed contribute knowledge towards the broader community, yet holds little personal value as the art object.

This is not to say that writing prompts could *never* be considered an artistic practice. Any medium has the potential to be used artistically, and as software

---

[9] This "skill" maybe short-lived: models already exist to translate between descriptive text and "prompt text" [17].

art and creative coding demonstrate, the authoring of a process may embody significant artistic qualities. However for prompt writing, there appears to be a (low) ceiling in skill development, which is unusual for an artistic medium, and is likely due to the highly limited possibilities for interaction provided by current systems.

Finally, and perhaps most importantly, prompt writing is entirely confined to a screen. This does not preclude its potential as an artistic medium, but rather brings into question whether this is the kind of artistic practice we as a community wish to foster, particularly if that comes at the expense of others. TTI models themselves are not only divorced from the physical world but are in turn emancipating artists from it too. This is further analysed in the following sections.

4.1 Embodiment in AI

There is a vast amount of research on understanding and defining artistic practice within artificial environments. A significant component concerns discerning the meaning of "embodiment" for AI systems, something that we reflect upon in this section.

Although the prevalent view of embodiment has established that it cannot exist in an "abstract algorithm" [22] and has emphasised the importance of physical grounding [4], others have put forward the idea that a physical body is not a fundamental requirement, and argue that embodiment is given by an agent *being situated* in an environment. Hence a *virtual* environment that provides a (high-fidelity) simulation of a physical world can enable embodiment, provided the agents are *structurally coupled* with that environment (i.e. the agents and their actions are bounded to the structure of the environment) [9,11]. Systems like DALL-E, Midjourney and Stable Diffusion, and in general TTI systems, do not exist in a virtual environment that enables any kind of agent-environment interaction. They learn statistical patterns from the data and generate new images altering these patterns, but they do not independently grow or evolve over time, they do not employ any mechanisms of change.

Cognitive scientists have also expanded the notion of embodiment to not only refer to the physical body but also to other aspects of an agent's interaction with the environment. In a classification of different notions of embodiment, Ziemke [30] highlights the notion of *historical embodiment*, which refers to the "result or *reflection* of a history of agent-environment interactions". The main premise behind this, is that a system's state is shaped by both its present interactions with the environment and its experiences (or history) from the past. This definition has been highlighted by some as one that "does not exclude domains other than the physical domain" [25]. By this definition, we might say that TTI systems could be considered embodied as they are trained on digitized versions of historic pieces of artwork; however, a lot (if not most) of the information about the artist-environment interactions and the physical media itself, which were intrinsic in the creation of these artworks, are not part of the training data. TTI systems focus on the final artefact (the pixelated, visual representation), with little or no information about the cultural and social environment of the time, the materials, creative process, etc.

When creating artworks, the boundary between the environment and artist is transformed, allowing both to experience each other, to move within each other's features and constraints, and adjust to them. TTI systems reinforce that boundary, significantly limiting the interaction between the artist and the system, as well as any experience of the system with the environment. It can be said that TTI systems' "experience" of the world is limited to their training process, decoupling them from the environment and setting a hard boundary with their users during the generation process. As argued in [6], this makes it "difficult to estimate if the value of a particular AI artwork should depend on the technological complexity and innovation involved in its production, or only on the final visual manifestation and contextual novelty".

4.2 Exposing Process

The black-box nature of TTI generators (and Deep Learning models more generally) leads to a privileging of output over *process*. The algorithm does not reveal it's process in producing output, nor is the process contained within the resulting image. Each new generated image is entirely independent of the images that came before it. In this way, we see a privileging of the visual image over its means of production. When employed in the conventional fashion, TTI systems generate visual imagery that is entirely divorced from process, artistic craft, or conceptual foundation; they fetishise surface aesthetics while often lacking a narrative component.

All of this is not to say that TTI and generative AI systems lack a materiality entirely. Of course, these systems are ultimately comprised of physical materials. The material entanglements of TTI systems stretch back to include echoes of the dataset (as elucidated in Sect. 3), and forward in their generative capacities and influence on visual and computing culture.

4.3 Material Agency

In his seminal paper, Malafouris puts forth an argument for distributed and emergent agency with the example of a potter moulding clay [15]. He argues that the potter not only exerts an agency onto the clay, but that through the clay's resistance to being moulded it also exerts agency onto the potter. Hence, agency is not a pre-existing quality but emerges from within an interaction. This notion of material agency has commonly been interpreted in relation to artistic media and practice. The qualities and constraints of a given medium are central to the artistic process; to create is to operate at the boundary – as Malafouris calls it, the Brain Artefact Interface (BAI). The information at the heart of digital systems may be incorrectly viewed as immaterial, yet "just as molecular materials act as resistance and come to transform action upon material objects, so digital materiality comes to enable and transform creative practices upon computers" [23].

In considering TTI models as a creative medium, we can see that it is precisely the resistance of the medium to being *moulded* that leads to it's true creative potential. In a large majority of work created with TTI models we observe an attempt to conceal the qualities of the medium – to render it invisible – often by imitating more traditional materials like paint, photography, or with aesthetic conventions such as "cinematic lighting". According to a media-specific analysis popular to modernist thought [5], it is through the abandoning of imitation and illusion that the full artistic potential of a given medium can be achieved. For example, the abstract art movement saw surrendering to the 2D plane, an escape from emulating the conditions of 3D media. In the same way we argue that TTI systems can become creatively interesting once we embrace their unique properties, rather than work against them or try and mimic popular aesthetic conventions.

Fig. 5. Image generated by DALLE-2. Prompt: *a man covered in tattoos of English words, long hair, rings on fingers, cinematic lighting, 8k,*

4.4   A New Medium?

Whatever you find weird, ugly, uncomfortable and nasty about a new medium will surely become its signature...It's the sound of failure: so much modern art is the sound of things going out of control, of a medium pushing to its limits and breaking apart.

– Brian Eno

While we argue that simply providing a prompt to a TTI system and taking its output does not constitute a rich artistic practice, we appreciate the possibility and potential of deep learning systems as an artistic medium. Duchamp's "Fountain" demonstrates how easily our understanding of what art is can be disrupted.

Rodolfo Ocampo, an AI artist, recently had some of their work showcased at a gallery in Sydney, Australia [21]. This work involved a portrait, generated by DALLE-2, of a person that does not exist. The motivation of the work was to see if people could build a connection with a non-existent subject. Generating people that do not exist is not a new concept [14] (indeed it has been practised in painting and photography for centuries) and there may still be debate on whether this work constitutes art. However, the use of DALLE-2 in this setting is more nuanced than simple image generation as it is employed within a broader artistic concept – building connections with non-existent AI generated humans.

It is common for AI artists to try and hide or fix errors generated by TTI systems. But exploiting these limitations might lead to more interesting artistic possibilities. An example of these errors can be seen in Fig. 5. The image portrays a subject covered in tattoos and the prompt specifies that the tattoos are English words. However, while the tattoos do have word-like qualities to them, none of them are in English (or in any human language). This demonstrates a common deficiency with TTI systems – they are hopeless at generating meaningful text. Other "errors" in Fig. 5 include the misshapen hands (a common issue for TTI systems) and the subject facing away from us even though the hands suggest he is facing towards us. These "errors" highlight interesting properties about the medium, revealing its limitations and raising questions about the underlying technology.

5 Conclusion

Technology has repeatedly transformed the arts and culture. Throughout modernity, many lamented the automation of artistic processes and practices due to technology ("Why bother making a painting when a camera will give you an image instantly"). As we sacrifice direct control and defer human intelligence to machines, we promote a shift that downgrades technique but promotes new sensibilities. People still bother making paintings because painting offers things that photography cannot.

The title of our paper asked the question: "Is writing prompts really making art?". As we have discussed, prompt writing raises many important concerns: conceptual, ethical, legal and moral. These concerns should make us wary of blindly accepting generative machine learning systems as a positive thing for human creativity and art.

If you gave a designer or illustrator a verbal brief (and paid them) to create a specific illustration for you, you would not credit yourself as the creator or artist in such a scenario. The creator would be the artist who drew the work. With TTI systems there is a broader question of authorship, because the data used for training directly involves literally billions of examples of human authorship.

Hence, at this point it is difficult to see prompt writing as a significant art practice. However, as artists have capably demonstrated, anything can become a medium for art and artistic practice, so we expect to see artists making use of these new technologies for new creative purposes. As we have outlined, this will most likely be by subverting their intended use, embodying the agency and interaction for these systems or just embracing their flaws and quirks to expose more of the process of deep learning and machine creativity. Moreover, the form and modalities of current TTI systems will likely change rapidly and unpredictably through the widespread technological, engineering, political and cultural advances driving these systems.

Acknowledgements. This research was supported by an Australian Research Council grant DP220101223.

References

1. Alexander, C.: Notes on the synthesis of form. Harvard University Press, Cambridge, Mass (1964)
2. Baio, A.: AI data laundering: how academic and nonprofit researchers shield tech companies from accountability. (webpage) (2022). https://waxy.org/2022/09/ai-data-laundering-how-academic-and-nonprofit-researchers-shield-tech-companies-from-accountability/. Accessed 21 Jan 2023
3. Boden, M.A.: The creative mind: myths & mechanisms. Basic Books, New York, N.Y. (1991)
4. Brooks, R.A.: Elephants don't play chess. Robot. Auton. Syst. **6**(1–2), 3–15 (1990)
5. Carroll, N.: The specificity of media in the arts. J. Aesthetic Educ. **19**(4), 5–20 (1985)
6. Cetinic, E., She, J.: Understanding and creating art with AI: review and outlook. ACM Trans. Multimedia Comput. Commun. Appl. (TOMM) **18**(2), 1–22 (2022)
7. Close, C.: From Hines to Christo: how artists influence other artists. (website) (2022). https://www.artandobject.com/articles/hines-christo-how-artists-influence-other-artists. Accessed 21 Jan 2023
8. Epstein, Z., Levine, S., Rand, D.G., Rahwan, I.: Who gets credit for AI-generated art? Iscience **23**(9), 101515 (2020)
9. Franklin, S.: Autonomous agents as embodied AI. Cybern. Syst. **28**(6), 499–520 (1997)
10. Gombrich, E.H.: The story of art. Englewood Cliffs, New Jersey: Prentice-Hall, 16th, rev., expanded and redesigned edn. (1995)
11. Guckelsberger, C., Kantosalo, A., Negrete-Yankelevich, S., Takala, T.: Embodiment and Computational Creativity. In: Proceedings of the Twelve International Conference on Computational Creativity, 2021. Association for Computational Creativity (ACC) (2021)
12. Hisarciklilar, O., Boujut, J.-F.: Symbolic vs. iconic: How to support argumentative design discourse with 3D product representations. In: Proceedings of IDMME - Virtual Concept 2008, Beijing, China, October 8–10, pp. 1–8 (2008)
13. Homburg, C.: The copy turns original : Vincent van Gogh and a new approach to traditional art practice. Oculi, v. 6, John Benjamins, Amsterdam; Philadelphia (1996)
14. Karras, T., Laine, S., Aila, T.: A style-based generator architecture for generative adversarial networks. (arXiv preprint) (2018). https://doi.org/10.48550/ARXIV.1812.04948. https://arxiv.org/abs/1812.04948. Accessed 21 Jan 2023

15. Malafouris, L.: At the Potter's wheel: an argument *for* material agency. In: Knappett, C., Malafouris, L. (eds.) Material Agency, pp. 19–36. Springer, Boston, MA (2008). https://doi.org/10.1007/978-0-387-74711-8_2
16. McNeill, N.B.: Colour and colour terminology. J. Linguist. **8**(1), 21–33 (1972)
17. Microsoft: Promptist. (Software on Huggingface.co) (2023). https://huggingface.co/spaces/microsoft/Promptist. Accessed 19 Jan 2023
18. Notaro, A.: State-of-the-art: AI through the (artificial) artist's eye. EVA London 2020: Electronic Visualisation and the Arts, pp. 322–328 (2020)
19. O'Hear, A.: Art and technology: an old tension. R. Inst. Philos. Suppl. **38**, 143–158 (1995)
20. Okada, T., Ishibashi, K.: Imitation, inspiration, and creation: cognitive process of creative drawing by copying others' artworks. Cogn. Sci. **41**(7), 1804–1837 (2017)
21. Panagopoulos, J.: This man does not exist. (website) (2022). https://www.theaustralian.com.au/the-oz/internet/this-man-does-not-exist/news-story/bff7ed91626562be57f248dafcc92128. Accessed 21 Jan 2023
22. Pfeifer, R., Scheier, C.: Understanding intelligence. MIT press (2001)
23. Poulsgaard, K.S., Malafouris, L.: Understanding the hermeneutics of digital materiality in contemporary architectural modelling: a material engagement perspective. AI & SOCIETY, pp. 1–11 (2020)
24. Ramesh, A., Dhariwal, P., Nichol, A., Chu, C., Chen, M.: Hierarchical text-conditional image generation with clip latents. (arXiv preprint) (2022). https://doi.org/10.48550/ARXIV.2204.06125. https://arxiv.org/abs/2204.06125. Accessed 21 Jan 2023
25. Riegler, A.: When is a cognitive system embodied? Cogn. Syst. Res. **3**(3), 339–348 (2002)
26. Schon, D.A.: The reflective practitioner: how professionals think in action, vol. 5126. Basic books (1984)
27. Shanahan, M.: Talking about large language models. (arXiv preprint) (2022). https://doi.org/10.48550/ARXIV.2212.03551. https://arxiv.org/abs/2212.03551. Accessed 21 Jan 2023
28. Singer, U., et al.: Make-a-video: Text-to-video generation without text-video data. arXiv preprint arXiv:2209.14792 (2022)
29. Vaswani, A., et al.: Attention is all you need. In: Advances in Neural Information Processing Systems 30 (2017)
30. Ziemke, T.: What's that thing called embodiment? In: Proceedings of the Annual Meeting of the Cognitive Science Society, vol. 25. Lawrence Erlbaum (2003)

Using GPT-3 to Achieve Semantically Relevant Data Sonificiation for an Art Installation

Rodolfo Ocampo[1]([⊠]), Josh Andres[2], Adrian Schmidt[2], Caroline Pegram[3], Justin Shave[3], Charlton Hill[3], Brendan Wright[1,3], and Oliver Bown[1]

[1] The University of New South Wales, Kensington, Australia
r.ocampo_blanco@unsw.edu.au
[2] The Australian National University, Canberra, Australia
[3] Uncanny Valley, Canberra, Australia

Abstract. Large Language Models such as GPT-3 exhibit generative language capabilities with multiple potential applications in creative practice. In this paper, we present a method for data sonification that employs the GPT-3 model to create semantically relevant mappings between artificial intelligence-generated natural language descriptions of data, and human-generated descriptions of sounds. We implemented this method in a public art installation to generate a soundscape based on data from different systems. While common sonification approaches rely on arbitrary mappings between data values and sonic values, our approach explores the use of language models to achieve a mapping not via values but via meaning. We find our approach is a useful tool for musification practice and demonstrates a new application of generative language models in creative new media arts practice. We show how different prompts influence data to sound mappings, and highlight that matching the embeddings of texts of different lengths produces undesired behavior.

Keywords: data sonification · machine learning · generative music · generative AI · large language models · word emdeddings

1 Introduction

Data sonification transforms data into sound to convey information sonically for data exploration, entertainment, or artistic expression [17] [10] [6].

The authors of this paper were commissioned with a data sonification art installation to produce a generative soundscape for the foyer of the building of The School of Cybernetics at The Australian National University. The brief requested that the sonified data represented the complex systems that the building is using a Cybernetic lens, where humans, technology and the environment adaptively interact [1,7].

Sonification approaches often involve a numeric mapping of data values to sonic values [6]. This mapping determines the extent to which the sonic properties convey data properties and the extent to which the sonified data results

C. Johnson et al. (Eds.): EvoMUSART 2023, LNCS 13988, pp. 212–227, 2023.
https://doi.org/10.1007/978-3-031-29956-8_14

in musical outputs acceptable for the purpose of the data sonification project. Usually, the mappings involve either continuous or discreet symbolic mappings. In a continuous approach, numerical data is mapped to sonic qualities such as sound frequency, attack, release or effects like delay and reverb. In a symbolic approach, categorical data or discreetly segmented data is mapped to symbolic music representations such as specific notes, MIDI values or instrument types. Sonification approaches involving more than one data source commonly employ continuous and symbolic approaches. A common challenge of these approaches is the potential arbitrariness or lack of meaning of such value mappings [6].

To address the installation's brief, we experimented with the creative potential of generative language models to enable a mapping not via value mapping but via semantic matching between data interpretations and sound descriptions. In other words: mapping data to sound via language.

In a preliminary examination, we found that GPT-3 could generate natural language interpretations of alphanumeric data and then match these to human-labeled sounds using word embeddings. Sounds were labeled according to what they evoked in the listener, thus our approach's objective is to match machine-generated subjective interpretations of data with human-generated subjective interpretations of sounds.

[10] proposed a three-level distinction for sonificiation approaches that is useful to situate our approach. The first level is generative, where the contents of the data are not relevant: it is merely used to achieve compositions that would be impossible or difficult manually. The second level, allusive sonification, aims to create a bridge between complex information with a more accessible artistic domain that appeals to sensory and emotional experiences. Lastly, curatorial sonification is mostly a scientific tool that aims to accurately convey salient features of the data through sound in a meaningful way.

Our approach can be placed within the Allusive level: our intention is to transform complex data into a sensory and emotional experience through the use of a semantic matching that captures expressed emotional information.

For this project, the authors, University of New South Wales academics, partnered with Uncanny Valley, music technology company and built upon their existing generative music engine, MEMU (https://memu.live/), which has been deployed previously in different contexts.

By developing a novel creative use of one of the most prominent machine learning models to date, GPT-3, we provide what we believe is a useful new tool for the data sonification and creative new media practice communities. We outline the challenges faced and the strategies used to address them. Lastly, we discuss limitations and qualitative and quantitative learnings that can be useful for future work.

2   Related Work

The sonification of data, particularly environmental and atmospheric, has fascinated researchers for decades to extend beyond text, tables, and visualisations on how humans can engage with data. For example, [15] explored an artistic approach to newly collected data on the earth's Ice Core, Seismic and Solar Wind

Data to map data attributes to musical elements for a one-hour pre-recorded performance. This work is an example of data sonification in a non-audio generative context because it uses static multimodal time series data where the musical elements are controlled, tweaked and finessed to achieve second by second a desired musical expression. Interestingly, in this creative decision-making process to select data attributes to musical elements, researchers can emphasise temporal data aspects for listeners that are often non-appreciated using numbers [4] [5] [14]. [18] developed a sonification approach rooted in Cybernetics by integrating urban open data streams in live musical performances, where live data was matched to audio effects and sonic parameters affecting music improvised by a human.

Recent data sonification explorations benefit from computational architectures that use real-time data streams and generative machine learning models to create audio experiences. From using atmospheric environmental data with generative Japanese music melodies data sets to query varying data periods [8] to experimental approaches combining multiple generative models, such as large language text models, to query creative music composition decisions for composer-to-machine dialogue opportunities [9]. Mubert is a recently developed text-to-music system that takes a user prompt and matches it to labeled pieces of sounds composed by humans via latent embeddings in order to build a finished track. Our approach to text-to-music matching is highly similar in the sense that sounds are composed by humans, and combined to create music tracks, and that matchings between text and sound are achieved via prompt-label word embedding similarity.

Our work contributes to the recent efforts in exploring the creative potential of Transformer and large language models, and provides a new tool to the creative new media and data soniciation communities.

Next, we describe the presented architecture and report on the audio generative learnings.

3   Implementation

3.1   Conceptual Development

To create a continuous data sonification soundscape of the building as a complex cybernetic system, we drew on the concept of Pace Layers by Stewart Brand [2], which describes complex systems as composed of six layers with distinct paces of change: nature, culture, infrastructure, governance, commerce and fashion/media. We carefully selected a data stream to each layer as follows:

- **Nature layer:** Global CO_2 concentration in the atmosphere (ppm) per month for the last 36 months, updated monthly.
- **Culture layer:** The most read articles of the day from Wikipedia, updated daily.
- **Infrastructure layer:** Latest indicator values from the Infrastructure topic from the WorldBank open data API, for Australia, updated daily.

– **Governance layer:** Latest indicator values from the Public Sector topic from the WorldBank open data API, for Australia, updated daily.
– **Commerce layer:** Latest indicator values indicator values from the Economy and Growth topic from the WorldBank open data API, for Australia, updated daily.
– **Fashion/media layer:** The latest tweet from the ANU School of Cybernetics Twitter handle, updated daily.

Each data source is mapped to a sonic layer in the soundscape that reflects a similar rate of change. For example, atmospheric sounds are the slowest changing sounds, thus they are mapped to Nature. On the other hand, accents are the fastest-changing, thus they are mapped to fashion. Therefore, we chose the mapping so that the soundscape, comprised of stacked layers of differently paced sounds, reflects the stacking of the pace layers diagram.

– Fashion layer: Accents
– Commerce layer: Bass
– Infrastructure layer: Timekeepers
– Governance layer: Pads and chords
– Culture layer: Scales
– Nature layer: Atmospheric

3.2   Human Labeling of Sounds by Audio Elicitation

We invited the building occupants, composed of academics, students, professional staff, and also friends, family, and the installation developers, to label a library of 1050 sounds created by Uncanny Valley, totalling approximately 80 labellers. Labellers were asked to enter a label description of multiple words based on what the sound elicited from them via a custom-built interface showing an audio player and a text field. The sound labelling interface randomly loaded sounds to each labeller as a strategy to ensure all sounds were labelled. From the 1050 sounds, 219 sounds received more than one label. The labels provided were succinct, averaging 2.9 words per label. Examples of labels include:

– A blue light traveling through a vacuum chamber, serious
– The dopplered siren fades to silence as it merges with the horizon.
– bladerunner electric android dreams

3.3   Generating Natural Language Interpretations of Data with GPT-3

The next step of our pipeline involves retrieving each data source at regular interval periods and then passing the data to GPT-3 as part of a prompt to generate natural language interpretations of the data.

A Python script retrieves each data source via web APIs. The data is converted into a string to be appended to the GPT-3 prompt, which has the following structure.

```
Prompt = data explanation + string representation of data + request.
```

Our intention was to use GPT-3 to produce an interpretation of the data that contained subjective descriptors and emotional language, since this is the type of language contained in the sound labels we aimed to match to. Thus one of our approaches involved requesting the generation of poems from the data. We tried different prompting approaches and the effects of each are shown in the results section.

An example prompt and GPT-3 response are provided below:

```
This data contains the last 36 months of CO2 measurements in the atmosphere.
The first column is the year, the second the month and the third is the CO2
level in parts per million.

Data:
[['2019', '11', '410.48'],
  ['2019', '12', '411.98'],
  ['2020', '1', '413.61'],
[...]
  ['2022', '8', '417.19'],
  ['2022', '9', '415.95'],
  ['2022', '10', '415.78']]

Write a poem about the data:
```

GPT-3 response:

```
The CO2 in the atmosphere
Is rising day by day
The levels are getting higher
And higher all the time
[...]
```

To obtain the GPT-3 generations, we used the OpenAI GPT-3 Python API [12] with the following settings:

- **Engine:** text-davinci-002
- **Temperature:** 0.9
- **Max Tokens:** 256
- **Top P:** 1
- **Frequency penalty:** 0
- **Presence penalty:** 0

3.4   Matching GPT-3's Data Interpretations to Sound Labels for Soundscape Generation

For the soundscape generation, we extended an existing music engine developed by Uncanny Valley called MEMU. This engine contains a music library of multiple thousands of sounds, and it is able to combine them in musically coherent ways to produce nearly infinite musical variations. This engine works by selecting and stacking different types of sounds (pads, atmospheric). The selected sounds in the engine are constrained by a globally defined chroma value as well as valence, and then sounds matching these contraints are selected at random. Our work consisted on modifying the engine so that instead of selecting sounds at random, sounds are selected based on the incoming data and the associated sound labels.

At regular intervals, the engine queries the generated GPT-3 text for each data source, including each embedding representation, which is a single 1024-valued vector for each text. The engine then matches these embeddings to pre-computed sound label embeddings via cosine similarity, a common measure used to calculate semantic similarity between word embeddings [19].

The word embeddings are generated using the "text-similarity-ada-001" model provided through the OpenAI embeddings Python API [13], a Transformer word embedding model derived from GPT [11].

The sounds that each data source can match to are constrained by three factors: 1) the layer-to-sound type mapping described in Sect. 3.1., 2) the global musical key, determined by the engine at random and 3) the musical mode, determined by the local weather outlook obtained from a weather API, according to the following mapping.

- Sunny outlook: Ionian mode
- Partly cloudy outlook: Dorian mode
- Cloudy or rainy outlook: Aeolian mode

4   Results

In order to evaluate our approach to semantic data musification, we are interested in understanding the following:

- **Interpretation quality:** how well does the GPT-3 generated interpretation reflect characteristics from the data
- **Matching quality:** can a clear semantic connection be drawn between the GPT-3 generated natural language interpretations of the data and the matched sound label?
- **Breadth:** how many different sounds are selected?

4.1   Experiment 1: Preliminary Testing of Matching GPT-3 Data Interpretations to Sound Labels

As described above, the prompt we passed to GPT-3 has the following structure:

```
Prompt = explanation + string version of data + generation request.
```

An example prompt with these three components, for the Culture layer data would be as follows.

```
"These data contains the titles for the most read wikipedia articles today.

Data:
Aaron Carter
Crown Jewel (2022)
"ICC Mens T20 World Cup"
Leslie Carter Cleopatra
"Weird Al" Yankovic
Elon Musk
Guy Fawkes Night
Nick Carter (singer)
Sally McNeil

Interpret these data."
```

We evaluated the effect of different prompts by varying the request component. We wrote six different requests:

- Request 1: Describe the data in only three words
- Request 2: Describe how these data make you feel in only three words
- Request 3: Describe in three words how these data make you feel and why
- Request 4: Write in one sentence how these data make you feel and why:
- Request 5: Write a poem about the data
- Request 6: Interpret the data

To test them, we isolated one data source: the most-read Wikipedia articles of the day corresponding to the Culture layer. We generated a hundred variants for each request, one for each of the last hundred days of data, thus obtaining six hundred generations and corresponding matchings.

We qualitatively evaluated the interpretation quality by observing if the generated text reflects the characteristics of the data. We evaluated the the matching quality qualitatively by examining if a clear semantic connection could be drawn by the GPT-3 generated text and the matched label. Lastly, we evaluated the sound selection breadth by calculating how many unique labels were selected for each request and obtaining the sound selection distribution of each selected sound.

A representative example of text generation and corresponding matching is provided in Table 1.

For each request, we calculate the number of unique sound labels matched, shown below:

- Request 1 (Describe the data in only three words): 21
- Request 2 (Describe how these data make you feel in only three words): 19
- Request 3 (Describe in three words how these data make you feel and why): 11
- Request 4 (Write in one sentence how these data make you feel and why:): 10

Table 1. Generated text by GPT-3 when asked to generate poems from the data and resulting sound label matchings

Request	Data	GPT-3 generated request	Matched Sound Label
Describe the data in only three words:	["Manti Te'o", 'Omegle', 'Casemiro' [...]]	Titles of Wikipedia articles.	exotic and interesting
Describe how these data make you feel in only three words:	["Manti Te'o", 'Omegle', 'Casemiro' [...]]	Curious, intrigued, satisfied.	cold, curious,Interested
Describe in three words how these data make you feel and why:	["Manti Te'o", 'Omegle', 'Casemiro' [...]]	Curious, because these are all topics that I am not familiar with. Intrigued, because I am curious to learn more about them. And entertained, because it is always interesting to learn new things.	Interesting, curious, Spatial, night, Open, airy, blue, whimsical, otherwordly, calling me
Write in one sentence how these data make you feel and why:	["Manti Te'o", 'Omegle', 'Casemiro' [...]]	The data make me feel curious about why Manti Te'o, Omegle, and Casemiro are the most read Wikipedia articles today.	eerie, suspenseful, sexy,inquisitive, reflective
Write a poem about the data:	["Manti Te'o", 'Omegle', 'Casemiro' [...]]	These are the titles for the most read wikipedia articles today. Manti Te'o is a football player who was caught in a hoax. Omegle is a website where you can chat with strangers. Casemiro is a Brazilian footballer. Sanna Marin is the Prime Minister of Finland. [...]	death and dying,death and dying,death and dying,fear death and dying,fear death and dying,fear death and dying,fear death and dying,fear death and dying
Interpret the data:	["Manti Te'o", 'Omegle', 'Casemiro' [...]]	The data shows the titles of the most read Wikipedia articles as of today. The top article is about Manti Te'o, followed by Omegle, Casemiro, Sanna Marin, and Andrew Tate.	death and dying,death and dying,death and dying,fear death and dying,fear death and dying,fear death and dying,fear death and dying,fear death and dying

- Request 5 (Write a poem about the data): 6
- Request 6 (Interpret the data): 6

We observed that requests 5 and 6 were producing text that was being matched to only six unique sounds, out of the 100 generations, thus displaying a degree of sound selection breadth that was undesirable given the installation's objectives. In fact, most times, these were being matched to a single label "death and dying [...]", as shown in Fig. 1 and Table 1. A very similar distribution was observed for request 6.

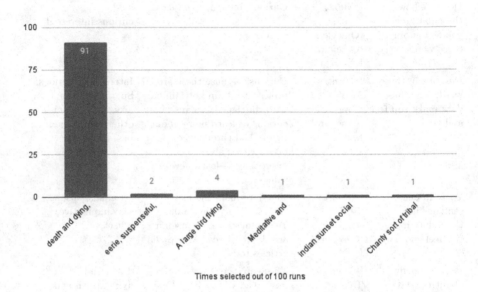

Times selected out of 100 runs

Fig. 1. Sound labels selected from Request 5 ("Write a poem about the data").

While some of the generated text had references to death, not all of them did but they were still being matched to the same label, and it was not clear why this was the case.

Meanwhile, requests 1 and 2 were producing text that matched to a larger number of unique labels, which also had clearer semantic relationships.

Matching Text of Different Lengths Produced Undesired Behavior. Upon closer examination, we found that requests 5 and 6 were producing the longest text out of all requests. The "death and dying [...]" label was the longest label in the library. While most labels in the library were around three words long, the death and dying label is 29 words long. Thus, we hypothesised that length was becoming a salient feature, perhaps due to word embedding averaging effects. While short GPT-3 generated text was producing semantically relevant matchings to sound labels, longer text seemed to only be match to longer text.

We sought to test this further by testing 6 hand-crafted test texts and evaluate what they matched to. We tested four short test texts and one long test text, corresponding to the first three paragraphs of the song "Stayin' Alive". The matching results are shown in Table 2.

Table 2. Testing sound label matchings using human written test inputs

Test input text	Matched sound label
A happy day	happy and warm
Sad and gloomy	depressing
Exciting and adventurous	exotic and interesting
Ethereal and spacey	exotic and ethereal
Well, you can tell by the way I use my walk I'm a woman's man [...] You're stayin' alive, stayin' alive Feel the city breakin' and everybody shakin' And we're stayin' alive, stayin' alive Ah, ha, ha, ha, stayin' alive, stayin' alive [...]	death and dying,death and dying,death and dying,fear death and dying,fear death and dying,fear death and dying,fear death and dying,fear death and dying

The results above provide supporting evidence that short text was producing clear semantic matchings, while long text was just matching to the longest label, even if the meaning was radically different. Thus using the text-similarity-ada-001 produced embeddings to match text of largely different lengths produces undesired behavior.

Repetitive Generation with Low Data Interpretation Quality. While the shorter text produced by Requests 1 and 2 produced clearer semantic matchings, when examining the generated text from these requests for interpretation quality, we found they did not reveal properties of the data, and instead were fixating around a small number of three-word descriptions. A subset of 10 generations for Request 2 showing this behavior is provided below.

1. Curious, intrigued, motivated.
2. Informative, curious, excited.
3. Curious, Engaged, Intrigued.
4. Curious, Engaged, Intrigued.
5. hopeful, inspired, curious.
6. Curious, interested, engaged.
7. Curious, intrigued, excited.

8. Intrigued, curious, interested.
9. Curious, intrigued, excited.
10. Happy, curious, and excited.

This repetive generation constrains the variety of matched labels, as shown in Fig. 2, where we plot the distribution of each of the selected labels for the 100 runs using Request 2 ("Describe how these data make you feel in only three words"). Out of the 19 unique sounds matched, 64 were matched to just one label: "question everything". A similar behaviour was observed for Request 1 ("Describe the data in only three words"), where out of the 21 unique labels selected for the 100 trials, 35 times it was matched to just one label: "cold, curious interested".

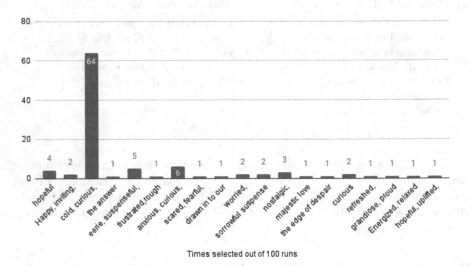

Times selected out of 100 runs

Fig. 2. Sound labels selected from Request 2 ("Describe how these data make you feel in only three words").

We hypothesised this was the result of asking the model to describe how the data made it "feel". We experimented with producing three-word descriptions that leverage the observed improvements of short text on matching quality and sound selection breadth while generating text that more closely captures characteristics of the data (better data interpretation quality). We tested this in experiment 2.

4.2 Experiment 2: Extracting Emotional Qualities of the Data for Better Matching

In the previous experiment, we found that short interpretations provide more varied and clearer sound label matchings than longer generations. However, these short texts mostly fixate around a few descriptors that don't reflect the data accurately. In order to address this, instead of how the data makes the model "feel", we experimented with requesting interpretations from GPT-3 that extract emotional qualities of the data. We trialled the following three phrasings:

– Request 1: Write three words that capture the emotional quality of the data
– Request 2: Write three words that summarise these titles
– Request 3: Write three words that summarise the vibe of these titles

Again, we generated 100 prompts for each request, one for each of the last 100 d of data for the Wikipedia data. We found these requests yield short data interpretations that reflect the data more closely, while still providing highly relevant semantic matchings, as shown in the examples in Table 3. Moreover, these generations more closely reflect qualities in the data.

We found this prompting approach yields text that is more varied than the approach in experiment 1, which fixated around curiosity related descriptors, A few generations are provided below.

– Readership, Popularity, Notability
– Sorrowful, triumphant, celebratory.
– Famous, Joyful, Controversial
– Gory, Terrifying, Shocking

We also we found that this approach yields a higher breadth in sound selection compared to the prompting approach in Experiment 1. Below, we list the unique sounds selected for each request, after generating 100 times for each.

– Write three words that capture the emotional quality of the data: 43
– Write three words that summarise these titles: 43
– Write three words that summarise the vibe of these titles: 45

These requests produce GPT-3 data interpretations that yield more than double the amount of unique sound matchings than the approach used in Experiment 1. Moreover, they provide flatter sound label selection distributions that are less dominated by particular labels, as shown in Fig. 3.

This prompting approach proved to be the best approach in terms of our musification objectives: matching quality, interpretation quality and sound selection breadth, thus it became our adopted approach in the installation.

In the supplementary materials, this approach is applied to all data sources.

Lastly, in the following link: https://github.com/rodolfoocampo/EvoMusArt 2023-SemanticSonification, we provide a repository data for the experiments, example code, a link to a live demo in the browser of the application and a link to an interface to play the selected sounds alongside with labels and matched GPT-3 generated text so that the reader gets a sense of the data-to-sound mappings. In the supplementary materials, we provide a video of the working installation.

Table 3. Comparison of generated text for different requests to GPT-3 and resulting sound label matchings

Request	Data	GPT-3 response	Matched Label
Write three words that capture the emotional quality of the data:	['Aaron Carter', 'Nick Carter (singer)', 'Leslie Carter' [...]]	Readership, Popularity, Notability	attention grabbing empowering
Write three words that capture the emotional quality of the data:	["Manti Te'o", 'Omegle', 'Casemiro' [...]]	humorous, lighthearted, fun	fun playful
Write three words that summarise these titles:	['Aaron Carter', 'Nick Carter (singer)', 'Leslie Carter' [...]]	Musicians, death, crime.	broken jazz band
Write three words that summarise these titles:	["Manti Te'o", 'Omegle', 'Casemiro' [...]]	1. People 2. Culture 3. Technology	confused people
Write three words that summarise the vibe of these titles:	['Aaron Carter', 'Nick Carter (singer)', 'Leslie Carter' [...]]	celebrities, murder, internet	extremely disturbing psycho
Write three words that summarise the vibe of these titles:	["Manti Te'o", 'Omegle', 'Casemiro' [...]]	humorous, chatty, lighthearted	fun playful

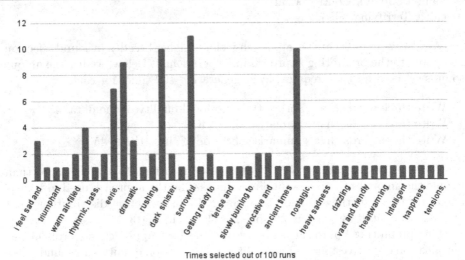

Times selected out of 100 runs

Fig. 3. Sound labels selected from Request 1b ("Write three words that capture the emotional quality of the data").

5   Discussion

Our sonification approach did not focus on precisely conveying properties of the data (curatorial scientific sonification [10]). Instead, our approach focused on allusive sonificiation [10] with the intention of transforming complex data into a more accessible artistic representation that appeals to emotional responses by matching data and sound through language interpretations.

For this, we set three objectives: interpretation quality, matching quality, and breadth of sound selection. We found that interpretation quality is highly influenced by prompting approaches as well as the model's own capabilities. While GPT-3 sometimes can produce factually accurate interpretations of data, such as recognising when values have gone up or down, or are positive or negative, this was not always the case, particularly when recognising more complex numerical relationships. We believe this is due to the well-documented weakness in GPT-3's numerical reasoning capabilities [3]. Musification or text-to-sound mappings that require higher objective descriptiveness might benefit from chain-of-thought few short prompting or fine-tuning. In our case, the intention was to produce short interpretations of data using subjective and emotional language that could be semantically mapped to sounds described in similar terms. We found that prompting approaches that require the model to extract the emotional qualities of the data best served this purpose. Moreover, the shortness of the generated text was important since we found that matching text of different lengths via word embedding cosine similarity produced undesired behavior, such that text appeared to be matched on the basis of length and not meaning.

We found that breadth of sound selection was influenced by the prompting approach. Initially, we wanted to create text that reflected a first person subjective evaluation of the data, so we asked the model to describe how the data made it"feel". However, this produced repetitive responses that limited the breadth of sound selection. As mentioned above, this was addressed by requesting the model to instead extract emotional qualities of the data, which yielded responses that more closely reflected the data and its variability. From a sonification perspective, breadth of sound selection was important for the soundscape to be interesting. However, maximum breadth was not the objective, but a degree that reflected the data's variability.

This paper does not involve any evaluation of the musicality of the MEMU engine. MEMU's sounds were created by professional musicians, and uses generative techniques to mix and arrange music. Our approach simply selects elements that MEMU will play, and could be easily adapted to work with other generative music engines.

Lastly, we believe this approach is relevant in the context of multimodal generative AI, which maps different modes of data, like text, image and music via latent embeddings. These approaches usually rely on datasets labelled in natural language [16]. While training text-to-image multimodal models benefit from image datasets with highly descriptive captions, music is often described in more subjective, abstract and emotional terms. A future direction for our work could be to use our labeled library to train a generative text-to-music model. This would require expanding the dataset and more standardised labelling.

6   Conclusion

We presented a novel machine learning pipeline built to create a continuous data sonification soundscape using large language models. Our work shows that large language models enable data sonification by semantic mapping between data interpretations and sound descriptions. We demonstrated via two experiments the potentials and limitations of the presented pipeline to support creative endeavours in data sonification practices.

Acknowledgements. This research was made possible by a commission from the School of Cybernetics at the Australian National University for music studio Uncanny Valley (UV). The development of the novel concept for a semantically relevant sonification using Large Language Models is an original contribution from Rodolfo Ocampo, who also led the technical development of the system, in collaboration with members of the UV team. The artwork uses UV's MEMU generative music system, developed by Justin Shave and Brendan Wright. The design and development of the visual user interface were led by Adrian Schmidt and Josh Andres. Oliver Bown and Rodolfo Ocampo's research is supported by an Australian Research Council Discovery Project (DP200101059).

References

1. Andres, J.: Adaptive human bodies & adaptive built environments for enriching futures. In: Frontiers in Computer Science, Special Issue Inbodied Interaction (2022)
2. Brand, S.: Pace layering: how complex systems learn and keep learning. J. Design Sci. (Jan 2018)
3. Floridi, L., Chiriatti, M.: GPT-3: Its nature, scope, limits, and consequences. Minds Mach. **30**(4), 681–694 (2020)
4. Flowers, J.H., Whitwer, L.E., Grafel, D.C., Kotan, C.A.: Sonification of daily weather records: issues of perception, attention and memory in design choices. Faculty Publications, Department of Psychology, p. 432 (2001)
5. Hermann, T., Drees, J.M., Ritter, H.: Broadcasting auditory weather reports-a pilot project (2003)
6. Hermann, T., Hunt, A., Neuhoff, J.G.: The sonification handbook. Logos Verlag Berlin (2011)
7. Andres, J., et al.: Cybernetic lenses for designing and living in a complex world. In: In Extended Abstracts of the 2022 OzCHI Conference on Human Factors in Computing Systems (OZCHI EA 2022). Association for Computing Machinery, New York, NY, USA. (2022)
8. Kalonaris, S.: Tokyo Kion-On: query-Based generative sonification of atmospheric data (Aug 2022)
9. Krol, S.J., Llano, M.T., McCormack, J.: Towards the generation of musical explanations with GPT-3. In: Artificial Intelligence in Music, Sound, Art and Design: 11th International Conference, EvoMUSART 2022, Held as Part of EvoStar 2022, Madrid, Spain, April 20–22, 2022, Proceedings, pp. 131–147. Springer-Verlag, Berlin, Heidelberg (Apr 2022)
10. Mardakheh, M.K., Wilson, S.: A strata-based approach to discussing artistic data sonification. Leonardo **55**(5), 516–520 (2022)

11. Neelakantan, A., et al.: Text and code embeddings by contrastive Pre-Training (Jan 2022)
12. OpenAI: OpenAI API. https://beta.openai.com/docs/introduction (2022). Accessed 17 Nov 2022
13. OpenAI: OpenAI API. https://beta.openai.com/docs/guides/embeddings (2022). Accessed 21 Aug 2022
14. Polli, A.: Atmospherics/weather works: a multi-channel storm sonification project (2004)
15. Quinn, M.: Research set to music: the climate symphony and other sonifications of ice core, radar, DNA, seismic and solar wind data (2001)
16. Ramesh, A., et al.: Zero-shot text-to-image generation (Feb 2021)
17. Rocchesso, D., et al.: Sonic interaction design: sound, information and experience. In: CHI'08 Extended Abstracts on Human Factors in Computing Systems, pp. 3969–3972 (2008)
18. Roddy, S.: Signal to noise loops: a cybernetic approach to musical performance with smart city data and generative music techniques. Leonardo, pp. 525–532 (2022)
19. Singhal, A.: Modern information retrieval: a brief overview. http://160592857366.free.fr/joe/ebooks/ShareData/Modern%20Information%20Retrieval%20-%20A%20Brief%20Overview.pdf (2001). Accessed 17 Nov 2022

Using Autoencoders to Generate Skeleton-Based Typography

Jéssica Parente(✉)[iD], Luís Gonçalo(✉)[iD], Tiago Martins[iD],
João Miguel Cunha[iD], João Bicker[iD], and Penousal Machado[iD]

CISUC, Department of Informatics Engineering, University of Coimbra,
Coimbra, Portugal
{jparente,lgoncalo,tiagofm,jmacunha,bicker,machado}@dei.uc.pt

Abstract. Type Design is a domain that multiple times has profited from the emergence of new tools and technologies. The transformation of type from physical to digital, the dissemination of font design software and the adoption of web typography make type design better known and more accessible. This domain has received an even greater push with the increasing adoption of generative tools to create more diverse and experimental fonts. Nowadays, with the application of Machine Learning to various domains, typography has also been influenced by it. In this work, we produce a dataset by extracting letter skeletons from a collection of existing fonts. Then we trained a Variational Autoencoder and a Sketch Decoder to learn to create these skeletons that can be used to generate new ones by exploring the latent space. This process also allows us to control the style of the resulting skeletons and interpolate between different characters. Finally, we developed new glyphs by filling the generated skeletons based on the original letters' stroke width and showing some applications of the results.

Keywords: Type Design · Variational Autoencoder · Skeleton-basis Typography

1   Introduction

The design of type has undergone numerous changes over time [4]. In the early years, typography was seen as a system made up of a series of rules. The artistic movements that arrived at the beginning of the twentieth century rejected the historical forms and transformed outdated aspects of visual language and expression. However, projects that combined software, arts and design only appeared a few years later with the proliferation of personal computers, allowing programming to reach a wider audience. Thanks to all these changes, the tools to design type changed, and new possibilities for typographic experimentation appeared, resulting in *(i)* grammar-based techniques that explore the principle of database amplification (*e.g.* [2]); *(ii)* evolutionary systems that breed design

---

J. Parente and L. Gonçalo—These authors contributed equally to this work.

C. Johnson et al. (Eds.): EvoMUSART 2023, LNCS 13988, pp. 228–243, 2023.
https://doi.org/10.1007/978-3-031-29956-8_15

Fig. 1. Interpolation of the skeleton and stroke width from two existing A's (light blue and red), resulting in a new A (dark blue). (Color figure online)

solutions under the direction of a designer (*e.g.* [15,22]); *(iii)* or even, even, Machine Learning (ML) systems that learn the glyphs features to build new ones (*e.g.* [14]) [18]. These computational approaches can also be helpful as a starting point of inspiration.

Most emerging fonts continue to be developed by type designers who study the shape of each letter and its design with great precision, despite the emergence of these new possibilities. Type design is a hugely complex discipline, and its expertise ensures typography quality [28]. Moreover, with the proliferation of web typography and online reading, the use of variable and dynamic fonts has increased, allowing more options for font designers and font users. Additionally, visual identities created nowadays are becoming more dynamic [17]. Museums, institutions, organisations, events and media increasingly rely on this type of identity. Consequently, designers should adapt their work to these new possibilities by creating dynamic identities with animations and mutations. Even though new computer systems create expressive and out-of-the-box results, they do not have the knowledge of an expert. But this is also an advantage, allowing non-arbitrary exploitation that extends the range of possibilities. It is necessary to create a balance to take advantage of the computational systems and the expert labour. Moreover, most generative systems that design type focus on the letters' filling and don't see the structure of a glyph as a variation parameter.

To overcome these limitations, we propose an Autoregressive model [9] that creates new glyph skeletons by the interpolation of existing ones. Our skeleton-based approach uses glyphs skeletons of existing fonts as input to ensure the quality of the generated results. The division of the structure and the filling of the glyphs add variability to the results. Different glyphs can be created by just changing the structure or the filling. The proposed approach enables the exploration of a continuous range of font styles by navigating on the Autoencoder (AE) learnt latent space. With the results of this approach, it is also possible to apply different filling methods that use the stroke width of the original letters to produce new glyphs (see Fig. 1).

The remainder of this paper is divided into three sections. The following section, Related Work, analyzes related projects in the domain of computational typography with Artificial Neural Networks (ANNs). The second section, Approach, describes the construction of the used dataset and explains the training process. Then, in the Results section, we present and discuss the different experimentations performed and the obtained results. In this section, we also present a set of different possible applications of the outputs of our system. In the final section, Conclusion and Discussion, we draw some conclusions and lay out future work.

2   Related Work

Over time, the methods and technologies available for type design have improved and designers have to evolve and adapt their process of thinking in accordance. Generative Adversarial Networks (GANs) have revealed impressive advances, presenting high-resolution images nearly indistinguishable from the real ones. In the typographic field, they are helpful when one wishes to obtain coherent glyphs in a typeface. When designing a typeface, one has to simultaneously seek an aesthetically appealing result and coherence among the different glyphs. This can be facilitated by exploring the similarities between the same letter present across diverse fonts, and the transferred stylistic elements within the same font [5]. Balashova et al. [2] develop a stroke-based geometric model for glyphs, a fitting procedure to re-parametrise arbitrary fonts to capture these correlations. The framework uses a manifold learning technique that allows for interactively improving the fit quality and interpolating, adding or removing stylistic elements in existing fonts. Campbell and Kautz [3] develop a similar contour-based framework allowing the editing of a glyph and the propagation of stylistic elements across the entire alphabet. Phan et al. [19] and Suveeranont and Igarashi [26] present two different frameworks that give one or more outline-based glyphs of several characters as input, producing a complete typeface that bears a similar style to the inputs. Rehling and Hofstadter [21] use one or more grid-based lowercase letters to generate the rest of the Roman alphabet, creating glyphs that share different style features. Azadi et al. [1] develop an end-to-end stacked conditional GAN model to generate a set of highly-stylised glyph images following a consistent style from very few examples.

We can also imitate the behaviour of a variable font using Recurrent Neural Networks (RNNs) and interpolate to obtain intermediate results. Lopes et al. [14] model the drawing process of fonts by building sequential generative models of vector graphics. Their model provides a scale-invariant representation of imagery. The latent representation may be systematically exploited to achieve style propagation. Shamir and Rappoport [24] present a parametric feature-based font design approach. The development of a visual design system and the use of constraints for preserving the designer's intentions create a more natural environment in which high-level parametric behaviours can be defined. By changing the glyph parameters they create several family instances. Also, outside the typographic field, there are some good examples exploring the latent space. Sketch-RNN [7] is an RNN able to construct stroke-based drawings. The network produces sketches of common objects in a vector format and explores the latent space interpolation of various vector images. There is also increased attention to these networks and their application to facilitate the use and combination of fonts. A usual way to combine different fonts is by using fonts from the same family or created by the same designer. Another way is to find fonts that match x-height and ascenders/descenders. Fontjoy [20] is another tool to facilitate the process of mixing and matching typefaces and choosing fonts to use side by side. FontMap [8] and Font-VAE [10] are tools developed with the goal of discovering alternative fonts with the same aesthetics.

3   Approach

In this section, we present the developed model that generates new letter skeletons by interpolating existing ones. This process allows us to control the style of the resulting font by navigating the latent space. We explain all the steps taken, from the data collection and editing, passing through the development of the network architecture until the experimentation and analysis of the results.

3.1   Data

One of the most important aspects of our approach is the collection and preprocessing of the dataset. We compile a collection of fonts in TTF font format with different weights from Google Fonts [6]. This dataset is composed of five different font styles, Serif, Sans Serif, Display, Handwriting and Monospace. We opted not to use handwriting and display fonts because they were largely distinct from the rest, which is not desirable for our approach. Their ornamental component, sometimes not even filled, complicates the extraction of a representative skeleton. We only worked with 26 characters (A-Z) of the Latin alphabet in their capital format. We believed that, as a work in progress, it would be best to create a dataset with a few characters. By just using capital letters, we are reducing the complexity of the approach.

After selecting the fonts, we remained with 2623 TTF files. Then, we use the library Skelefont [16][1] to extract the skeleton of a font file. It applies the Zhang-Suen Thinning Algorithm [29] to derive the structural lines of a binary image. This library also allows the extraction of the points of the skeletons as well as the connections between them. It can also calculate the distance between the points and their closest borderline pixel, returning the stroke width of the original glyph at each of these points.

For each font, we rasterise the vectors that compose the skeleton of each glyph into a 64 × 64 px black and white image. We also save all points' positions

---

[1] <https://github.com/tiagofmartins/skelefont>.

and stroke width of the original glyph in a file to use later to generate the filling of the glyphs. Then, we repeat the process for the 26 letters of the alphabet (capital letters of the Latin alphabet only). This process is shown in the first three images of the diagram of Fig. 2.

3.2   Network Architecture

The proposed model consists of a Conditional Conditional Variational Autoencoder (VAE) [11] and an Autoregressive sketch decoder. We used a VAE instead of a regular AE to allow us to manipulate the latent vectors more easily. The output of the VAE are parameters of distribution instead of vectors in the latent space. Moreover, the VAE imposes a constraint on this latent distribution forcing it to be a normal distribution which makes sure that the latent space is regularised. Therefore, we can create smoother transitions between different fonts when we sample the latent space moving from one cluster to the other. The Conditional part of the model allows us to input which letter we are encoding and decoding allowing us to manipulate better which letter we are creating. Finally, as all the letters share the same latent space we can also explore the skeletons between different letters.

Figure 2 shows a diagram of the architecture used. In summary, the encoder employs a Convolutional Neural Network (CNN) that processes the greyscale images and encodes them into two 64-D latent vectors which consist of a set of means (μ) and standard deviations (σ) of a Gaussian representation. Through experimentation, we found that size 64 for the latent code presents the best results for our approach as it is a good trade-off, allowing us to compress all the characteristics of the letter while keeping its tractability. Then, using the mean and standard deviation we take a sample from the Gaussian representation z to be used as input for both decoders, the image decoder and the sketch decoder. The image decoder consists of a set of convolutional transpose layers that receive the z vector and decodes it into a greyscale image which is compared with the original input. The sketch decoder consists of an LSTM [9] with dropout [23, 25] that transforms the z vector into a sequence of 30 points creating a single continuous path. This path is rasterised using a differentiable vector graphics library [13] to produce an output image. This library allows converting vector data to a raster representation while facilitating backpropagation between the two domains. In the rasterisation process, we take the sequence of 30 x and y values and transform them to canvas coordinates. Then, we create a line that connects all points following the same order they are returned from the sketch decoder. The width of this path needs to be carefully selected to match the width of the original skeleton. If the width of the path is thinner than in the original images, at some part of the training process, the network stops trying to compose the whole letter and starts to fill the width of the letter in a zig-zag manner. However, if the line is thicker than in the original images we lose detail in the final skeleton.

Finally, we render the produced path in a canvas as greyscale image that is compared with the original image. Although the standard VAE works at the

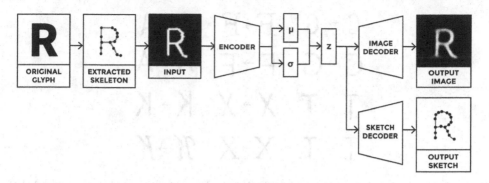

Fig. 2. Diagram of the architecture of our approach.

pixel level, the output of our sketch decoder is a sequence of points, thus allowing the generation of scalable vector graphics that allow easier manipulation of the generated skeletons without losing quality. The loss value is calculated in a similar way as in the standard VAEs. We calculate the Binary Cross Entropy between the output images of the image decoder and the original inputs. We also calculate the Kullback-Leibler Divergence [12] to allow a regularised distribution of the latent space. Finally, we compute the Binary Cross Entropy between the original inputs and the output of the sketch decoder. To obtain the final loss value we add the three values together.

4   Results

The VAE and sketch decoder trained for 50 epochs with a learning rate of 0.001 and a batch size of 256. As mentioned before, we use 2623 64×64 px black and white images of skeletons for each capital letter of the Latin alphabet, so our dataset is constituted of 68 198 images.

4.1   Reconstruction of Skeletons

As mentioned before, the model returns a sequence of points that, when connected, create a reconstruction of the skeleton image used as input. In most cases, the generated strokes reconstruct the basic features of the skeleton. For example, in the case of the letter "A", the network first creates one stem, then the crossbar connects both stems, and finally draws the second stem. Even though there is nothing to control the distance between points or to enforce them to be close, the network learns that it needs to connect both stems at the beginning and the end of the sequence. Another interesting feature observable in the reconstruction is related to how the ANN handles the letter "T". This letter presents one of the simplest skeletons of the alphabet, so the network can learn how to generate the whole structure of the letter very quickly in comparison with others.

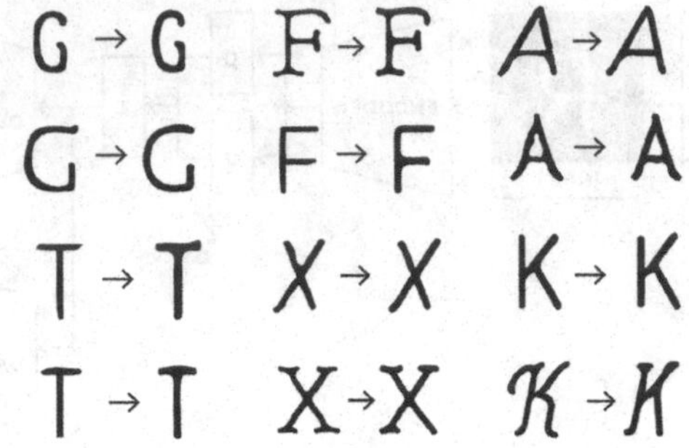

G → G   F → F   A → A
G → G   F → F   A → A
T → T   X → X   K → K
T → T   X → X   𝒦 → 𝒦

Fig. 3. Comparison between the originals (left) and the reconstructed skeletons (right).

Figure 3 presents a comparison between the original inputs and the reconstructed skeletons using a single stroke. The reconstructions of "G", "T" or "A", for example, are very similar. The letters "A", "X" and "K" present a more complex challenge to the network as it needs to create a path that overlaps itself to draw the whole letter structure with only one line. Sometimes, the serif is lost in the reconstruction due to the same issue. The line must overlap itself multiple times to create the small parts without messing with the overall structure of the letter. But the other reason for this could be that the number of letters with serif is lower than the number of letters without it.

In summary, even though the small details of the letters might be lost, our network is able to create the minimal structure of the letter, generating skeletons that cannot be confused with any other letter.

4.2   Latent Representation of Font Style

To understand if the trained model can learn a latent representation for the different letters that is smooth and interpretable, we need to visualise the 64-dimensional z vectors for the dataset. So we take all the images of the dataset (68198 images) and encode them using our network. Then, using the means and standard deviations of each encoded image we took a sample from the distribution. Finally, we took all the z vectors and reduced their dimensionality using the t-SNE algorithm [27]. This allows us to reduce the z vectors from a size of 64 to two dimensions which can be translated to positions in a two-dimensional domain. For each position of a two-dimensional grid, we place the image of the best candidate. We select this candidate by finding the two-dimensional encoding closest to that position. Figure 4 presents the visualisation of the results. In general, the model can separate the different letters into clusters. In some cases, it is also possible to observe that similar letters are placed near each other, for example in the case of the letters "B", "R" and "P". These three letters present similar anatomical characteristics, they share a top bowl and they all have a vertical stem, thus they are placed near each other. The same happens for the letters "T" and "I" which are placed more separated from the rest but near each other. Even though the majority of the skeletons for the letter "I" is represented with a single stem, in some cases, when they have serif, they are similar to the letter "T" but with a cross stroke on the top and bottom part of the letter. This leads to both letters having a strong similarity between each other, therefore they are placed together in the latent space.

We also create a similar representation contemplating the skeleton images of a single letter (2623 images). To understand if the trained model was able to smoothly change styles within the same letter we created a similar visualisation as in Fig. 4. Figure 5 presents the visualisation of the results for the letter "R". As it is possible to observe, the model is able to separate the different font weights across the latent space, creating different regions. The zoom-in boxes show four separate locations where we notice a concentration of specific font styles. In (A) it is presented a region where the condensed fonts are, while the opposite corner (D) represents the most extended fonts. It is also possible to observe that (B) represents the italic, and finally (C) presents most of the fonts with serifs. Local changes within these regions are also visible, where the font width increases when distancing from the region (A) and approximating to the region (D). It is also possible to observe a slight increase in the font height in the top-bottom direction.

4.3   Exploring the Latent Space

After analysing whether the latent space translates font characteristics for meaningful latent representation, we explore linear interpolations between pairs of skeletons for a given glyph. First, we encode two randomly selected fonts from the dataset into their corresponding z vectors. Then, we perform a linear interpolation between the two vectors and, using the trained sketch decoder, we reconstruct the skeletons for these vectors. Figure 6 shows some results of this exploration. The first and last glyph of each row are the original skeletons, and in the middle are the interpolations between them two. The interpolation percentage starts at 0% and ends at 100%, which means that the second skeleton is a reconstruction of the glyph on the left side, and the penultimate skeleton is a reconstruction of the glyph on the right.

The results show that the model is not only able to decode meaningful skeletons but it is also able to control several characteristics of it. In the example of the letter "N", not only the model can control the width of the letter, but it also controls its height.

As it is possible to observe in the interpolations presented in Fig. 6, not only the model is able to decode meaningful skeletons but it is able to control several characteristics of it.[2] In the example of the letter "H", the width of the letter is slightly changed until it matches the width of each skeleton input image. In the case of the letter "N", not only the model is able to control the width of the letter, but it also controls its height. At the same time that the width of the

---

[2] An example video containing multiple skeleton interpolations can be seen at https://imgur.com/a/qf1m2Da.

Fig. 4. t-SNE visualisation of the learned latent space z for all the capital letters of the Latin alphabet.

letter changes, its height is also modified to match its parents, which allows wider control over the skeleton that can be created. In the case of the letter "T", it is possible to observe that the model can also control how much the letter is italic. As we go from the left input skeleton image to the right, the stem of the letter gets closer to a vertical position. This not only shows that the model is capable of perceiving different angles but it can also transition between them gradually. Therefore, we might be able to control all these stylisations of the skeletons by navigating the latent space. This can be observed in the visualisation shown in Fig. 5. There are certain regions dedicated to different letter styles. So, we can navigate this space in order to create fonts that demonstrate a set of desired styles.

We also interpolate between skeletons of different letters. By observing the resulting skeletons present in Fig. 7, we observe that the model is able to pass from one skeleton to another from different letters. Sometimes the morphings are not even expected to be smooth, because some letters have anatomical parts completely different, like for instance the "Z" and "T". The generated skeleton

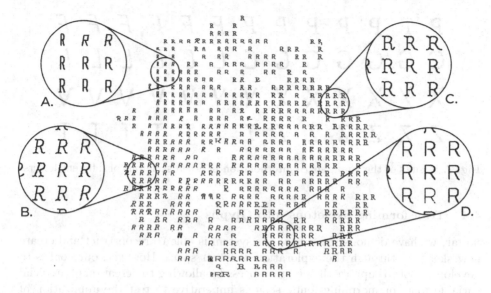

Fig. 5. t-SNE visualisation of the learned latent space z for a single letter.

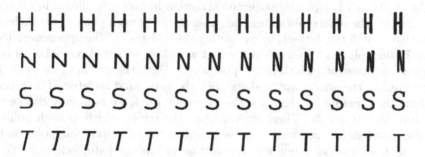

Fig. 6. Results of the latent space interpolation between different skeletons of the same letter. An example video of multiple interpolations can be found at https://imgur.com/a/qf1m2Da.

starts as "Z" but over time it loses its bottom cross stroke. Moreover, its diagonal stroke slightly changes its angle and transforms itself into the stem of a "T". There are also other transformations that are expected, such as the case of "P" and "F", which share a stem. Over the line, the generated skeleton goes opening its bowl to create the arms of the "F" and at the same time slightly inclines the stem to create an italic glyph according to the inclination of the "F". Another information that we can obtain is that sometimes we start to visualise intermediate skeletons that look like other existing letter's skeletons. For example, when we explore the latent space between "G" and "L" in some intermediary steps we can observe some resemblance with the letter "C".

Fig. 7. Results of the latent space interpolation between skeletons of different letters.

4.4 Transforming Skeletons into Glyphs

So far, we have demonstrated how our system is able to reconstruct and create new skeletons through the exploration of latent space. However, our goal is to develop a tool to support the design process by allowing the creation of artificial variable fonts or morphing fonts, so it is imperative to test the application of the generated skeletons.

As mentioned before, the skeleton extraction library [16] allows, in addition to extracting the points, obtaining the stroke width at each point of the skeleton. When we created the dataset, by extracting the skeletons of the uppercase letters of the Latin alphabet for each font file that we select, we saved the points of each skeleton and its stroke width to use posteriorly. With these values, we were able to interpolate the stroke width along with the generated skeleton. The process of filling the generated skeletons is the following. First, we randomly choose two skeletons to interpolate. Then, we calculate the stroke width at each point of the generated skeletons. To do this, we calculate the corresponding point on the skeletons that serve as input for the creation of intermediate skeletons. We do this calculation by overlapping the input skeletons and the generated skeleton and calculating the closest match. The stroke width at each point is a result of combining the interpolation of the widths of the input skeletons. Figure 8 shows some results in which each row represents a different interpolation. Looking at the generated glyphs, we can see that they look similar to a regular font. With a few adjustments, we could use them as a variable font. Now, with interpolated fill, the contrast between variations is more visible, because we had another parameter to the glyph design. By splitting the skeleton and the filler we have more visual possibilities because we are not stuck with a filler. In these tests, we use filling in the original fonts to fill in the intermediate ones, but it is not mandatory. We can even use some fonts to create the skeleton and others to create the filling or even use a fixed value along the skeleton. By applying the filling, the interpolated glyphs become more unique, by suffering more alterations when moving between the two input glyphs. For example, in the "S" (Fig. 8) we can observe that besides the axis alteration, the glyphs also change in contrast. The generated "S" near the left is styled more like a modern font, with high contrast and serifs. From left to right the contrast inside the generated glyphs turns almost nil and they lost the serifs.

D D D D D D D D D
N N N N N N N N N
S S S S S S S S S
R R R R R R R R R

Fig. 8. Results of the latent space interpolation filling the skeleton with an interpolated stroke width.

Fig. 9. First example of application of the generated skeletons into glyphs to create a typographic identity. The glyphs present in the images are composed of the two input glyphs, in red and light blue, and the interpolated glyph, in dark blue. An example video of multiple interpolations can be found at https://imgur.com/3XTecg5 (Color figure online).

As mentioned before, our system provides a tool to facilitate the process of building these dynamic identities with a typographic component. With this tool, designers can generate skeletons and develop a filling to create their versions of glyphs. To demonstrate the application of our system we made a series of experimentations with different ways of using the obtained skeletons by our model (see Figs. 9 and 10).

In the first application (Fig. 9), we present the interpolation[3] between two input glyphs. The input glyphs are represented in red and light blue while the generated one is in dark blue. To visualise the three superimposed glyphs, we apply the multiply effect, thus obtaining another colour that represents the common parts between the generated and the original ones. The generated glyphs are very diverse on a visual level, enabling the design of a dynamic visual identity with the use of only two fonts. We believe that the mutating factor of these results provides an identity that is easily placed side by side with the dynamic visual identities and variable fonts that are made these days. In the second application (Fig. 10), the generated glyphs use just the interpolated skeletons. The stroke width is also calculated based on the input glyphs. However, the filling is

Fig. 10. Second example of application of the generated skeletons into glyphs to create a typographic identity. The glyphs present in the image are the result of the interpolation of two input skeleton glyphs.

---

[3] A video showing multiple skeleton and stroke width interpolations can be seen at https://imgur.com/3XTecg5.

further away from the traditional typographic visual aspect. Along the skeleton line, we draw a series of crosswise line segments to define the width of the glyph' stroke. The density changes to accommodate the same number of line segments between each pair of points.

5   Conclusion and Discussion

Since its emergence, type design has been adapting to technological advances. Nowadays, most typefaces are developed by type designers, who studied the design and anatomy of each character with great precision. Type design is a difficult and time-consuming process. Our approach takes advantage of the knowledge present in a design of a typeface and the computational possibilities that ANNs provide. We propose a VAE combined with an Autoregressive model to generate glyphs' skeletons by interpolating existing ones. Our contributions are the following, a sketch decoder capable of *(i)* reconstructing images of glyphs' skeletons using a single stroke, *(ii)* controlling font styles by navigating the latent space, *(iii)* interpolating between two skeletons to create new ones. By creating interpolations between existing fonts we develop a method to help designers in making their artificial variable fonts, easing the usual glyph production. We also explored a feature of a skeleton extraction library, which calculates the stroke width at each point of the letter skeleton, to produce a fill for the generated skeletons. By interpolating between skeletons of different letters we are creating new glyph forms that resemble other existing glyphs. This opens up new exploration possibilities for the future. We envision that our approach can find use as a tool for graphic designers to facilitate font design. We can employ this system to generate new skeletons, which the designer can fill with the desired style, but also be used as inspiration seed to create new glyphs.

We expect to make several future contributions. First, we want to change the architecture of the sketch decoder to be able to use multiple strokes. In some cases, our approach was able to draw skeleton letters that require more than one line by overlapping them. However, if the sketch decoder had access to multiple strokes, this problem could be solved more easily. Finally, we intend to change the input of the network so it can receive a vector version of the skeletons instead of a pixel-based image. This way we can work with an end-to-end architecture focused on vector format leading to better quality skeletons without any loss of information.

Acknowledgments. This work is partially funded by national funds through the FCT - Foundation for Science and Technology, I.P., within the scope of the project CISUC - UID/CEC/00326/2020 and by European Social Fund, through the Regional Operational Program Centro 2020, and under the grant SFRH/BD/148706/2019.

References

1. Azadi, S., Fisher, M., Kim, V.G., Wang, Z., Shechtman, E., Darrell, T.: Multi-Content GAN for Few-Shot Font Style Transfer. CoRR, abs/1712.00516 (2017)
2. Balashova, E., Bermano, A.H., Kim, V.G., DiVerdi, S., Hertzmann, A., Funkhouser, T.A.: Learning a stroke-based representation for fonts. Comput. Graph. Forum **38**(1), 429–442 (2019)
3. Campbell, N.D.F., Kautz, J.: Learning a manifold of fonts. ACM Trans. Graph. **33**(4), Jul 2014. ISSN 0730–0301. https://doi.org/10.1145/2601097.2601212
4. Cheng, K.: Designing type. Yale University Press (2020)
5. Cunha, J.M., Martins, T., Martins, P., Bicker, J., Machado, P.: Typeadviser: a type design aiding-tool. In: C3GI@ ESSLLI (2016)
6. Google. Google Web Fonts (2012). http://www.google.com/webfonts/v2/. visited 2022-01-02
7. Ha, D., Eck, D.: A neural representation of sketch drawings. In: ICLR (2018). https://openreview.net/forum?id=Hy6GHpkCW
8. Ho, K.: Organizing the World of Fonts with AI (2017). https://medium.com/ideo-stories/organizing-the-world-of-fonts-with-ai-7d9e49ff2b25, visited 03/01/2022
9. Hochreiter, S., Schmidhuber, J.: Long short-term memory. Neural Comput. **9**(8), 1735–1780 (1997)
10. Hong, S.: Font-VAE (2019). https://github.com/hngskj/Font-VAE, visited 2022-01-02
11. Kingma, D.P., Welling, M.: Auto-encoding variational bayes. In: 2nd International Conference on Learning Representations, ICLR 2014, Banff, AB, Canada, April 14–16, 2014, Conference Track Proceedings (2014)
12. Kullback, S., Leibler, R.A.: On information and sufficiency. Ann. Math. Statist. **22**(1), 79–86 (1951)
13. Li, T.-M., Lukáč, M., Gharbi, M., Ragan-Kelley, J.: Differentiable vector graphics rasterization for editing and learning. ACM Trans. Graph. (TOG) **39**(6), 1–15 (2020)
14. Lopes, R.G., Ha, D., Eck, D., Shlens, J.: A learned representation for scalable vector graphics. In: DGS@ICLR. OpenReview.net (2019)
15. Martins, T., Correia, J., Costa, E., Machado, P.: Evotype: evolutionary type design. In: Johnson, C., Carballal, A., Correia, J. (eds.) EvoMUSART 2015. LNCS, vol. 9027, pp. 136–147. Springer, Cham (2015). https://doi.org/10.1007/978-3-319-16498-4_13
16. Martins, T., Parente, J., Bicker, J.: Skelefont (2018). https://github.com/tiagofmartins/skelefont, visited 2022-02-01
17. Martins, T., Cunha, J.M., Bicker, J., Machado, P.: Dynamic visual identities: from a survey of the state-of-the-art to a model of features and mechanisms. Visible Lang. **53**(2) (2019)
18. McCormack, J.P., Dorin, A., Christopher, T.: Innocent. Generative design: a paradigm for design research. In: Redmond, J., Durling, D., de Bono, A. (eds.) Futureground, vol. 2, Monash University, 2005. ISBN 0975606050
19. Phan, Q.H., Fu, H., Chan, A.B.: FlexyFont: learning transferring rules for flexible typeface synthesis. Comput. Graph. Forum **34**(7), 245–256 (2015)
20. Qiao, J.: Fontjoy - Generate font pairings in one click. http://fontjoy.com/, visited 2022-01-02
21. Rehling, J., Hofstadter, D.: Letter spirit: a model of visual creativity. In: ICCM, pp. 249–254 (2004)

22. Schmitz, M.: genoTyp, an experiment about genetic typography. In: Proceedings of Generative Art 2004 (2004)
23. Semeniuta, S., Severyn, A., Barth, E.: Recurrent dropout without memory loss. In: Calzolari, N., Matsumoto, Y., Prasad, R. (eds.) COLING, pp. 1757–1766. ACL (2016). ISBN 978-4-87974-702-0
24. Shamir, A., Rappoport, A.: Feature-based design of fonts using constraints. In: Hersch, R.D., André, J., Brown, H. (eds.) EP/RIDT -1998. LNCS, vol. 1375, pp. 93–108. Springer, Heidelberg (1998). https://doi.org/10.1007/BFb0053265
25. Srivastava, N., Hinton, G.E., Krizhevsky, A., Sutskever, I., Salakhutdinov, R.: Dropout: a simple way to prevent neural networks from overfitting. J. Mach. Learn. Res. 15(1), 1929–1958 (2014)
26. Suveeranont, R., Igarashi, T.: Example-based automatic font generation. In: Taylor, R., Boulanger, P., Krüger, A., Olivier, P. (eds.) SG 2010. LNCS, vol. 6133, pp. 127–138. Springer, Heidelberg (2010). https://doi.org/10.1007/978-3-642-13544-6_12
27. van der Maaten, L., Hinton, G.: Visualizing data using t-SNE. J. Mach. Learn. Res. 9, 2579–2605 (2008)
28. Willen, B., Strals, N.: Lettering & type: creating letters and designing typefaces. Princeton Architectural Press (2009)
29. Zhang, T.Y., Suen, C.Y.: A fast parallel algorithm for thinning digital patterns. Commun. ACM 27(3), 236–239 (1984)

Visual Representation of the Internet Consumption in the European Union

Telma Rodrigues⬤, Catarina Maçãs(✉)⬤, and Ana Rodrigues⬤

Centre for Informatics and Systems of the University of Coimbra,
Department of Informatics Engineering, University of Coimbra, Coimbra, Portugal
trodrigues@student.dei.uc.pt, {cmacas,anatr}@dei.uc.pt

Abstract. The impact of internet usage on the environment is a contradictory topic. While it can help reduce carbon emissions, with smart grids or the automation of services and resources, it can also increase e-waste that end up affecting the environment. To draw attention to the impact of energy consumption on the environment, we proposed and developed a computational artifact that unites the areas of Data Aesthetics and Interaction Design. The artifact, displayed in an interactive installation, was divided into three panels: (i) the left panel, which represents the countries—from the European Union (EU)—with the lowest energy consumption impact on the environment; (ii) the central panel, which use swarming boids to represent the internet usage at the installation site and its impact; and (iii) the right panel, which represents the EU countries with the highest energy impact on the environment. The arrangement of the three panels in a single interactive installation aims to establish a visual connection between the energy consumption in the EU and the energy consumption in the installation's site and to promote awareness of its impact on the environment.

Keywords: Data Aesthetics · Swarming System · Environment · internet usage · Awareness

1 Introduction

The impact of the increasing use of internet resources—and consequent increase in energy consumption—on the environment can be seen in two ways, through its positive and negative impacts [1]. On the positive side, there was an improvement of energy efficiency, with the optimization of energy resources and the use of renewable energies, and a reduction of unnecessary travel, through e-commerce and remote work. On the negative side, there was an increase of energy consumption and pollution related to the production of electronic infrastructures and devices.

This work is funded by the FCT - Foundation for Science and Technology, I.P./MCTES through national funds (PIDDAC), within the scope of CISUC R&D Unit - UIDB/00326/2020.

In recent years, there has been a growing interest in exploring data representations of environmental problems. Climate change harms the entire population and therefore there have been several ways of alerting, raising awareness, and emphasizing information on this topic [2,5,12]. Nonetheless, there is still data that needs to be correlated, explored, and represented, such as energy consumption, internet use, and their impact on the environment. To communicate such data in a more aesthetic way, one can resort to Data Aesthetics. Data Aesthetics is a sub-area of Information Visualization and its main focus is the visual exploration of data rather than functionality [6,7,9]. In this context, data is represented to draw the public's attention and communicate a message, regardless of its readability. However, and due to the effectiveness of visual mappings, in many cases they can still be understood by the viewer [6].

To promote awareness on the impacts of energy consumption and internet usage but also to highlight the adoption of positive measures in recent years by EU countries [4], we propose a computational artifact that combines Data Aesthetics and Interaction Design. Our artifact represents both the impacts of energy consumption and internet usage on the environment, such as the increase in greenhouse gas (ghg) and fossil fuels, and the application of positive actions, such as the increasing adoption of renewable energy by EU countries. More specifically, we visualize data regarding internet usage, energy consumption, energy sources, and impact on the environment for every EU country.

This project aims to analyze and visually represent the aforementioned data in an aesthetic way as well as to explore visual metaphors that highlight the impact of internet usage at the installation site. Through the visual representation of the energy consumption in the EU countries—left and right panels of the installation—our intent is to make the viewers aware of the high consumption values in different countries and the measures that each country is applying to reduce the impacts on the environment. Through the visual artifact based on swarming boids—central panel, which represents the impact of internet users logged in the installation site in multiple perspectives—we intend to draw the viewers' attention and make them conscious of the impacts of internet usage.

The remainder of this article is organized as follows. Section 2, presents other projects that use Data Aesthetics and Information Visualization to represent the energy's consumption impact on the environment. Section 3, presents the left and right panels—which show the countries with the lowest and highest impacts of energy consumption on the environment—the central panel—representing the ecosystem of internet networks—and the interaction mechanisms. The third Sect. (4) presents the user tests regarding interaction and visual perception. In Sect. 5, we discuss the results obtained and possible improvements. Finally, in Sect. 6, we present the conclusions and future work.

2   Related Work

The related work was chosen due to its relation to the theme and application of Information Visualization and Data Aesthetics.

"Earth Bits—Sensing the Planetary" is an installation developed by Dot-dotdot that represents the carbon footprint of humanity [2]. The installation is divided into sub-projects, each one visualizing different data. Below, we present the sub-projects more related to ours.

The subproject "Power Rings—Energy Consumption in Portugal (2019–2020)" consists of a visualization of the changing patterns of daily electricity consumption in Portugal, between the years 2019 and 2020. In addition, other representations are made concerning specific Portuguese cities, such as Faro, Viana do Castelo, and Porto. The interactive console with multi-user capability, called "CO2 mixer—Identifying Human Impact", measures the individual ecological footprint, according to the users choices in categories such as nutrition, housing and mobility. "CO2 mixer" also features a sonification that reflects the measured data. Finally, the sub-project "Planet Calls—Imagining Climate Change", offers, through images and satellite data, a historical correlation between emissions and the increase of occurrences of environmental phenomena on Earth.

The "7000 Oaks and Counting" project, by Tiffany Holmes, reinterprets ecological data through new technological and artistic means with the aim of educating and changing consumer behaviors [5]. This visualization aims to explore the individual public commitment to carbon footprint reduction and consists of an animated sequence of images of trees representing carbon emissions. The data was collected from the NCSA building and refers to steam, chilled water, electricity flow, among others. Then, the carbon footprint is calculated in real time and is converted into the number of trees that would be needed to offset the carbon emitted. This project uses metaphor to facilitate the understanding of the concept and create a relationship between the viewer and the data.

"Waves to waves to waves" is an interactive audiovisual installation that uses visualization and sonification to represent the electromagnetic energy generated by humans that is imperceptible to the human senses [12]. This project aims to reflect on the growing dependency with technology. The changes detected in the environment by Wi-Fi devices, television, and mobile phones are converted into electrical signals that generate sounds and abstract shapes in real time.

The related works were chosen due to their focus on the impacts of energy and pollution on the environment and due to the variety of representations. The projects that best relate to the concept of our project is the "Earth Bits-Sensing the Planetary", specifically the subprojects "Power Rings-Energy Consumption in Portugal", that addresses the changing patterns of electricity consumption in Portugal, and the sub-project "Power Rings-Energy Consumption in Portugal", that represents the energy consumption in different cities. The project "Planet Calls-Imagining Climate Change" reveals an historical correlation between CO2 emissions and the increase in the occurrence of environmental phenomena on Earth. The visual metaphor is best applied in "7000 oaks and counting" project, as it creates a direct link between the way data is presented and the topic.

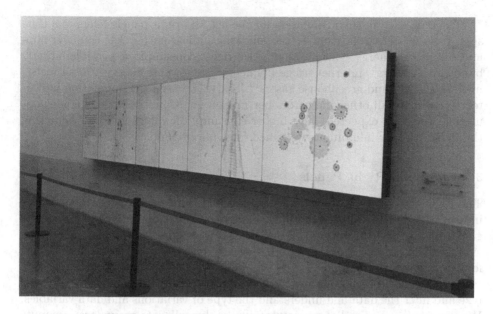

Fig. 1. Picture of the installation on the site.

3   Impacts of Consumption on the Environment

Our project consists of an interactive installation divided into three panels
(Fig. 1). The installation was developed in Processing, which is an open-source
tool developed by artists and designers that uses a simplified language built
on the Java language. We chose this tool as its main focus is the creation of
visual and interactive media. In the first and third panels, left and right sides
of the installation, respectively, we represent the EU countries. This focus is
due to the response of the 27 countries in the fight against climate change and
environmental degradation.

In the second panel, central panel, we visually represent the energy consumption of the EU countries and data from the internet networks at the installation
site through an ecosystem. This ecosystem aims to explore the creation of a
narrative of boids that evolve and change to characterize the EU energy consumption impacts on the environment.

3.1   Data

The first step in developing the installation was to acquire the data through open
source platforms. We selected datasets according to: reliable source and required
time span. The selected open source databases were: Eurostat, Organization for
Economic Co-operation and Development, World Bank, International Telecom-
munication Union, European Environment Agency, and Institute National of
Statistics.

The data used in the installation for the left and right panels refers to a time span from 2007 to 2020, and to the following indicators per EU country: population[1], total energy consumption[2], electricity consumption per inhabitant[3], bandwidth[4], users[5], internet usage frequency[6], renewable energy[7], fossil fuels[8], nuclear energy[9] and greenhouse gasses[10]. Lastly, we used the population indicator to calculate all other indicators per inhabitant. Countries with higher population would, as expected, have higher consumption values and by calculating the indicators per inhabitant, we aimed for a fairer comparison.

3.2   Left and Right Panels

The EU countries were divided into the left and right panels according to their indicators' average values, regarding energy consumption, energy sources, and ghg from energy consumption. We ranked all countries by the aforementioned averages and placed countries with higher ranking on the right panel, and with lower ranking on the left.

To understand the appropriate visual representation, it was necessary to analyze and filter the data and understand the type of variations and data variables. We chose to focus on radial representations, as they allow to group large amounts of information and distinguish pieces of information in a reduced space [3]. Also, this representation format allowed us to visually highlight countries through their size and position in space. The countries are then placed on the panels according to their geographic positions, without overlapping.

To map the data values to the visual variables (e.g., color, size, length), we used a simple normalization between the minimum and maximum values by indicator (regardless of the country or year). This way, we can compare the indicators among countries and over the different years. The visual mappings of each indicator in each country is presented in Fig. 2.

As we aimed to represent all 10 indicators per country and this could lead to a complex and clustered artifact, we decided to create two views on the same data. The first, the overview, aims to give a more general view of the data. The second, the country view, aims to give more detailed view with all the consumption indicators per country.

The overview only includes four indicators: (i) internet users; (ii) frequency of internet use; (iii) internet energy consumption; and (iv) internet ghg. In the overview (Fig. 3), users are represented by the radius of the yellow and blue

---

1   shorturl.at/tzF15.
2   shorturl.at/EIN05.
3   shorturl.at/jopr0.
4   shorturl.at/orAVX.
5   shorturl.at/aCOZ0.
6   shorturl.at/lptz0.
7   shorturl.at/DKOPV.
8   shorturl.at/dNR69.
9   shorturl.at/bILX4.
10   shorturl.at/kmO12.

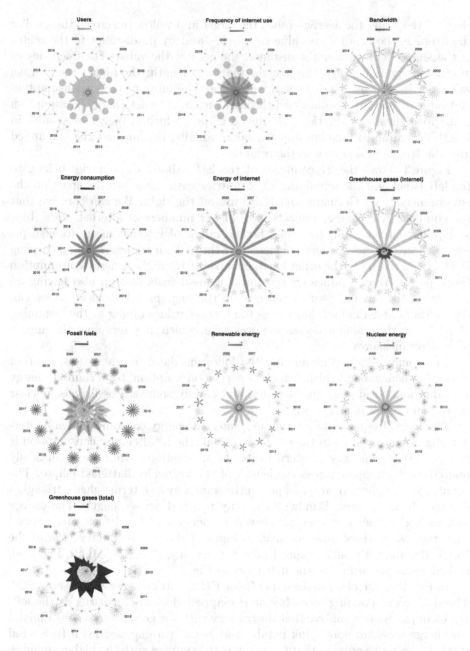

Fig. 2. Visual mapping of left and right panel. Here we highlight the shapes resulting in the visual mapping of each indicator.

circles—the larger the average value, the larger and yellow the circle, the smaller the average, the smaller and blue the circle—and by its distance to the center of the shape—the greater the distance, the greater the value. The frequency of internet usage is mapped through the pie slices length: the higher the average value per inhabitant, the longer the pie slice. The internet power consumption and internet ghg are also mapped through length. Internet energy consumption is mapped through a vertically flattened ellipse, which is bigger or smaller in length according to the consumption value. Finally, the internet ghg is mapped into the triangle's distance to the center.

Figure 3 shows the 27 countries of the EU. About 14 countries belong to the left panel and the remaining EU countries with the greatest impact on the environment (e.g., Germany and Italy) are on the right. We can perceive that the countries with higher impacts have higher numbers of internet users (blue and yellow circles) and, therefore, higher values of internet usage (orange pie slices). Note that both values are increasing throughout the years (i.e., increasing clockwise in size). On the other hand, countries with higher energy consumption (pink petals) that produce energy through fossil fuels end up also having an impact on the environment as reflected in the ghg (pink triangles), as we can see in Malta and Finland. Malta has the highest values of ghg at the beginning of 2007, but this value is decreasing over time, which may represent the impact of positive measures.

The country view contemplates the following data: internet users, internet usage, frequency, bandwidth, internet energy consumption, total country energy consumption, total ghg, internet ghg, and energy sources (fossil fuels, nuclear energy and renewable energy).

The users, frequency of internet use, internet energy consumption, and internet ghg are mapped as in the overview. Then, the total energy consumption is mapped in the same way as internet energy consumption as they are semantically related—it is mapped across the length of the vertically flattened ellipse. The country's total ghgs are also mapped in the same way as internet ghgs—triangle's distance to the center. Bandwidth is only mapped across length. The energy sources such as nuclear energy, renewable energy and fossil fuels are mapped with the size of three different shapes. Figure 4 shows the country view of the Czech Republic, Estonia, Spain, Finland, Germany, Bulgaria, Malta, Portugal, and Slovenia, according to the indicators in Fig. 2.

In Fig. 4 we can observe the data of nine EU countries between 2007 and 2020. The data corresponding to each year is mapped clockwise, starting at the left. For example, we can analyze that the country with the greatest impact is Finland (in energy consumption—pink petals) and Malta (in ghg and fossil fuels—red star). We can also observe that Germany is the country with the highest number of internet users (yellow circles), and Slovenia and Finland the countries with the highest values in terms of nuclear energy (yellow stars). In terms of highest values of renewable energy (green flower), Malta, Slovenia, Finland, and Estonia are at the top.

Impact of energy consumption on the environment

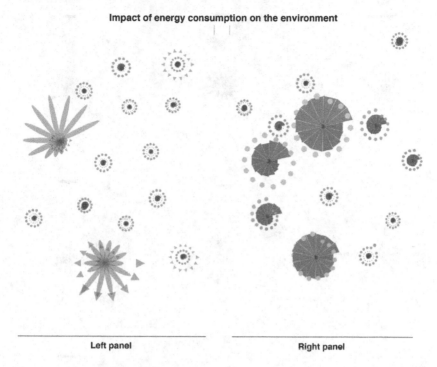

Left panel   Right panel

Fig. 3. Overview of the left and right panel. On the left, countries with lowest impacts on the environment. On the right, countries with highest impacts on the environment.

3.3   Central Panel

The main goal of the visual metaphor for the central panel is to translate the impact of energy consumption represented on the left and right panels, into an aesthetic experience.

In the central panel, we use an ecosystem of swarming boids—also seen in other projects [8,10,13]—that behave as an interpretation of the data represented in the left and right panels. In our representation, the higher the values of energy consumption in the EU countries, the higher the noise and the more cumulative the ecosystem becomes. Our intent is to create awareness on our consumption behaviors by creating a more complex and visually cluttered artifact.

The very unpredictability of the visual results of the ecosystem allows space for imagination. The perception of the artifact is subjective and differs according to several factors: (i) experience/knowledge of ecosystems functioning; and (ii) association of color and shape to data. Next, we discuss how the swarming rules are applied, how we define the number of organisms (i.e., boids) in the ecosystem, and how we visually represent them.

The ecosystem is based on the swarm system proposed by Craig Reynold [11] and it represents the number of users of 5 internet networks available at the installation site. The behavior of the boids is based on three fundamental rules:

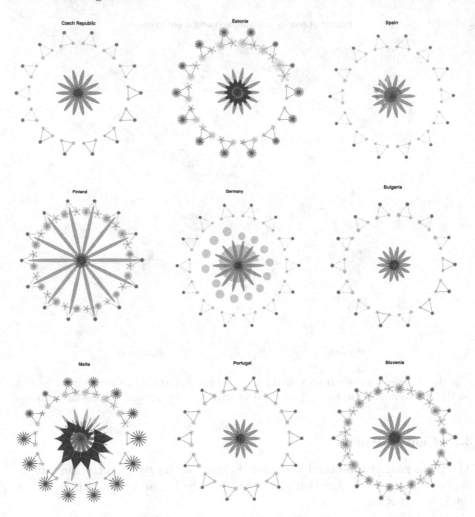

Fig. 4. Country view of different EU countries.

separation, alignment, and cohesion. Each boid represents a user from a certain internet network and swarms together with other boids of the same type.

To enable the viewers to create a visual connection between panels, we change the boids' behaviors—and therefore representations—, according to the left and right panels. When no country is selected, the behaviors are adjusted to the average consumption values of the EU countries, and when a country is selected in the left/right panel, the behaviors are adjusted to the consumption values of the selected country. We chose to use only three indicators: internet usage frequency, internet energy consumption, and internet ghg. This choice is due to the fact that they are represented in both overview and country view of the left and right panel.

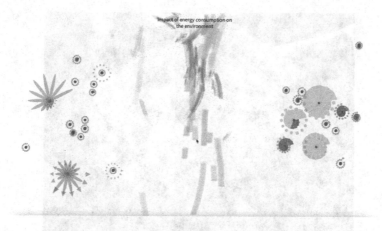

Fig. 5. Central panel with the final visual representation.

To define the changes in the boids' behaviors, we compute the average indicator value of all years and divide it by the average by the maximum value (of the countries and indicator between 2007 and 2020). This final value defines the behavior/representation of each boid. Thus, the behavior/representation of the boid is defined and redefined as the user interacts with the panel and selects different countries or the overview.

Regarding the boids' visual representation, they are colored according to the internet network they represent and their shape and size represent the indicator and the indicator value, respectively. Regarding the shape, internet usage frequency is represented by an arch, internet energy is represented with petals and internet ghg is represented by triangles. The bigger these shapes, the higher the indicator value.

We tested the change of background, the trail to give continuity to the positions of the boids, and addition of particles. However, we opted for a cleaner and more organized ecosystem to facilitate the perception of the different representations for the different countries' values (Fig. 5). The final result of the ecosystem can be seen at the following url: https://drive.google.com/file/d/1hPyZK_z2J-sTNKN6PhiYKZtL8Bt_dmEG/view.

3.4   Interaction

The interaction with the installation was developed with the aid of the Arduino IDE tool and other devices considered necessary to test the approach we intended to follow. We used the Serial library, to allow communication between Processing and Arduino, and used other two Arduino libraries: the EspRotary library and the Button 2 library. The main objective of the interaction is data filtering and therefore we opted for the following solutions: (i) use a rotary encoder to change

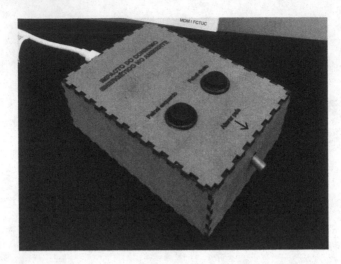

Fig. 6. Box created for the interaction with the left and right panel.

the country selected by the user; and (ii) use on/off buttons to trigger specific events in the system. To interact with the installation the users can use the rotary encoder to change countries and the buttons to change views—press once to show the Country View and press twice to show the Overview.

We created a box that allows interaction in any location of the installation. For the development of the box (Fig. 6), we opted for laser cutting and used laser printing for additional information (i.e., title and identification of the buttons). Finally, all elements were arranged on the box to facilitate the handle and interaction with the box.

4 User Testing

We defined a set of user tests to analyze and understand the participants' interaction with the artifact, detecting main errors, improvements, and suggestions. To carry out the tests, we defined a set of objectives: (i) understand the interpretation of the visual representations; (ii) perceive if the interaction is efficient; (iii) analyze if the colors and shapes aided the visual interpretation of the artifact; (iv) analyze if the exploratory ecosystem of the central panel conveyed the connection between panels, improving the communication with the user; and (v) evaluate whether the artifact raised awareness on energy consumption from internet use.

We performed two tests: an interaction test, to analyze the interaction with the artifact, and a perception test, to analyze the representations' visual impact.

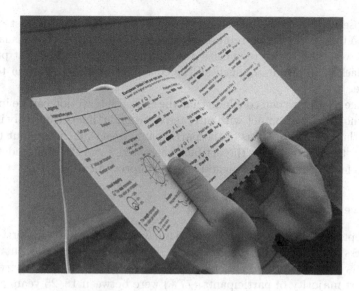

Fig. 7. Leaflet with the visual mappings.

4.1 Interaction Test

The interactive tests were carried out at the Department of Informatics Engineering of the University of Coimbra with six participants. The participants where from distinct areas: two from the area of management and administration, one Master's student in Design and Multimedia, two Master's students in Data Science and a researcher in the field of computer engineering.

All interactive tests followed the same procedure: introduction to the artifact, completion of a set of interactive tasks, and final questionnaire. The tasks covered the left and right panel and aimed to understand the effectiveness of the interaction. Participants interacted spontaneously with the materials provided: box and information leaflet. The leaflet (Fig. 7) contains the description of the visualizations and the project's synopsis. The box provides interaction with the interactive panel.

The participants were asked to perform the following tasks: (i) explore the left and right panel overview; (ii) select a desired country; and (iii) explore the resulting artifact of the selected country.

Some participants felt an initial difficulty in interacting with the artifact due to, for example, not knowing that the leaflet contained information regarding the interaction possibilities.

According to the time needed to accomplish the tasks, we could see that participants with greater knowledge about Information Visualization had a more active participation and made more comments about what could be changed in the panels to be better perceived by the participants (e.g., adding a caption with percentages to highlight relevant data in each country).

The observations given by the participants at the end of the test centered around two topics: the visual representations and the caption. The visual representations were difficult to memorize and, in the caption, the biggest problems were the lack of information, such as the consumption by country in text and percentage.

With the results we can conclude that the chosen interaction can be improved so those who have less contact with interactive visual artifacts can easily interact. However, through the analysis of the participants' reaction throughout the test, we can say that, in general, the artifact generated curiosity and lead them to explore its features and make suggestions for improvement.

4.2   Perception Test

The perception test consists of a questionnaire that aims to understand the effectiveness of the visual representations. The tests were carried out with 31 participants with different experiences and areas of professional activity.

The vast majority of participants (77%) were between 18–25 years old, with the others being between 26 and 41 years of age. Of the 31 participants, 16 (52%) are male and the remaining 15 (48%) are female. The professional area of the participants differs, but three main areas of activity stand out: graphic design (29%), management and administration (16%) and marketing and advertising (16%).

We used google forms[11] to make a quantitative and qualitative analysis of the answers. The tests were not supervised, they were sent to the participants who performed the tests in the environment that was most convenient for them.

The perception test was divided into four sections. In the first section, the project and the questionnaire were contextualized. Subsequently, we present a set of questions related to the participants' profile, to understand their activity, experience, and knowledge in the area under study. In the second section, we first defined single choice questions to understand how the reader perceived the data presented in the visual representations. Then, the questions are related to the artifact problems, ecosystem functioning, and data insight. In the third section, the questions were aimed at analyzing the participant's perception of the experience with the artifact through first person premises and selection responses. Finally, a visual and global appreciation of the artifact was requested.

According to the answers from the perception test, the best way to raise community awareness of environmental problems is through dissemination in the media (38%), exposure of data on consumption and ecological footprint (25%), or through environmental manifestations (16%).

We could perceive that the academic training of the participants had some impact on the reading of the artifact, and the participants with greater technological knowledge showed less difficulties. This was also verified during the interaction test in which a Data Science participant (with previous knowledge in the area of Data Visualization) was able to easily understand how the data were

---

[11] https://forms.gle/EGfCQ3ooQjJaMGcGA.

mapped and mentioned intervals of data values. However, the artifact managed to catch the attention of all participants and arouse curiosity, even without prior knowledge in the area of Data Visualization.

In the central panel, we analyzed how the swarm system was perceived by the user, in which we obtain the following results: 61% of the participants were able to distinguish the networks by colors; 58% of the participants were able to associate the shape of the boids with the data and 67% of the participants were able to understand the developed system.

Most participants gave suggestions for improvement, regarding the way the caption was presented and the way the data was mapped into visual representations. The suggested changes in the perception test focus on the following: remove overlaying shapes, addition of detail and information, and improve the relationship between panels.

5   Discussion

In the left and right panels, we used color, shape, and size which, together, created higher visual emphasis on the countries' indicators. In this way, we intended to make the users aware of the issue of environmental impact and the influence of energy and the internet on the environment. Through color, we aimed to distinguish different values. For example, high values of energy or internet consumption, greenhouse gases, and fossil fuels are colored with reddish colors and low values are colored with bluish colors. In the other hand, high values of renewable energy and nuclear energy are colored in green and yellow, respectively. These colors were chosen due to their semantic connotations: green is can be associated to nature, red can be associated to an alert state, yellow can be associated to energy, and blue to calmness. Also, by using size the aim was to better distinguish indicators with higher values from the ones with lower values.

With the results obtained from the user tests, we noticed that the participants had some difficulties in understanding the visual representations of the indicators. This may be due to the high number of indicators represented and to the caption not being fully explicit and objective. Additionally, the high number of indicators also produced other constraints such as the memorization of each representation and overlapping figures. However, much of the representations were perceived by the participants, as well as its main message. The majority of indicators accurately perceived by the participants were internet related.

In the central panel, the visual metaphor involves the symbolic use of spatially arranged boids that metaphorically express the concept (impact of internet usage on the environment) through visual transformations that occur according to the selections of the left or right panels. When boids represent a greater value of consumption, more boids are added, creating a cumulative impact of consequences on the environment. The ecosystem metaphor suggests the relationship between the use of internet networks (central panel) and the energy consumption impacts of the EU (left and right panel), through the demonstration of the consequences in the way boids are represented in the ecosystem.

We believe that to speed up the reading of the artifact for different people with different knowledge and experiences, we need, for example, to create an artifact that incorporates contextual indications to the visual representations. These indications can be textual or visual (e.g., the display of rankings to help people contextualize a country in relation to the EU).

6   Conclusion

The main focus of this project was the analysis and visual representation of the impact of energy consumption on the environment. The first phase of the work process was directed towards the identification of the problem and the definition of target audience and visual metaphor. The second phase focused on the development and implementation in Processing and Arduino. Finally, in the last phase of the development process, user tests were carried out to understand how the participants interact with the installation (interaction test) and how they interpret the visual representations (perception test).

The conclusions drawn from the results of the tests clarify that there is room to improve the artifact, especially regarding the information leaflet and the visual representations. Nonetheless, participants considered the project important for the community and mostly appreciated the visual representations. The main improvements in our project are: (i) use of the most current data, preferably up to the current year; (ii) improve the relationship between the interactive panel and the information leaflet; (iii) add rankings and informative captions on the panel; and (iv) improve the box handling (i.e., wireless).

As a future work, we intend to create a personalized experience for each user, through the calculation and analysis of their carbon footprint data in real time. By introducing and analyzing these variables in detail, we aim to help understand the human footprint derived from energy consumption in the environment.

Finally, the development of this project made it possible to recognize the impacts of European Union countries on the environment, energy consumption, and the relationship with internet use. The impact of energy consumption and internet usage on the environment is an open field for new explorations in the area of data visualization due to the relevance of the topic and available data related to energy consumption and the environment.

References

1. Berkhout, F., Hertin, J.: Impacts of information and communication technologies on environmental sustainability: speculations and evidence. Report for the OECD (2001). https://www.oecd.org/sti/inno/1897156.pdf. Accessed 25 Jun 2022
2. DotDotDot: Earth bits, sensing the planetary (2021). https://ext.maat.pt/cinema/earth-bits-sensing-planetary. Accessed 15 Jun 2022
3. Draper, G.M., Livnat, Y., Riesenfeld, R.F.: A survey of radial methods for information visualization. IEEE Trans. Visual Comput. Graphics 15(5), 759–776 (2009). https://doi.org/10.1109/TVCG.2009.23

4. EuropeanCommission: In focus: The digital transformation of our energy system (2021). https://ec.europa.eu/info/news/focus-digital-transformation-our-energy-system-2021-dic-16_en. Accessed 10 Jul 2022
5. Holmes, T.G.: Eco-visualization: combining art and technology to reduce energy consumption. In: Proceedings of the 6th ACM SIGCHI Conference on Creativity & Cognition, pp. 153–162. C&C 2007, Association for Computing Machinery, New York, NY, USA (2007). https://doi.org/10.1145/1254960.1254982
6. Kosara, R.: Visualization criticism - the missing link between information visualization and art. In: 2013 17th International Conference on Information Visualisation. vol. 1, pp. 631–636. IEEE Computer Society, Los Alamitos, CA, USA (2007). https://doi.org/10.1109/IV.2007.130
7. Lowe, R.K.: Diagrammatic information: techniques for exploring its mental representation and processing. Inf. Design J. **7**(1), 3–17 (1993). https://doi.org/10.1075/idj.7.1.01low
8. Maçãs, C., Cruz, P., Martins, P., Machado, P.: Swarm systems in the visualization of consumption patterns. In: Proceedings of the 24th International Conference on Artificial Intelligence, pp. 2466–2472. IJCAI2015, AAAI Press, Buenos Aires, Argentina (2015)
9. Manovich, L.: 17. Introduction to Info-Aesthetics, pp. 333–344. Duke University Press, New York, USA (2008). https://doi.org/10.1515/9780822389330-021
10. Ogawa, M., Ma, K.L.: code_swarm: a design study in organic software visualization. IEEE Trans. Visual Comput. Graphics **15**(6), 1097–1104 (2009). https://doi.org/10.1109/TVCG.2009.123
11. Reynolds, C.W.: Flocks, herds and schools: a distributed behavioral model. In: Proceedings of the 14th Annual Conference on Computer Graphics and Interactive Techniques, pp. 25–34. SIGGRAPH 1987, Association for Computing Machinery, New York, NY, USA (1987). https://doi.org/10.1145/37401.37406
12. Sugrue, C., Stewart, D.: Waves to waves to waves (2008). https://csugrue.com/waves/. Accessed 20 Jul 2022
13. Vande Moere, A., Lau, A.: In-formation flocking: an approach to data visualization using multi-agent formation behavior. In: Randall, M., Abbass, H.A., Wiles, J. (eds.) ACAL 2007. LNCS (LNAI), vol. 4828, pp. 292–304. Springer, Heidelberg (2007). https://doi.org/10.1007/978-3-540-76931-6_26

GTR-CTRL: Instrument and Genre Conditioning for Guitar-Focused Music Generation with Transformers

Pedro Sarmento[1]([✉])[iD], Adarsh Kumar[2], Yu-Hua Chen[3], CJ Carr[4],
Zack Zukowski[4], and Mathieu Barthet[1][iD]

[1] Queen Mary University of London, London, UK
{p.p.sarmento,m.barthet}@qmul.ac.uk
[2] Indian Institute of Technology Kharagpur, Kharagpur, India
[3] National Taiwan University, New Taipei, Taiwan
[4] Dadabots, Boston, USA
https://dadabots.com/

Abstract. Recently, symbolic music generation with deep learning techniques has witnessed steady improvements. Most works on this topic focus on MIDI representations, but less attention has been paid to symbolic music generation using guitar tablatures (tabs) which can be used to encode multiple instruments. Tabs include information on expressive techniques and fingerings for fretted string instruments in addition to rhythm and pitch. In this work, we use the DadaGP dataset for guitar tab music generation, a corpus of over 26k songs in GuitarPro and token formats. We introduce methods to condition a Transformer-XL deep learning model to generate guitar tabs (GTR-CTRL) based on desired instrumentation (inst-CTRL) and genre (genre-CTRL). Special control tokens are appended at the beginning of each song in the training corpus. We assess the performance of the model with and without conditioning. We propose instrument presence metrics to assess the inst-CTRL model's response to a given instrumentation prompt. We trained a BERT model for downstream genre classification and used it to assess the results obtained with the genre-CTRL model. Statistical analyses evidence significant differences between the conditioned and unconditioned models. Overall, results indicate that the GTR-CTRL methods provide more flexibility and control for guitar-focused symbolic music generation than an unconditioned model.

Keywords: Controllable Music Generation · Deep Learning · Conditioning · Transformers · Interactive Music AI · Guitar Tablatures

1 Introduction

The field of music generation encompasses a diverse set of approaches and has seen progressive improvements towards automation: from musical dice games of the 18th century used by composers to create full compositions [20], to the earliest instances of computer music making [10], and the following explorations

C. Johnson et al. (Eds.): EvoMUSART 2023, LNCS 13988, pp. 260–275, 2023.
https://doi.org/10.1007/978-3-031-29956-8_17

of techniques such as rule-based models [14], Markov Models [22], or genetic algorithms [23]. Lately, state of the art in machine music creation has been reached with the use of deep learning approaches [3,4].

This work focuses on automatic music generation using symbolic notations which digitally encode musical score attributes [25]. We address the generation of *guitar-focused music*, referring here to music for which the melodic and harmonic arrangements are predominantly conveyed by fretted string instruments, namely guitars and bass guitars. Although fretted instruments are central to tabs, the generation techniques proposed in this work support multiple instruments (e.g. guitar, keyboard, bass, drums). Prior work on symbolic music generation often relies on datasets using formats such as MIDI, MusicXML, and ABC [8]. We use the DadaGP dataset which, in contrast, is built upon the GuitarPro (GP)[1] format suitable for *guitar-focused symbolic music*. For string instruments, such tab format enables to specify not only *what* music to play but also *how* it should be played [17]. The GP tab format can support expressive renderings of *guitar-focused music* providing information on fingerings and guitar-specific techniques, which are currently not supported by MIDI.

[28] introduced CTRL, a conditional Transformer [31] language model that can control style and content structure for text generation tasks. *Control codes* are special tokens appended to text sequences in order to categorize them and can be used at inference stage as prompts to condition generation features. Inspired by this work, we followed a similar approach by appending special control codes to condition tab generation using a set of instruments and musical genres.

The main contributions of this work are: (1) GTR-CTRL, methods to control a Transformer-based model for automatic tab music generation trained on the DadaGP dataset. The main goal of GTR-CTRL is to provide more flexibility and control at the time of inference. GTR-CTRL is able to be conditioned on instrumentation (inst-CTRL) and musical genre (genre-CTRL); (2) GPBERT, a model to classify songs into distinct musical genres using the DadaGP dataset, which can be used to assess if the generated music fits expected genres. We hope this work will foster guitar-focused symbolic music and AI research.

2   Background

2.1   Symbolic Music Generation with Deep Learning

Symbolic music generation techniques with deep learning can be categorized according to the architecture used, namely Variational Autoencoder (VAEs) models [30], Generative Adversarial Networks (GANs) [9], and models that closely stem from natural language processing (NLP) field, such as Recurrent Neural Networks (RNNs) [19], Long Short-Term Memory (LSTMs) [29], or Transformers [31]. The Transformer is a sequence-to-sequence model that is able

---

[1] Available at: https://www.guitar-pro.com/; alternative software available at: https://sourceforge.net/projects/tuxguitar/, http://www.power-tab.net/guitar. php.

to learn the dependencies and patterns among elements of a given sequence by incorporating the notion of self-attention. The Music Transformer by Huang et al. [12] was first to apply a self-attention mechanism to generate longer sequences of symbolic piano music. Similar seminal works include Musenet [24], in which a large-scale Transformer model, GPT-2, was used to generate symbolic multi-instrument music from different musical genres, the Pop Music Transformer [13], which uses Transformer-XL [7] as a backbone architecture and is able to generate pop piano symbolic music with a better rhythmic structure, and the Compound Word Transformer [11], that explores novel and more efficient ways of tokenizing symbolic music for training purposes.

Not much work on guitar-focused symbolic music generation has been done to date despite the proliferation and abundance of tablatures [2,16]. [18] presented an automatic guitar solo generator in tablature format, dependent on both input chord and key sequence, by exploring probabilistic models. Regarding guitar tab music generation with deep learning, Chen et al. [5] presented a fingerstyle guitar generator, trained on a dataset of 333 examples (not using the GuitarPro format), using a Transformer-XL model as a backbone.

2.2   Controllable Symbolic Music Generation

Despite the compelling results of deep learning models for automatic symbolic music generation, difficulties to interpret and control models have persisted. This has fostered research into ways of conditioning and guiding the generation process [32]. Wang et al. [32] proposed a VAE model able to generate short piano compositions that could be conditioned on chord structure and style. Similarly, in Music FaderNets [30], a VAE architecture is used to generate piano pieces conditioned on rhythm and note density. Regarding genre conditioning, in [15], a VAE framework was used to generate MIDI pieces in the style of either Bach chorales or Western folk tunes. Closely related to our work as it used a Transformer architecture [27] for conditional symbolic music generation, the Theme Transformer is a novel architecture to generate MIDI scores conditioned on musical *motifs* or themes [27].

3   Motivations

The main motivation of this work is to devise AI techniques for the production of expressive guitar-focused music which give producers some agency. The adaptation of models proposed in this paper could lead to the development of music making tools for artists/bands, allowing them to stir the generation into different creative directions. To the best of our knowledge, there are no prior guitar-focused symbolic music generation models supporting multiple instruments and that are controllable. The initial experiments we conducted with an *unconditional* model indicate that some degree of control can be achieved through the use of a prompt. For example, by defining one note for each instrument from a desired instrument combination, it is possible to stir the output into a particular instrument arrangement. However, defining said initial notes would often force the key of the composition (which can be a desirable feature or not). By using an excerpt of a song from a given genre as a prompt, the generated music is likely to fit the source musical genre but the content often mimics the initial prompt *motif* which limits creative possibilities. In an attempt to eliminate these constraints and give more creative agency to the user, we investigate alternative token representations facilitating a finer degree of control for tab music generation.

4   Conditioning Experiments

4.1   DadaGP Dataset

The DadaGP dataset [26] comprises 26,181 songs in two representations: *token format*, which resembles a text representation, and *GuitarPro format*, named after the GuitarPro software for tablature edition and playback. Conversion between these two file formats is possible thanks to a dedicated encoding/decoding tool using PyGuitarPro [1], a Python library which manipulates GuitarPro files. We used DadaGP as a guitar-focused symbolic music dataset since it represents notes in a syntactic format that is best suited for fretted instruments such as the guitar.

In the DadaGP *token format*, songs start with `artist`, `downtune`, `tempo` and `start` tokens; these *header tokens* are essential for the decoding process from token to GuitarPro format (or vice-versa). Notes from pitched instruments are represented by tokens of the form `instrument:note:string:fret`. This syntax is not only suitable for string instruments since the string/fret combination is eventually mapped to a MIDI note. It can be used for any other pitched instrument supported by DadaGP. Percussive instruments, namely the drumkit, are represented using tokens in the form `drums:note:type`. In order to quantify note duration or rest, the `wait:ticks` token is used. A resolution of 960 ticks per quarter note is used, as in most digital audio workstations (DAWs), and the decoder infers note duration from the tick value. For a more detailed description of DadaGP, please refer to [26].

4.2   Model Description

We used a Transformer-XL model [7] as backbone architecture, for it expands on the vanilla Transformer [31] by modifying the positional encoding scheme and introducing the concept of recurrence, enabling it to use information from tokens that occurred before the current segment. To address guitar-focused music generation, we employed the Pop Music Transformer model [13] which uses a similar architecture to generate piano compositions in MIDI. We trained three variants of a Transformer-XL model configuration, consisting of 12 self-attention layers with 8 multi-attention heads: (i) an *unconditional* model without any possibility of control (apart from side effects of prompting), (ii) a model to control instrumentation (inst-CTRL), (iii) and a model to control musical genre (genre-CTRL). All models were trained for 300 epochs, with a learning rate of $1e-4$ and a batch size of 8 samples. Model parameters were heuristically tuned based on prior experiments.

4.3   Instrument Conditioning Experiment

Instrumentation Control Tokens: Following the introduction of control codes presented in [28], we devised a list of tokens for the instrumentation, i.e. which instruments play in the piece. Before training, instrumentation information is appended at the start of each song as part of header tokens. Tokens for each instrument in a given song are inserted between `inst_start` and `inst_end` tokens. At the time of inference, the model is forced to produce music in the instrumentation given by the tokens. As stated in Sect. 3, it is possible to control to an extent instrumentation with the *unconditional* model using prompts specifying initial notes for each desired instrument (e.g. `distorted0:note:s6:f0`). However, this indirectly pushes generation towards a specific key, based on the note specified in the prompt (in this example, E minor or major). The proposed inst-CTRL method aims to provide more flexibility.

Instrumentation Inference Prompts: In order to assess the ability of inst-CTRL to control instrumentation, we devised three distinct initial prompts for the inference stage: *full-prompt*, describing both the instrumentation control codes and an initial note for every selected instrument, *partial-prompt*, which includes instrumentation control codes and a note from one instrument, and *empty-prompt*, containing just the instrumentation control codes. To a varied set of instruments, we selected eight distinct instrument combinations, dividing them into combinations of two, three and four instruments: bass and drums (b-d); distorted guitar and drums (dg-d); distorted guitar, bass and drums (dg-b-d); clean guitar, bass and drums (cg-b-d); distorted guitar, piano and drums (dg-p-d); two distorted guitars, bass and drums (dg-dg-b-d); clean guitar, distorted guitar, bass and drums (cg-dg-b-d); distorted guitar, piano, bass and drums (dg-p-b-d). To compare results from the conditioned and unconditioned models, the *unconditional* case uses a prompt with an initial note for every selected instrument. We generated 1,200 examples for each model/prompt, comprising a total of 150 examples for each of the eight instrument combinations. We defined a limit of 1,024 generated tokens per song using a temperature-controlled stochastic sampling method with *top-k* truncation [28], as employed in [13].

4.4   Genre Conditioning Experiment

Genre Control Tokens: Similarly to the procedure used for instrumentation control (Sect. 4.3), for the genre conditioning experiment we created a list of tokens with musical genre information and placed it at the start of every song in the DadaGP dataset. In order to discard genres with fewer songs, we only selected genres with more than 200 examples in the training corpus. This resulted in around 100 distinct musical genres, with a predominance for genres and subgenres of *Rock* and *Metal*, but also including other genres like *Jazz*, *Folk* and *Classical*.

Genre Inference Prompts: To assess the genre-CTRL model, we conducted inferences for the following five genres: *Metal*, *Rock*, *Punk*, *Folk*, and *Classical*. Prompts used at inference were chosen similarly to the instrument conditioning experiment. The *full-prompt* provides two measures plus the corresponding genre token. We sampled the first two measures of randomly selected songs from the training corpus in each genre. The *partial-prompt* provides genre tokens, but only uses the first note of every two measure-long snippet. The *empty-prompt* only describes the genre token. The prompt in the *unconditional* case specifies the first two measures but omits the genre token. We generated 20 songs for each of the 5 distinct song prompts, for each of the 5 different genres, resulting in a total of 500 songs. The number of generated tokens and sampling hyper-parameters were kept the same as in the instrument conditioning experiment.

4.5   Examples

Following the procedures described in Sects. 4.3 and 4.4, we cherry-picked examples of generated songs for the instrument and genre conditioning models. These examples without any post-processing are made available for listening[2].

5   Evaluation Methods

5.1   Overall Pitch and Rhythmic Metrics

In order to assess the overall consistency of the generated results, from both a melodic/harmonic and rhythmic perspective, we utilized two metrics, namely **pitch class histogram entropy** (PCE) and **groove consistency** (GC) [33]. These metrics were computed on examples generated using both the instrument and genre conditioning methods by using the MusPy library, a toolkit for symbolic music generation and evaluation [8]. We also used metrics specific to each conditioning as outlined in the following sections.

The PCE computes the degree of tonality within a song. Low entropy is obtained when specific pitch classes dominate (e.g. the first and fifth degree of the key), whereas high entropy points towards tonality instability. For inst-CTRL, we computed a combined pitch class histogram entropy figure for every melodic instrument (i.e. drums excluded) in every selected instrument combination and averaged across the different instrument combinations. For genre-CTRL, averaging was done across musical genres.

---

[2] Currently   available   at: https://drive.google.com/drive/folders/1ds5D01YW-8PAkIf-KxbyACOIBEYqk-he?usp=share_link.

The GC as described in [8], can be seen as a metric that measures rhythmicity. In compositions where there is a clearly defined rhythm, the grooving patterns between bars remains identical, thus yielding a high score. In opposition, this metric scores low for songs in which there is no rhythmic consistency across measures.

5.2   Instrumentation Metrics

As described in Sect. 4.3, during inference time we prompted the model testing different levels of prompt completeness, in order to investigate the influence of the control codes for instrumentation. To objectively measure the effects, we introduce two new metrics. The **prompted instrument presence (PIP) score** provides insight on how well the model is capable of generating the instruments that were selected in the control codes and instrumentation prompt. For a given instrument, an empty measure consists of rests throughout the whole duration of the measure. For a given song, the PIP metric indicates the percentage of measures that contain the prompted instruments with respect to all the instruments, as expressed by Eq. 1:

$$PIP = \frac{\sum\limits_{i \in P} M_i - E_i}{\sum\limits_{i \in A} M_i} \tag{1}$$

where A is the set of all instruments that appear in that song, P is the set of instruments prompted for a song, hence a subset of A, M_i represents the total number of measures with respect to instrument i (including empty measures), and E_i represents the total number of empty measures assigned to instrument i. In order to make the metric more robust, we considered as valid measures, the measures, M_i, that contain at least one note; we subtracted empty measures, E_i, to prevent high PIP scores to occur for instrumental parts with many empty measures.

Given that the model may generate parts for instruments that are not in the instrumentation control codes, we implemented an **unprompted instrument presence (UIP) score**, which measures the percentage of measures with unprompted instruments with respect to the total number of non-empty measures, as expressed by Eq. 2:

$$UIP = \frac{\sum\limits_{i \in (A \backslash P)} M_i - E_i}{\sum\limits_{i \in A} M_i - E_i} \tag{2}$$

Based on observations from previous experiments, when the model generates only a few notes for a given instrument, the decoder procedures in DadaGP automatically fill the rest of the parts of that instrument with rests, so that the number of measures for that instrument matches the number of measures for the remaining instruments. For example in a 20-measure long composition, if an instrument has only one measure with note information, it will get 19 empty measures assigned to it. As we focus on active musical content (i.e. notes) for unprompted instruments, we discarded empty measures in the UIP formulation.

5.3 Genre Metrics

Aware of the difficulties regarding genre classification, often leading to a lack of consensus even amongst human experts [21] and given the sheer volume of generated data, we used a machine-based classifier to assess the genre-CTRL model. Genre recognition is complex considering music in the symbolic domain, which is the domain in which the network outputs music in this work, since acoustic information convey cues for genre identification (e.g. timbre). It is possible to render audio from the generated symbolic notation but the choice and quality of the virtual instruments may affect genre identification. The authors in [6] introduced MIDIBERT, a Bidirectional Encoder Representations from Transformers (BERT)-based masked language model trained on polyphonic piano MIDI pieces, able to be configured for downstream classification tasks. Following a similar approach, we propose GPBERT, a variant model which was first pre-trained on the DadaGP dataset for a total of 50 epochs, and later fine-tuned for the task of genre classification for 10 epochs. For the latter, we selected a corpus of 800 songs from each of the five genres described in Scct. 4.4, with a 80/10/10 split between training, validation, and test sets. Achieving an overall test accuracy of 90.67%, we deemed this model to be suitable for the assessment of the genre-CTRL model. However, this type of assessment present limitations since an AI system is used to classify outputs from another AI system, without human feedback. However, to the credit of the method, the musical genre metadata used for the training of GPBERT were gathered via the Spotify Web API, which provides human-produced genre labels from Spotify's curation teams.

6 Quantitative Analysis and Discussion

6.1 Pitch and Rhythm Metrics Results

We computed the **pitch class histogram entropy** and **groove consistency** on the pieces generated for the instrument (1,200 examples in total) and genre (500 examples in total) conditioning experiments. To benchmark the proposed conditioning methods, we used the training corpus as groundtruth data (theoretical best case scenario), and corpi produced by randomizing certain musical attributes (theoretical worst case scenario). For PCE (GC, respectively), the randomized corpus was obtained by randomizing the notes' pitch (rhythm, respectively) in DadaGP songs. We report in this section statistical analyses investigating the effects of the music source (4 different prompts, groundtruth and random) on the PCE and GC dependent variables, with a Type I error α of .05. Notched boxplots of pitch class entropy can be seen on the left plots in Fig. 1 for instrumentation (top) and genre (bottom) conditioning, respectively.

We performed a Kruskal-Wallis rank sum test which yielded a highly significant effect of the music source (4 prompts, groundtruth and random) on pitch class entropy ($H(5) = 3169.6$, $p < .001$). Pairwise comparisons were conducted using the Wilcoxon rank sum test and significant differences between conditions are reported using brackets in Fig. 1. These tests indicate that the three conditional models as well as the unconditional model yield significantly different PCEs compared to the groundtruth and random conditions, for both the instrumentation (top left) and genre conditioning (bottom left).

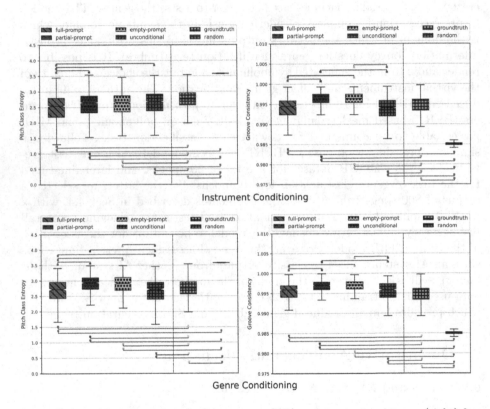

Fig. 1. Notched boxplots of pitch class entropy (left) and groove consistency (right) for the instrument (top) and genre (bottom) conditioning experiments by comparison with groundtruth and random groups. Brackets indicate significant differences ($p < .001$) in pairwise Wilcoxon signed-rank tests using a Bonferroni-adjusted α level of .008 (.05/6).

However, the model-generated examples have PCEs that are much closer to the groundtruth than the random condition. The *full-prompt* obtains PCEs that are significantly different from the other prompts and present the lowest median (the median is slightly below that of the groundtruth). This means that the *full-prompt* condition produces the highest tonal stability on average. However, a low PCE is not a sufficient condition to ensure musical quality given that less tonal variations could be for example due to repetitive structures in the generated music. On average, groundtruth examples comprise 84.6±57.0 measures, whereas generated examples have on average of 34.4 ± 24.6 measures. We hypothesize that the slightly higher tonal stability compared to the training corpus is due to the shorter length of the generated snippets; shorter length often translates into compositions that stay within a given key and without or with only little harmonic modulation which would increase the PCE score. We highlight that it is difficult to precisely define what "better" results are in terms of PCE due the subjective nature of music. A lower PCE may not necessarily translate into better sounding examples, rather it indicates that the music stays more strongly within a given tonality, on overall. The impact of PCE differences on the musical expectations of listeners should be further assessed in perceptual tests.

We conducted a Kruskal-Wallis rank sum test to assess the effect of the music source on GC (right plots in Fig. 1). A highly significant effect can be observed ($H(5) = 3666.9$, $p < .001$). For the instrument conditioning experiment (top right), the *partial-prompt* and *empty-prompt* configurations obtained a slightly significantly higher GC median compared to the groundtruth indicating more rhythmic stability. More rhythmic variations tend to occur in longer musical excerpts such as in the groundtruth corpus. The more consistent rhythm obtained for the generated pieces is likely due to the shorter length of the pieces, but it could also be due to the presence of very repetitive structures. Similar conclusions can be made for the genre conditioning experiment.

6.2   Instrumentation Metrics Results

For the instrumentation metrics, no comparisons were made with the grountruth and random cases since the metrics are self-explanatory.

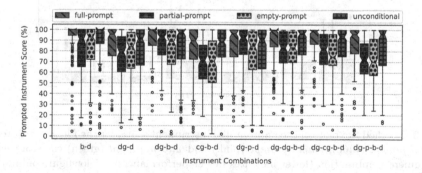

Fig. 2. Notched boxplots of the prompted instrument score, for the *full-prompt* (red), *partial-prompt* (blue), *empty-prompt* (pink) and *unconditional* (green) cases, for each instrument combination (higher indicates better performance). (Color figure online)

Figure 2 shows notched boxplots for the **prompted instrument score**. All the prompt configurations achieved a median PIP score above 70%. The *full-prompt* achieves the best PIP median (78.2%, closer to 100%) for every instrument combination. The *unconditional* outperforms the *partial-prompt* and *empty-prompt* for every instrumentation, with the exception of *dg-p-b-d* (distorted guitar, piano, bass and drums). We believe this is due to a lack of songs with this instrumentation in the dataset yielding less training examples (see instrument distribution in DadaGP in Fig. 4(h) [26]). A Kruskal-Wallis rank sum test yielded a significant effect of the prompt configuration on the PIP score ($H(3) = 304.93$, $p < .001$). Table 1 reports mean PIP and significant differences occurring between the three conditional models and the unconditional model (Wilcoxon rank sum test with Bonferroni correction). These results show that *full-prompt* obtains significantly better results than the *unconditional*.

Table 1. Mean values of prompted and unprompted instrument score for the *full-prompt*, *partial-prompt*, *empty-prompt* and *unconditional* conditions. Stars indicate significant differences compared to the unconditional model based on pairwise Wilcoxon rank sum test with Bonferroni correction ($p < .001$ for all cases). Best results in **bold**.

	Full-prompt	Partial-prompt	Empty-prompt	Unconditional
Prompted Inst. Score	**87.54*****	78.23***	77.26***	83.62
Unprompted Inst. Score	4.09	4.19	**4.01**	5.2

In Fig. 3 we present the notched boxplots for the unprompted instrument score.

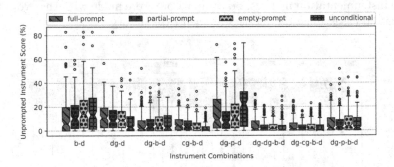

Fig. 3. Notched boxplots of the unprompted instrument score, for the *full-prompt* (red), *partial-prompt* (blue), *empty-prompt* (pink) and *unconditional* (green) cases, for each instrument combination (lower indicates better performance). (Color figure online)

It seems that there is a progressive improvement in avoiding unprompted instruments (lower UIP score) as we move from combinations with less instruments to configurations with more instruments. This may be related to the smaller amount of songs with two or three instruments in DadaGP compared to songs with four or more instruments. However, the *dg-p-d* and *dg-p-b-d* instrumentations obtained worst UIP scores on overall which may be related to the fact that DadaGP only contains four and 25 examples for these configurations, respectively (compared to 4,930 in total for the six remaining selected instrument combinations, the range being [98-3,058]). We conducted a Kruskal-Wallis rank sum test to assess the effect of the instrument combinations on UIP, grouping by number of instruments (respectively 2, 3, and 4 instruments). A significant effect of the number of instruments in instrument combinations can be observed on the UIP score ($H(2) = 352.6$, $p < .001$). A pairwise Wilcoxon rank sum test with Bonferroni correction shows significant differences ($p < .001$) between the three instrument combination groups. Combinations with 2, 3 and 4 instruments achieved an average UIP of 6.9%, 4.35% and 2.69%, respectively.

6.3 Genre Metrics Results

Genre classification scores from GPBERT are shown in Table 2, reporting the average softmax score for 100 examples generated for each genre and prompt.

Table 2. Genre classification softmax scores from GPBERT, for the *full-prompt* (F-P), *partial-prompt* (P-P), *empty-prompt* (E-P) and *unconditional* (UNC), when conditioned on the genres of *Metal, Punk, Rock, Classical* and *Folk*. Highest results for each genre per prompt in **bold**.

		Genre Classification Score				
		Metal	Punk	Rock	Classical	Folk
Metal	F-P	**0.6680**	0.0925	0.2028	0.0263	0.0103
	P-P	0.2217	0.1929	**0.4537**	0.1162	0.0155
	E-P	0.2509	0.1412	**0.4322**	0.1577	0.0180
	UNC	**0.7189**	0.0675	0.1697	0.0404	0.0035
Punk	F-P	0.0041	**0.8126**	0.1727	0.0012	0.0094
	P-P	0 .1264	0.3406	**0.4807**	0.0434	0.0088
	E-P	0.0997	0.2176	**0.5043**	0.1342	0.0441
	UNC	0.0290	**0.6954**	0.2464	0.0113	0.0179
Rock	F-P	0.0236	0.3325	**0.6403**	0.0021	0.0015
	P-P	0.1069	0.1661	**0.6401**	0.0757	0.0112
	E-P	0.0863	0.0608	**0.5763**	0.1994	0.0772
	UNC	0.0266	0.2207	**0.7487**	0.0030	0.0010
Classical	F-P	0.0338	0.0128	0.2293	**0.6753**	0.0488
	P-P	0.0300	0.0126	0.1883	**0.7035**	0.0657
	E-P	0.0877	0.0644	0.3175	**0.4767**	0.0538
	UNC	0.0415	0.0227	0.2638	**0.5940**	0.0780
Folk	F-P	0.0692	0.0451	**0.4338**	0.2751	0.1768
	P-P	0.0751	0.0533	0.2904	**0.4323**	0.1489
	E-P	0.0822	0.1141	0.3192	**0.3900**	0.0946
	UNC	0.0885	0.0430	**0.4849**	0.2901	0.0934

The *unconditional* and *full-prompt* achieved the highest genre classification results for *Metal, Punk* and *Rock*. Whilst the *unconditional* model was able to perform better for *Rock* and *Metal*, the *full-prompt* attains higher scores for *Punk*. It is interesting to note that GPBERT often confuses genres that would also be difficult to discriminate for a human listener (e.g. *Metal/Punk* being judged as *Rock*). Overall, the results from the *Folk* genre were often classified as either *Rock* (*full-prompt* and *unconditional*) or *Classical* (*partial-prompt* and *empty-prompt*). For both the *partial-prompt* and *empty-prompt*, when the intended genre is *Metal* or *Punk*, the generated pieces tend to get classified as *Rock*. We believe this is due to the high *Rock/Metal* bias in DadaGP, with more than 50% of tracks in these two genres (see Fig. 3 in [26]).

7   Subjective Analysis and Discussion

We conducted a subjective analysis of the cherry-picked generated examples. For genre control, particularly in the *empty-prompt* and *partial-prompt* cases, we noted that the model struggles to force a given multi-instrument configuration, frequently producing only one or two instrumental parts. This behavior influences GPBERT genre classification scores which expects multi-instrument compositions for correct classifications. To circumvent this, an interesting approach could be to combine inst-CTRL and genre-CTRL into one single model, accounting for instrumentation and genre controllability.

Fig. 4. First four measures in <u>id-139</u> (prompt from "Immigrant Song" by Led Zeppelin). Prompt highlighted in red (distorted guitar, bass and drums are visible).

Although not a design objective, we found that some of the proposed models with the *full-prompt* genre conditioning could effectively be used as *motif/song continuators*. For example, in song <u>id-139</u>[3], the model maintains the overall rhythmic structure of the initial *motif*, while adapting it through different chord

---
[3] Underlined song ids are hyperlinked to facilitate listening.

progressions (see Fig. 4). In <u>id-67</u> (prompt from the first two measures of "Sea of Lies" by Symphony X), both guitars perform lead and rhythmic functions, alternating between phrases with guitar solos or chord structures. Furthermore, <u>id-459</u> (prompt from the first two measures of "Canon in D" by Johann Pachelbel) · stands out for its unusual harmonic exploration that maintains the characteristics of the initial idea, and <u>id-103</u> (see Fig. 5) showcases the model's ability to generate novel guitar phrases that fit the key and style of the prompt.

Fig. 5. First measures in <u>id-103</u> guitar track (prompt from "Back In Black" by AC/DC). Measures used as prompt are highlighted in red (only distorted guitar is visible). (Color figure online)

8 Conclusion and Future Work

In this paper we presented the conditioning of two Transformer-based models, inst-CTRL and genre-CTRL, for the task of guitar-focused symbolic music generation. We studied the performance of these models against an unconditional model, and explored the use of three different prompting strategies at the time of inference. Results show that both models succeed on the respective conditioning tasks: the conditional prompts in inst-CTRL achieve a better performance than the unconditional case in all the metrics, whilst for genre-CTRL, the *full-prompt* outperforms the unconditional model on the *Punk*, *Classical* and *Folk* genres. We believe that the highest scores for the *unconditional* model in *Rock* and *Metal* are due to the large Rock and Metal bias in the dataset, making the *unconditional* model perform well at generating songs in these genres. Despite slightly poorer results, the *empty-prompt* model, by bypassing the need of prompting the model with specific instrument notes that often condition the key/melodic nature of the composition, accounts for an increased level of flexibility, and a completely uninfluenced choice of notes and keys.

In future work, we plan to combine both models to control instrumentation and genre simultaneously. We also intend to explore GPBERT for different downstream tasks, such as music inpainting and artist/composer classification. One interesting research avenue would be to develop a model that is able to generate guitar sections on the style of a particular guitar player, once again using a variant of GPBERT to classify its output, supported by listening tests with expert guitar players. Finally, we envision collaborations with artists/bands willing to engage in human-AI driven compositions.

Acknowledgments. This work is supported by the EPSRC UKRI Centre for Doctoral Training in Artificial Intelligence and Music (Grant no. EP/S022694/1).

References

1. Abalumov, S.: PyGuitarPro (2014). https://github.com/Perlence/PyGuitarPro. Accessed 3 Nov 2022
2. Barthet, M., Anglade, A., Fazekas, G., Kolozali, S., Macrae, R.: Music Recommendation for Music Learning: Hotttabs, a Multimedia Guitar Tutor. In: Workshop on Music Recommendation and Discovery pp. 7–13. Chicago, IL, USA (2011)
3. Briot, J.P., Hadjeres, G., Pachet, F.D.: Deep Learning Techniques for Music Generation. Computational Synthesis and Creative Systems Series. Springer (2019)
4. Carnovalini, F., Rodà, A.: Computational Creativity and Music Generation Systems: An Introduction to the State of the Art. Frontiers in AI 3 (2020)
5. Chen, Y.H., Huang, Y.H., Hsiao, W.Y., Yang, Y.H.: Automatic Composition of Guitar Tabs by Transformers and Groove Modelling. In: Proceedings of the 21st International Soc. for Music Information Retrieval Conference, pp. 756–763 (2020)
6. Chou, Y.H., Chen, I.C., Chang, C.J., Ching, J., Yang, Y.H.: MidiBERT-Piano: Large-scale Pre-training for Symbolic Music Understanding. Tech. rep. (2021)
7. Dai, Z., Yang, Z., Yang, Y., Carbonell, J., Le, Q.V., Salakhutdinov, R.: Transformer-XL: attentive language models beyond a fixed-length context. In: Proceedings of the 57th Annual Meeting of the Ass. for Computational Linguistics, pp. 2978–2989. Florence, Italy (2019)
8. Dong, H.W., Chen, K., McAuley, J., Berg-Kirkpatrick, T.: MusPY: A Toolkit for Symbolic Music Generation. In: Proceedings of the 21th International Society for Music Information Retrieval, pp. 101–108. Montréal, Canada (2020)
9. Dong, H.W., Yang, Y.H.: Convolutional Generative Adversarial Networks with Binary Neurons for Polyphonic Music Generation. In: Proc. of the 19th Int. Soc. for Music Information Retrieval Conf. pp. 190–198. Paris, France (2018)
10. Hiller, L.A., Isaacson, L.M.: Experimental Music. Composition with an Electronic Computer. Greenwood Publishing Group Inc., USA (1979)
11. Hsiao, W.Y., Liu, J.Y., Yeh, Y.C., Yang, Y.H.: Compound word transformer: learning to compose full-song music over dynamic directed hypergraphs. In: Proceedings of the AAAI Conference on Artificial Intelligence, pp. 178–187 (2021)
12. Huang, C.Z.A., et al.: Music transformer: generating music with long-term structure. In: Proceedings of the 7th International Conference on Learning Representations, New Orleans, LA, USA (2019)
13. Huang, Y.S., Yang, Y.H.: Pop music transformer: beat-based modeling and generation of expressive pop piano compositions. In: Proceedings of the 28th ACM International Conference on Multimedia, Seattle, WA, USA, pp. 1180–1188 (2020)
14. Johnson-Laird, P.N.: How Jazz Musicians Improvise. In: Music Perception, vol. 19 (3), pp. 415–442. University of California Press (2002)
15. Lim, Y.Q., Chan, C.S., Loo, F.Y.: Style-Conditioned music generation. In: Proceedings of IEEE International Conference on Multimedia and Expo. London, UK (2020)

16. Macrae, R., Dixon, S.: Guitar tab mining, analysis and ranking. In: Proceedings of the 12th International Society for Music Information Retrieval Conf, pp. 453–459, Miami, FL, USA (2011)
17. Magnusson, T.: Sonic Writing: Technologies of Material. Symbolic & Signal Inscriptions, Bloomsbury Academic (2019)
18. McVicar, M., Fukayama, S., Goto, M.: AutoLeadGuitar: automatic generation of guitar solo phrases in the tablature space. Int. Conf. on Signal Processing Proc, pp. 599–604 (2014)
19. Meade, N., Barreyre, N., Lowe, S.C., Oore, S.: Exploring Conditioning for Generative Music Systems with Human-Interpretable Controls. Tech. rep. (2019)
20. Nierhaus, G.: Algorithmic Composition: Paradigms of Automated Music Generation. Springer Vienna (2009)
21. Oramas, S., Barbieri, F., Nieto, O., Serra, X.: Multimodal deep learning for music genre classification. Trans. Int. Soc. Music Inf. Retriev. **1**(1), 4–21 (2018)
22. Pachet, F.: The continuator: musical interaction with style. In: International Computer Music Conference, pp. 333–341, Gothenborg, Sweden (2002)
23. Papadopoulos, G., Wiggins, G.: A genetic algorithm for the generation of jazz melodies. In: Human and Artificial Information Processing: Finnish Conference on Artificial Intelligence, pp. 7–9. Jyväskylä, Finland (1998)
24. Payne, C.: Musenet (2019). https://openai.com/blog/musenet. Accessed 12 Jun 2022
25. Raffel, C., Ellis, D.P.W.: Extracting ground truth information from MIDI files: a MIDIfesto. In: Proceedings of the 17th International Society for Music Information Retrieval Conference, pp. 796–803. New York City, USA (2016)
26. Sarmento, P., Kumar, A., Carr, C., Zukowski, Z., Barthet, M., Yang, Y.H.: DadaGP: a Dataset of Tokenized GuitarPro Songs for Sequence Models. In: Proc. of the 22nd Int. Soc. for Music Information Retrieval Conf. pp. 610–618 (2021)
27. Shih, Y.J., Wu, S.L., Zalkow, F., Müller, M., Yang, Y.H.: Theme Transformer: Symbolic Music Generation with Theme-Conditioned Transformer. Tech. rep. (2021)
28. Shirish Keskar, N., Mccann, B., Varshney, L.R., Xiong, C., Socher, R., Research, S.: CTRL: A Conditional Transformer Language Model for Controllable Generation. Tech. rep. (2019)
29. Sturm, B.L., Santos, J.F., Ben-Tal, O., Korshunova, I.: Music transcription modelling and composition using deep learning. In: Proceedings on the 1st Conference on Computer Simulation of Musical Creativity (2016)
30. Tan, H.H., Herremans, D.: Music FaderNets: controllable music generation based on high-level features via low-level feature modelling. In: Proceedings of the 21th International Society for Music Information Retrieval Conference, pp. 109–116. Montréal, Canada (2020)
31. Vaswani, A., Shazeer, N., Parmar, N., Uszkoreit, J., Jones, L., Gomez, A.N., Kaiser, Ł., Polosukhin, I.: Attention Is All You Need. In: Proceedings of the 31st Conference on Neural Information Processing Systems. Long Beach, CA, USA (2017)
32. Wang, Z., Wang, D., Zhang, Y., Xia, G.: Learning interpretable representation for controllable polyphonic music generation. In: Proceedings of the 21st International Society for Music Information Retrieval Conference, pp. 662–669. Montréal, Canada (2020)
33. Wu, S.L., Yang, Y.H.: The jazz transformer on the front line: exploring the shortcomings of AI-composed music through quantitative measures. In: Proceedings of the 21th International Society for Music Information Retrieval Conference, pp. 142–149. Montréal, Canada (2020)

Artistic Curve Steganography Carried by Musical Audio

Christopher J. Tralie(✉) ⓘD

Ursinus College, Collegeville, PA 19426, USA
ctralie@alumni.princeton.edu

Abstract. In this work, we create artistic closed loop curves that trace out images and 3D shapes, which we then hide in musical audio as a form of steganography. We use traveling salesperson art to create artistic plane loops to trace out image contours, and we use Hamiltonian cycles on triangle meshes to create artistic space loops that fill out 3D surfaces. Our embedding scheme is designed to faithfully preserve the geometry of these loops after lossy compression, while keeping their presence undetectable to the audio listener. To accomplish this, we hide each dimension of the curve in a different frequency, and we perturb a sliding window sum of the magnitude of that frequency to best match the target curve at that dimension, while hiding scale information in that frequency's phase. In the process, we exploit geometric properties of the curves to help to more effectively hide and recover them. Our scheme is simple and encoding happens efficiently with a nonnegative least squares framework, while decoding is trivial. We validate our technique quantitatively on large datasets of images and audio, and we show results of a crowd sourced listening test that validate that the hidden information is indeed unobtrusive.

Keywords: Steganography · Traveling Salesperson Art · Hamiltonian Cycles · Audio Processing · Hidden Signals

1 Introduction/Background

Steganography is the process of hiding one data stream "in plain sight" within another data stream known as a "carrier." In this work, we are interested in using audio as a carrier to store images, which has been relatively unexplored in the academic literature (Sect. 1.1). As with any stegonographic technique, one can use this as a general means of covert communication, artists can also use it to hide images and to watermark their tunes. One such example occurred when Aphex Twin embedded a face image inside the spectrogram of their "Windowlicker" song [30], though the technique yields very loud, inharmonic, scratchy sounds. By contrast, we seek an embedding of the image that is undetectable. Like Aphex Twin, however, *we would like the image to survive lossy audio compression* so that people can communicate their images via social media, which makes the problem significantly more challenging.

To make our audio steganography system as robust and as usable as possible, we have the following design goals:

1. The hidden data should be audibly imperceptible
2. The hidden data should be faithfully preserved *under lossy compression*
3. The hidden data should be robust to frame misalignment, or it should be possible to recover a frame alignment without any prior information
4. No metadata should be required to retrieve the hidden data; (lossily compressed) audio alone should suffice
5. It should be possible to recover the data partially from partial audio chunks; that is, we don't need to wait for the entire data stream to recover the signal

Goals 1 and 2 are at odds with each other, and satisfying them simultaneously is the biggest challenge of this work. We formulate an objective function in Sect. 2.1 to trade off both of these goals.

Rather than storing images directly in audio, we constrain our problem to hiding artistic curves that trace out images (Sect. 1.2) or 3D surface shapes (Sect. 1.3). We hide each dimension of our curve in a different frequency, and we use sliding window sums of the frequency magnitudes to smooth them in time, which makes them more robust to compression, as we show in Sect. 3. Furthermore, since we know that the curves only move by a small amount between adjacent samples in time, we can exploit this fact to recover from frame misalignments to satisfy Goal 3, as explained in Sect. 2.4. In the end, all of the hidden information needed to reconstruct the curves is stored in the magnitudes and phases of frequencies in the audio, so no metadata is needed (Goal 4). Finally, the information needed to recover individual curves samples is localized in time, so the curves can be partially decoded from partial audio (Goal 5).

1.1 Prior Work in Audio Steganography

Before we proceed to describing the techniques for constructing artistic curves (Sect. 1.2, Sect. 1.3), and to ultimately hiding them in audio (Sect. 2), we briefly review adjacent work in the area of audio steganography to put our work in context (for more information, review to these survey articles [12,14]).

As is the case for most other steganographic carriers, the simplest techniques for audio steganography rely on changing the least significant bit of uncompressed encodings (e.g. [11]). Other early works rely on the "frequency masking" property of the human auditory system. Binary data is stored by controlling the relative amplitude of two frequencies that are masked in this way [17,18].

Beyond amplitude perturbations, it is also possible to keep the amplitudes fixed and to change the phases [36,37]; the technique of [29] was particularly successful at this by manipulating the poles of allpass filters to encode binary data. One can also hide binary data by adding and manipulating echoes of a signal [21], which is more robust than other techniques to lossy compression since it is akin to impulse responses of natural environments; however, the bit rate is quite low (on the order of dozens of bits per second). For another approach to compressed audio, some techniques are able to use the properties of mp3 files directly to hide information [2,32]. Other techniques adapt wireless transmissions schemes, such as and on-off keying (OOK) [28] (at a low bit rate of 8 bps) and orthogonal frequency-division multiplexing (OFDM) [15] (at a higher bit rate of closer to 1 kbps), to the audio domain.

Most of the hidden messages in the audio steganography literature are in binary format, but a few recent works have focused on hiding images pixels using deep learning to train a "hiding network" and a "reveal network" in tandem to encode an image in an audio stream and then to decode its addition into the carrier, respectively, while maximizing quality of the decoded image and the encoded audio [10,13,16,34]. Not only do we also attempt to hide images, but we have a similar perspective that rather than attempting to extract binary sequences exactly, we can tolerate progressive noise in the reconstructions. Challenges of deep-learning based approaches include the need to extensively train on examples, the difficulty of including a loss term that can model the effect of lossy audio compression, and the difficulty of training to be robust to frame misalignment (Goal 3). We sidestep these challenges in our work by using a simple model based on hiding in frequencies.

1.2   Traveling Salesperson Art

To devise 2D curves that stand in for images, we use Traveling Salesperson (TSP) art [4,6,23], which is an automated artistic line drawing pipeline which computes a *simple closed loop* to approximate a source image. "Simple" in this context does not imply a lack of complexity; rather, it means that a curve does not intersect itself (see, for example, the Jordan Curve Theorem [5]). To construct such a curve, we first place a collection of dots in the plane in a "stipple pattern" to approximate brightness in the image (e.g. more dots concentrate in darker regions), and then connect them in a loop via an approximate traveling salesperson (TSP) tour. Figure 1 shows an example. We follow the TSP Art technique of [23], with a few modifications, as described below.

Fig. 1. Our modified pipeline for creating TSP art. Color on the tour indicates phase along the loop.

The first step in TSP art is to generate a stipple pattern X to best approximate a grayscale image G. Following the authors of [23], we use Secord's technique for Voronoi stippling [33], which takes an initial sample of points according

Fig. 2. Since mp3 compression introduces noise to our embedded curves, we pre-smooth them using one iteration of curvature-shortening flow at $\sigma = 1$, which smooths the curves without introducing any crossings.

to some density weights W and then repeatedly moves them closer the weighted centroid of their Voronoi regions (an instance of Lloyd's algorithm) so that they are spread more evenly. Secord [33] takes the weight W_{ij} at a pixel to be inversely proportional to its brightness, using a weight of 0 above some brightness threshold b. This leads more dots to concentrate in darker regions; however, the algorithm may fail to sample any dots along important edges between brighter regions (the authors of [27] also observed this). To mitigate this, we run a Canny edge detector [8] on the original image and set the weight of any pixels along an edge to be 1, so that the final weights promote samples both in darker regions and along edges of any kind. This addition is particularly helpful for line drawings.

Once a stipple has been established, the next step in TSP art is to "connect the dots" with a closed loop that visits each stipple point exactly once, referred to as a "tour". A well known objective function for a tour that doesn't "jump too much" is the total distance traveled, or the sum of all edge lengths, and a tour that achieves the optimum is known as a *traveling salesperson (TSP) tour*. Since the TSP problem is NP-hard, the authors of [23] use the Concorde TSP solver [1] for an approximate solution. We opt for a simpler technique that first creates a 2-approximation of a TSP from a depth-first traversal through the minimum spanning tree of the stipple dots, which is a already a 2-approximation of the optimal tour. We then iteratively improve on this tour via a sequence of 2-opt relaxations [22]; that is, if for some $i > 1, i < j < N$ the distances between the 4 points X_i, X_{i+1}, X_j, and X_{j+1} satisfy

$$d(X_i, X_j) + d(X_{i+1}, X_{j+1}) < d(X_i, X_{i+1}) + d(X_j, X_{j+1}) \tag{1}$$

then it is possible to perform a swap to yield a new tour with a smaller distance. This amounts to reversing the indices in the tour between index $i + 1$ and j, inclusive. We repeat this step as long as such a swap is still possible. Though this is not guaranteed to yield an optimal TSP tour, it does produce aesthetically pleasing tours which are simple; that is, every crossing is removed.

Curvature Shortening Flow. Since mp3 compression introduces noise into our embedded curves, we smooth them before embedding to improve visual quality. To this end, we apply a numerical version of curvature shortening flow described by Mokhtarian and Mackworth [31] which applies to piecewise linear curves (like our TSP tours). The technique works by numerically by convolving coordinates of each curve with smoothed versions of Gaussians and their derivatives.

To approximately smooth a curve via one step of curvature-shortening flow, Mokhtarian and Mackworth [31] show that it suffices to first re-parameterize by the arc length, and then to smooth the curve by convolving with once with a Gaussian[1]. By the Gage-Hamilton-Grayson theorem, simple curves that undergo curvature shortening flow *remain simple* and eventually become convex, shrinking to a point under repeated applications of the flow. Figure 2 shows an example of one application of curvature shortening flow for different σ values.

1.3   Hamiltonian Cycles on Watertight Triangle Meshes

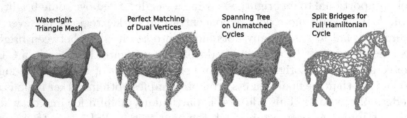

Fig. 3. An example of the algorithm of Gopi and Epstein [19] on a horse model from the Princeton mesh segmentation benchmark [9].

In addition to 2D TSP tours on stipple patterns on planar images, we also create artistic 3D space loops that fill out the surfaces of 3D shapes following the technique of Gopi and Epstein [19]. They observe that watertight triangle meshes (those with no boundary) have a dual graph in which a perfect matching exists. In other words, it is possible to partition the triangles into a set of adjacent pairs (second column of Fig. 3). In practice, we use the Blossom-5 algorithm [24] to find perfect matches of the dual graph. Then, adding edges between unpaired triangles leads to a set of disconnected cycles (third column of Fig. 3), which one can connect by a spanning set of edges between paired triangles. Finally, the spanning edges are split into bridges between the disconnected cycles, joining them together into one large cycle that covers the surface of the triangle mesh.

2   Curve Embedding in Audio

We now introduce our new algorithm for hiding artistic space curves in audio. Before we go into the details, we first define quantitative measurements for measuring the fit to the original audio (Goal 1) and the geometric quality of the

---

[1] The beauty of convolving with Gaussians as such is that γ does not even have to be differentiable, so this works on our piecewise linear TSP tours.

hidden curve (Goal 2). Let the original carrier audio be x and let the steganography audio be y, each with N samples. Then we define the steganographic signal to noise ratio in decibels (dB) as

$$\mathrm{snr}(y|x) = 10 \log_{10} \left(\sum_{j=1}^{N} x_j^2 \right) - 10 \log_{10} \left(\sum_{j=1}^{N} (y_j - x_j)^2 \right) \qquad (2)$$

For our geometric measurement of quality, let X_i be a sequence of M target curve points in $\mathbb{R}^d$ and Y_i be a sequence of M reconstructed points in $\mathbb{R}^d$. Then we define the *mean geometric distortion* is simply as the mean Euclidean distance between these points:

$$\mathrm{distortion}(Y|X) = \frac{1}{M} \sum_{j=1}^{M} ||X_j - Y_j||_2 \qquad (3)$$

2.1 Formulation of Least Squares Problem

We now formulate an objective function that trades off Eq. 2 and Eq. 3. Let $T = (T_1, T_2, \ldots, T_N)$ be the sequence of points of the target curve to hide, where each $T_i \in \mathbb{R}^d$, and let $T_{i,m}$ refer to m^{th} coordinate of T_i, and let x be a set of audio samples which will serve as a carrier. The goal is to perturb the samples of x so that some function of x matches T as closely as possible. The function we choose is based on a "time regularized" version of the magnitude Short-Time Fourier Transform (STFT) which we call a *sliding window sum* STFT (SWS-STFT). As a first step, we compute a *non-overlapping* STFT S based on a chosen window length w with a total of N frames:

$$S_{k,j} = \sum_{n=0}^{w} x_{jw+n} \left(e^{-i2\pi kn/w} \right) = M_{k,j} \left(e^{iP_{k,j}} \right) \qquad (4)$$

and we factor S it into its magnitude and phase components M and P, respectively. Next, we choose a subset of d frequencies $k_i, i \in \{1, 2, \ldots, d\}$, each of which will hide a different dimension of T. Given a second window length ℓ, we then define the following *sliding window sum* function, which we apply to each row k_i of the magnitudes of S that we wish to perturb to obtain the SWS-STFT

$$\mathrm{SWS}^{\ell}(M)_{k_i,j} = \sum_{n=0}^{\ell-1} M_{k_i,j+n} \qquad (5)$$

The effect of ℓ is to smooth out the noisy rows k_i of the magnitude spectrogram so that the rows in the SW-STFT match smoother transitions in the target curves. Each row of the SW-STFT has $N - \ell + 1$ samples. Let's assume momentarily that T has exactly this many samples; we will address the case where $\mathrm{length}(T) > N - \ell + 1$ in Sect. 2.2. We then seek a perturbed version of the magnitudes $\hat{M}$ so that each coordinate i is hidden in a single frequency index k_i of $\hat{M}$. To that end, we minimize the following objective function, one coordinate dimension $i = 1, ...d$ at a time:

$$f(\hat{M}_{k_i}) = \sum_{j=1}^{N-\ell+1} \left(\left(\sum_{n=0}^{\ell-1} \hat{M}_{k_i,j+n} \right) - T_{i,j} \right)^2 + \lambda \sum_{j=1}^{N} \left(M_{k_i,j} - \hat{M}_{k_i,j} \right)^2 \qquad (6)$$

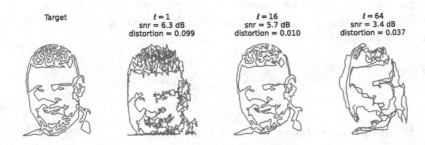

Fig. 4. Varying ℓ for a fixed $\lambda = 0.1$, using the lowest two non-DC frequencies to carry. A larger ℓ for the SW-STFT in Eq. 6 leads to smoother curves which are more likely to survive compression, but an ℓ that's too large may over-smooth.

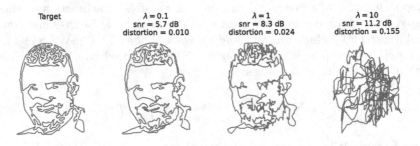

Fig. 5. Varying λ for a fixed $\ell = 16$. A smaller λ in Eq. 6 leads to higher geometric fidelity (Goal 2), at the cost of audio quality (Goal 1), as measured by SNR.

subject to $\hat{M}_{k_i,j} \geq 0$. In other words, we want the magnitude SW-STFT of a perturbed signal to match the target coordinate as well as possible (minimizing Eq. 3), while preserving the original audio as well as possible (minimizing Eq. 9), according to λ. A greater λ means that the signal $\hat{M}$ will fit the original audio better, at the cost of a noisier curve. Figure 5 shows an example.

After solving for $\hat{M}_{k_i,j}$, we replace all rows of M_{k_i} with $\hat{M}_{k_i}$, and we perform an inverse STFT of Me^{iP} using the original phases P (Sect. 2.3 will explain when it is necessary to modify the phases as well). We then save the resulting audio in a compressed format, and we recover the hidden signal by loading it and extracting the corresponding components of the magnitude SW-STFT.

Computational Complexity. Minimizing Eq. 6 can be formulated as a sparse nonnegative linear least squares problem, and there are myriad algorithms (e.g. [7]) for solving such systems efficiently via repeated evaluation of the linear system and its adjoint on iterative estimates of a solution. Furthermore, though Eq. 6 suggest an $O(N\ell)$ time complexity to evaluate the objective function at each iteration, we implement the linear operator and its adjoint with $O(N)$ operations only, independent of ℓ, using the 1D version of the "summed area tables" trick in computer vision [25]. For example, let $C_{k_i,j} = \sum_{i=0}^{j} \hat{M}_{k,i}$, and $C_{k_i,j<1} = 0$. Then Eq. 6 can be rewritten as

$$f_i(\hat{M}) = \sum_{j=1}^{N-\ell+1} \left((C_{k_i,j+\ell-1} - C_{k_i,j-1}) - T_{i,j}\right)^2 + \lambda \sum_{j=1}^{N} \left(M_{k_i,j} - \hat{M}_{k_i,j}\right)^2 \quad (7)$$

It is also worth noting that a higher λ in Eq. 6 leads to a lower condition number[2] of the matrix in the implied linear system, which leads to faster minimization of Eq. 6. But as Fig. 5 shows, it may still be worth it to use smaller λ values. In practice, we see it as a difference between an encoding in 30 s of audio that takes a few seconds for $\lambda = 0.1$ versus an encoding that takes a split second for $\lambda = 10$ on a CPU.

The Choice of Non-overlapping Windows. Though some other works use transforms with overlapping windows [16,37], changing different STFT bins independently leads to STFTs that do not correspond to any real signal due to discrepancies between overlapping windows. An algorithm like Griffin-Lim [20] can recover a signal whose spectrogram has a locally minimal distance to the perturbed STFT, but we find that this step introduces an unacceptable level of distortion both in the target and the carrier audio. This problem occurs even for real-valued transforms such as the Discrete Cosine Transform (which was used by [16]). Like [36], we sidestep this problem by using non-overlapping windows.

Aside from window effects, the main downside of non-overlapping windows is that they lead to fewer STFT frames. For instance, with a 30 s audio clip at 44100 hz using a window length of 1024, we are limited to about 1292 frames, which is half of what we would get using an overlapping STFT with a hop length of 512. However, a quick thought experiment shows that this is still a reasonable data rate. Suppose that we use a sliding window length $\ell = 16$, for a total of 1277 carrier samples. Suppose also that the precision of our embedding of a 2D plane curve is roughly on par with an 8 bit per coordinate quantization in a more conventional binary encoding scheme. Then the total number of bits transmitted over 30 s is $1277 \times 8 \times 2$, or about 681 bits/second. This number jumps to about 1022 bits/second for a 3D curve. These numbers are on par with other recent techniques that are designed to be robust to noise (e.g. 900 bits/second in the OFDM technique of [15]). Furthermore, our equivalent of a "bit error"

---

[2] The condition number of a matrix is defined as the ratio of the largest to smallest singular values, and lower condition numbers are more numerically desireable.

is additive coordinate noise, and the hidden signal degrades continuously with increased noise, rather than reaching a failure mode when a bit error is too high.

2.2   Shifting, Scaling and Re-parameterizing Targets for Better Fits

In this section, we explain how to modify the target curves to better match the given SW-STFT so that the STFT magnitudes don't have to be perturbed as much for the SW-STFT to match the target, leading to a less noticeable embedding of the hidden curve for the same quality geometry.

Vertical Translation/Scaling. A crucial step to keep the hidden signal imperceptible is to shift and rescale the target coordinates of the hidden curve to match the dynamic range of the SW-STFT components. We first choose a scale ratio a_i for each component as the ratio of the standard deviations $\text{stdev}_j \left(\text{SWS}^\ell(M)_{k_i,j} \right) / \text{stdev}_j(T_{i,j})$. Then, letting N be the length of the two signals, we compute the vertical shift b_i as

$$b_i = \frac{1}{N} \sum_{j=1}^{N} \text{SWS}^\ell(M)_{k_i,j} - a_i T_{i,j} \tag{8}$$

The rescaled target coordinate $\hat{T}_i$ is then defined as $\hat{T}_{i,j} = a_i T_{i,j} + b_i$. Unfortunately, we lose relative scale information between the components, but we will explain how to hide and recover this in Sect. 2.3

Viterbi Target Re-parameterization. Beyond shifting and scaling the targets coordinates vertically, we may also need to re-parameterize them in time, since, in general, the SWS-STFT sequence will not have the same number of samples as the shifted target $\hat{T}$. Furthermore, the target curves are cyclic, so the starting point is arbitrary, nor does it matter if we traverse the curve left or right, or at a constant speed. This gives us a lot of freedom in how we choose to re-parameterize the target to best match the signal even before we perturb the signal. Let N_M be the number of frames in the SW-STFT and N_T be the number of samples of the target, and assume that $N_T > N_M$ (we can always resample our target curves to make this true). Let $\Theta = \{\theta_1, \theta_2, \theta_3, \ldots, \theta_{N_M}\}$ be new indices into the target, and let $K > 0$ be a positive integer so that $1 \leq ((\theta_i - \theta_{i-1}) \mod N_T) \leq K$; that is, K is the maximum number of samples by which the target re-parameterization can jump in adjacent time steps. We seek a Θ minimizing the following objective function for a d-dimensional shifted target $\hat{T}$

$$g(\Theta = \{\theta_1, \theta_2, \ldots, \theta_{N_M}\}) = \sum_{j=1}^{N_M} \sum_{i=1}^{d} \left(\text{SWS}^\ell(M)_{k_i,j} - \hat{T}_{i,j} \right)^2 \tag{9}$$

Algorithm 1. Viterbi Target Re-Parameterization

1: **procedure** VITERBITARGETREPARAM($\hat{T}$, SWS$^\ell(M)$, K)
2:    $N_T \leftarrow \text{len}(\hat{T})$ ▷ Number of target points
3:    $N_M \leftarrow \text{len}(\text{SWS}^\ell(M))$ ▷ Number of SW-STFT frames
4:    $S[i > 1, j] \leftarrow 0$ ▷ $N_T \times N_M$ Cumulative cost matrix
5:    $S[1, j] \leftarrow \sum_{i=1}^{d}(\hat{T}_{i,1} - \text{SWS}^\ell(M)_{k_i,j})^2$
6:    $B[i, j] \leftarrow 0$ ▷ $N_T \times N_M$ backpointers to best preceding states
7:    **for** $j = 2 : N_M$ **do**
8:       **for** $t = 1 : N_T$ **do**
9:          $S[t, j] = \min_{k=(t-K) \bmod N_T}^{k=(t-1) \bmod N_T} S[k, j-1]$ ▷ Find best preceding state
10:         $S[t, j] \leftarrow S[t, j] + \sum_{i=1}^{d}(\hat{T}_{i,t} - \text{SWS}^\ell(M)_{k_i,j})^2$ ▷ Add on matching cost
11:         $B[t, j] = \arg\min_{k=(t-K) \bmod N_T}^{k=(t-1) \bmod N_T} S[k, j-1]$ ▷ Save reference to best
         preceding state
12:      **end for**
13:   **end for**
14:   Backtrace B to obtain the optimal sequence Θ
15:   **return** Θ
16: **end procedure**

For a fixed K, we use the Viterbi algorithm to obtain a Θ minimizing Eq. 9 in $O(MNK)$ time. Algorithm 1 gives more details.

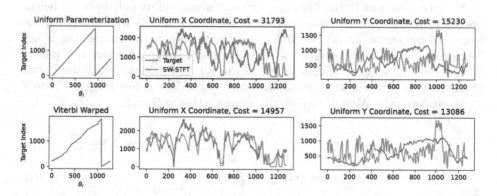

Fig. 6. Circularly shifting the target loop and traversing it at a non-uniform speed traces out the same shape, while matching the given SW-STFT better than a uniform parameterization starting at an artibrary place on the target.

In practice, we re-run the algorithm starting at $K = 1$, and we repeatedly increment K until the optimal Θ goes through at least one full loop on the target. Since clockwise or counter-clockwise traversal of the target is arbitrary, we then rerun this procedure again for a reversed version of the sequence and keep the result which minimizes Eq. 9. As a rule of thumb, we find that having a target curve with about 1.5–2x as many samples as there are SWS-STFT frames gives enough wiggle room for the Viterbi algorithm. In our experiments in Sect. 3, we will generate TSP and Hamiltonian sequences with 2000 samples for our 1200–1300 SWS-STFT frames. Figure 6 shows an example running this algorithm under these conditions on the Usher example in column 3 of Fig. 4.

2.3   Storing Component Scales in Phase

Since we intentionally rescale the dimensions of the target to match the dynamic range of each SW-STFT component, we lose the aspect ratio between the dimensions. But since we have only perturbed the magnitude components of the frequency indices k_i, we still have some freedom to perturb the phases to store additional information. To that end, we use the technique presented by the authors of [36] to store the relative scale of each dimension in the phase. We store the same scale in the phase of every STFT frame, and we take the scale to be the median of the phases upon decoding.

2.4   Recovering Frame Alignments

What we've described so far works for audio that is aligned to each window, but additional work needs to be done to address Goal 3 if the audio to decode comes in misaligned. To this end, we use the fact that the hidden curves move only slightly between adjacent samples; a TSP tour is defined as length-minimizing, and adjacent samples in Hamiltonian cycles on meshes move only between neighboring triangles on the original mesh. If the embedding is frame aligned, the length of the curve should be minimized. Conversely, if the embedding is not frame aligned, the curve becomes noisy and is more likely to jump around quickly from sample to sample. Therefore, we can pick the alignment which minimizes the length in all possible shifts from 0 to the STFT window length. Figure 7 shows an example. We will empirically evaluate this in Sect. 3.

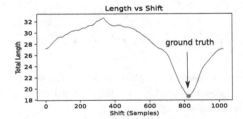
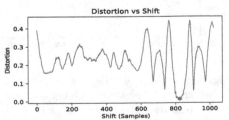

Fig. 7. A shift which minimizes curve length is most likely the shift needed to re-align audio to frame windows. In this example with the embedding using the lowest two non-DC frequencies with $\lambda = 0.1, \ell = 16$, the global mins exactly match ground truth.

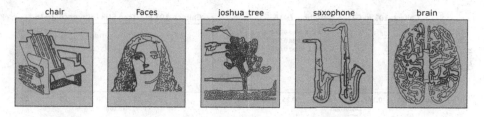

Fig. 8. Examples of TSP art on the Caltech-101 dataset [26].

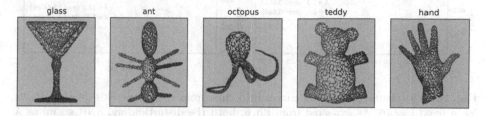

Fig. 9. Examples of Hamiltonian cycles on triangle meshes from the Princeton mesh segmentation benchmark [9].

3   Experiments

We now quantitatively assess the performance of our system. To generate a large set of curves, we generate TSP art on the roughly 10,000 images in the Caltech-101 dataset [26] (e.g. Fig. 8), and we generate Hamiltonian cycles on the 380 watertight triangle meshes in the Princeton mesh segmentation database [9] (e.g. Fig. 9). We then use the 1000 30 s audio clips from the Tzanetakis genre dataset [35] as carrier audio. For the Caltech-101 database, we partition the images into sets of 10, which are each encoded in one of the audio carriers varying λ and ℓ. Likewise, for the mesh segmentation dataset, we hide each Hamiltonian path in three different audio clips from the Tzanetakis dataset. In all cases, we use an STFT window length of 1024 at a sample rate of 44100 hz, and we encode the audio using lossy mp3 compression at 64 kbps.

Overall, we see slightly higher distortions and lower SNRs for 3D embeddings than 2D embeddings, which makes sense since there is one additional coordinate to hide in 3D. As expected, increasing λ increases both the SNR and distortion, as shown in Fig. 10. Also, as Fig. 11 shows, increasing the window length has a positive effect on geometric distortion, while moderate window lengths lead to the best SNR. We also see a positive effect of the Viterbi alignment from Sect. 2.2 on the SNR in Fig. 12. Though $\approx+1$dB may not seem significant, it can make a huge difference in audio, particularly in quiet regions. Finally, we run an experiment on the Caltech-101 dataset by choosing 4 random frame offsets per embedding, and we see in Fig. 13 that the algorithm of Sect. 2.4 recovers alignments well.

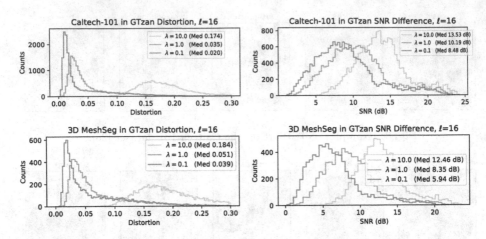

Fig. 10. The results embedding curves into clips from the Tzanetakis genre dataset, for a fixed $\ell = 16$. As expected from Eq. 6, both the distortion and SNR go up as λ increases.

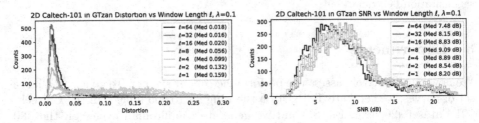

Fig. 11. Embedding curves into clips from the Tzanetakis genre dataset varying the window length ℓ for a fixed λ. Moderate window lengths are the best choices for both SNR and distortion. We recommend $\ell = 16$. (3D is similar; see supplementary)

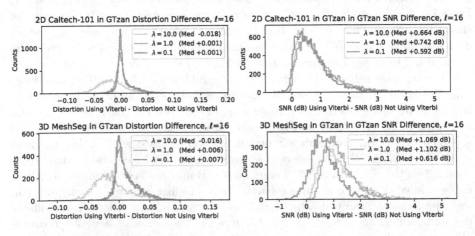

Fig. 12. Pre-warping the target with Viterbi alignment (Sect. 2.2) overall improves the resulting distortion and SNR.

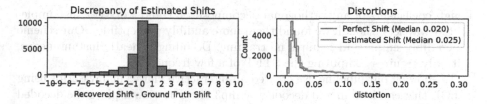

Fig. 13. Estimating the frame alignment by minimizing the length (Sect. 2.4) works nearly as well as perfect knowledge of the alignment. Using $\lambda = 0.1, \ell = 10$ over all images in Caltech-101, 93% of the shifts are within 10 of the ground truth for an STFT window length of 1024, which has hardly any effect on the geometry of the curve.

Subjective Listening Experiment. Though the experiments above are encouraging, SNR can be misleading; frequencies that are more audible may actually have a higher SNR due to psychoacoustic phenomena. To address this, we performed a crowd-sourced listening experiment on the Amazon Mechanical Turk where we embedded a random image from the Caltech-101 dataset in each of the Tzanetakis clips. We split them into 4 groups with no embedding (control), and with $\lambda = 0.1, 1, 10$. We asked the listeners to rate the quality of the noise on the 5 point impairment scale of [3] (5: imperceptible, 4: perceptible but not annoying, 3: slightly annoying, 2: annoying, 1: very annoying). In our experiment, we had 46 unique Turkers, 21 of whom participated in at least 40 rankings. Figure 14 shows the results. Mean opinion scores (MOS) are correlated with λ, but there is little difference between $\lambda = 0.1$ and $\lambda = 1$, which suggests using the former as a rule of thumb due to its lower geometric distortion.

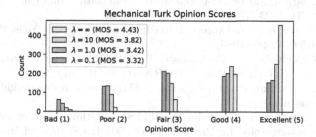

Fig. 14. Results of the listening experiment on the Amazon Mechanical Turk. A lower λ leads to a lower mean opinion score, as expected, though not to an intolerable degree.

4   Discussion/Supplementary Material

We have presented a model-based scheme for hiding artistic curves in audio, and the curves survive lossy compression while remaining reasonably imperceptible, as validated both with quantitative measurements and by humans. We hide the dimensions of curves in time regularized magnitudes of STFT frequencies, though coefficients of any orthogonal decomposition could work (we also implemented wavelets, though we found them more audibly perceptible). Our scheme is incredibly simple and requires no training. Decoding is nearly instantaneous, as it only requires computing the STFT of a few frequencies.

To show off our pipeline, we created an interactive viewer in Javascript using WebGL that can load in and decode any mp3 file. The viewer plays the decoded curve synchronized to the music. We provide a variety of 2D and 3D precomputed examples to demonstrate our capabilities. To view our supplementary material, source code for encoding and decoding, and live examples, please visit https://github.com/ctralie/AudioCurveSteganography.

References

1. Applegate, D.: Concorde tsp solver (2001). https://www.math.uwaterloo.ca/tsp/concorde.html. Accessed 12 Feb 2023
2. Atoum, M.S., Ibrahimn, S., Sulong, G., Zeki, A., Abubakar, A.: Exploring the challenges of mp3 audio steganography. In: 2013 International Conference on Advanced Computer Science Applications and Technologies, pp. 156–161. IEEE (2013)
3. Bassia, P., Pitas, I., Nikolaidis, N.: Robust audio watermarking in the time domain. IEEE Trans. Multimedia 3(2), 232–241 (2001)
4. Bosch, R.: Connecting the dots: the ins and outs of tsp art. In: Bridges Leeuwarden: Mathematics, Music, Art, Architecture, Culture, pp. 235–242 (2008)
5. Bosch, R.: Jordan as a jordan curve. Mathematical Wizardry for a Gardner, p. 175 (2009)
6. Bosch, R., Herman, A.: Continuous line drawings via the traveling salesman problem. Oper. Res. Lett. 32(4), 302–303 (2004)
7. Branch, M.A., Coleman, T.F., Li, Y.: A subspace, interior, and conjugate gradient method for large-scale bound-constrained minimization problems. SIAM J. Sci. Comput. 21(1), 1–23 (1999)
8. Canny, J.: A computational approach to edge detection. IEEE Trans. Pattern Anal. Mach. Intell., 679–698 (1986)
9. Chen, X., Golovinskiy, A., Funkhouser, T.: A benchmark for 3D mesh segmentation. ACM Trans. Graph. (Proc. SIGGRAPH) 28(3) (Aug 2009)
10. Cui, W., Liu, S., Jiang, F., Liu, Y., Zhao, D.: Multi-stage residual hiding for image-into-audio steganography
11. Cvejic, N., Seppanen, T.: A wavelet domain LSB insertion algorithm for high capacity audio steganography. In: Proceedings of 2002 IEEE 10th Digital Signal Processing Workshop, 2002 and the 2nd Signal Processing Education Workshop, pp. 53–55. IEEE (2002)
12. Djebbar, F., Ayad, B., Meraim, K.A., Hamam, H.: Comparative study of digital audio steganography techniques. EURASIP J. Audio Speech Music Process. 2012(1), 25 (2012)
13. Domènech Abelló, T.: Hiding images in their spoken narratives. Master's thesis, Universitat Politècnica de Catalunya (2022)
14. Dutta, H., Das, R.K., Nandi, S., Prasanna, S.R.M.: An overview of digital audio steganography. IETE Tech. Rev. 37(6), 632–650 (2020)
15. Eichelberger, M., Tanner, S., Voirol, G., Wattenhofer, R.: Receiving data hidden in music. In: Proceedings of the 20th International Workshop on Mobile Computing Systems and Applications, pp. 33–38. ACM (2019)

16. Geleta, M., Punti, C., McGuinness, K., Pons, J., Canton, C., Giro-i Nieto, X.: PixInWav: Residual steganography for hiding pixels in audio
17. Gopalan, K.: A unified audio and image steganography by spectrum modification. In: 2009 IEEE International Conference on Industrial Technology, pp. 1–5 (2009)
18. Gopalan, K., Wenndt, S.: Audio steganography for covert data transmission by imperceptible tone insertion. In: Proceedings the IASTED International Conference on Communication Systems and Applications (CSA 2004), Banff, Canada (2004)
19. Gopi, M., Eppstien, D.: Single-strip triangulation of manifolds with arbitrary topology. In: Computer Graphics Forum, vol. 23, pp. 371–379. Wiley Online Library (2004)
20. Griffin, D., Lim, J.: Signal estimation from modified short-time fourier transform. IEEE Trans. Acoust. Speech Signal Process. **32**(2), 236–243 (1984)
21. Gruhl, D., Lu, A., Bender, W.: Echo hiding. In: International Workshop on Information Hiding. pp. 295–315. Springer (1996)
22. Johnson, D.S., McGeoch, L.A.: The traveling salesman problem: a case study in local optimization. Local Search Comb. Optim. **1**(1), 215–310 (1997)
23. Kaplan, C.S., Bosch, R.: Tsp art. In: Renaissance Banff: Mathematics, music, art, culture, pp. 301–308 (2005)
24. Kolmogorov, V.: Blossom v: a new implementation of a minimum cost perfect matching algorithm. Math. Program. Comput. **1**(1), 43–67 (2009)
25. Lewis, J.: Fast template matching, vision interface 95. Canadian Image Processing and Pattern Recognition Society, pp. 15–19 (1995)
26. Li, A., Ranzato, P.: Caltech 101. Accessed 12 Feb 2023. doi:https://doi.org/10.22002/D1.20086
27. Li, H., Mould, D.: Structure-preserving stippling by priority-based error diffusion. In: Proceedings of Graphics Interface 2011, pp. 127–134 (2011)
28. Madhavapeddy, A., Scott, D., Tse, A., Sharp, R.: Audio networking: the forgotten wireless technology. IEEE Pervasive Comput. **4**(3), 55–60 (2005)
29. Malik, H.M.A., Ansari, R., Khokhar, A.A.: Robust data hiding in audio using allpass filters. IEEE Trans. Audio Speech Lang. Process. **15**(4), 1296–1304 (2007)
30. Mathews, P.D.: Music in his own image: The aphex twin face. Nebula **1**(1), 65–73 (2004)
31. Mokhtarian, F., Mackworth, A.K.: A theory of multiscale, curvature-based shape representation for planar curves. IEEE Trans. Pattern Anal. Mach. Intell. **14**(8), 789–805 (1992)
32. Qiao, M., Sung, A.H., Liu, Q.: Steganalysis of MP3stego. In: 2009 International Joint Conference on Neural Networks, pp. 2566–2571. IEEE (2009)
33. Secord, A.: Weighted voronoi stippling. In: Proceedings of the 2nd International Symposium on Non-Photorealistic Animation and Rendering, pp. 37–43 (2002)
34. Takahashi, N., Singh, M.K., Mitsufuji, Y.: Source mixing and separation robust audio steganography
35. Tzanetakis, G., Cook, P.: Musical genre classification of audio signals. IEEE Trans. Speech Audio Process. **10**(5), 293–302 (2002)
36. Xiaoxiao Dong, Bocko, M., Ignjatovic, Z.: Data hiding via phase manipulation of audio signals. In: 2004 IEEE International Conference on Acoustics, Speech, and Signal Processing, vol. 5, pp. V-377-80. IEEE (2004)
37. Yun, H.S., Cho, K., Kim, N.S.: Acoustic data transmission based on modulated complex lapped transform. IEEE Signal Process. Lett. **17**(1), 67–70 (2009)

LyricJam Sonic: A Generative System for Real-Time Composition and Musical Improvisation

Olga Vechtomova$^{(\boxtimes)}$ and Gaurav Sahu

University of Waterloo, Waterloo, Canada
{ovechtom,gsahu}@uwaterloo.ca

Abstract. Electronic music artists and sound designers have unique workflow practices that necessitate specialized approaches for developing music information retrieval and creativity support tools. Furthermore, electronic music instruments, such as modular synthesizers, have near-infinite possibilities for sound creation and can be combined to create unique and complex audio paths. The process of discovering interesting sounds is often serendipitous and impossible to replicate. For this reason, many musicians in electronic genres record audio output at all times while they work in the studio. Subsequently, it is difficult for artists to rediscover audio segments that might be suitable for use in their compositions from thousands of hours of recordings. In this paper, we describe LyricJam Sonic, a creative tool for musicians to rediscover their previous recordings, re-contextualize them with other recordings, and create original live music compositions in real-time. A bi-modal AI-driven approach uses generated lyric lines to find compatible audio clips from the artist's past studio recordings, and uses them to generate new lyric lines, which in turn are used to find other clips, thus creating a continuous and evolving stream of music and lyrics. The intent is to keep the artists in a state of creative flow conducive to music creation rather than taking them into an analytical/critical state of deliberately searching for past audio segments. The system can run in either a fully autonomous mode without user input, or in a live performance mode, where the artist plays live music, while the system "listens" and creates a continuous stream of music and lyrics in response (LyricJam Sonic: https://lyricjam.ai

Demo videos: https://sites.google.com/view/supplementary-material-for-evo/home).

Keywords: Generative Music · Lyrics Generation · Neural Network · Variational Autoencoder · Generative Adversarial Network

1 Introduction

Electronic music artists and sound designers can create complex and unique audio paths by patching multiple modules and standalone instruments to achieve a desired sound effect. Such audio paths typically include a combination of analog and digital sound processing. The process is inherently unpredictable, resulting in sounds that can be impossible to replicate, especially in complex audio paths. For this reason, many electronic artists record all of their studio sessions. Many artists in electronic genres also have an organic approach to composition, where they may start with an open mind and get inspiration by playing the instruments.[1] This is in part due to the experimental nature of the instruments and the importance of texture in electronic music. As a result of their music recording sessions, artists can accumulate many hours of recordings. Furthermore, since sampling has become one of the mainstream approaches to electronic music composition [8], many electronic artists create samples from their past recordings and integrate them into their new compositions, with some electronic compositions created entirely from samples. Finding a past recording to sample or use as overdub in a new composition is problematic, as sifting through and listening to hours of recordings is impractical and can take artists out of their creative flow.

In this work, we envisioned and built LyricJam Sonic, a AI-based creative tool for artists and composers to: 1) tap into their catalogue of studio recordings, 2) rediscover sounds, 3) re-contextualize them with other sounds and 4) interactively create new original music compositions or soundscapes with this tool.

An important consideration that such tool must have is to be conducive to creativity and not take the artist out of their creative flow. For this reason, our system was envisioned as facilitating serendipitous discovery, rather than enabling a deliberate search process controlled by the artist. The tool is therefore close to the philosophy of serendipitous discovery in electronic music composition.

In *live performance* mode, the system works by "listening" to the artist's live music performance, and playing a continuous and coherent musical accompaniment that it composes in real-time using the clips from the artist's own catalogue of studio recordings. In this mode, the system can be used by an artist as a partner in live performances.

In *autonomous* mode, the system creates a music composition or a soundscape without the artist's live music input. In this mode, the system relies on the previously played clip to predict the next one, and can be used by an artist as a composition tool in the studio to autonomously create new original compositions from the artist's catalogue of studio recordings. The artist can also switch between the interactive and autonomous modes to steer and guide the composition as needed.

In either of these modes, the system creates an original and coherent composition, re-contextualizing audio segments from past recordings. This process not only lets the artists hear clips of their previous recordings in new musical context, but also inspire them to combine these segments in new ways.

Concurrently with the next clip prediction, the system also generates lyric lines based on the audio it receives from the user (live mode) or the previous

---

[1] https://www.factmag.com/2020/01/08/alessandro-cortini-in-the-studio/.

clip (autonomous mode). The generated lyric line also influences the prediction of the next audio clip.

The system consists of two neural networks that interact with each other and generate continuous and evolving stream of music and lyric lines. One neural network generates lyric lines, while the other uses the generated line and the previously played clip to find a congruent audio clip from the artist's catalogue.

The rationale for using generated lyric lines in predicting the next audio clip, as opposed to just using the previously played clip (autonomous mode), is to introduce progression in the created music composition or soundscape, and also bring about elements of surprise. If the next sound clip was selected based solely on its degree of similarity to the previous clip, it would lead to monotonous compositions, effectively preventing any kind of musical transition and development. Past research has shown that conditional lyrics generation based on audio clips of instrumental music results in lyrics that are emotionally congruent with the music they are conditioned upon [20,21]. We expect that emotional valence of lyrics will similarly lead to the prediction of congruent audio clips. While lyric generation is not the main goal of this tool, the generated lyric lines can also be used for inspiration if an artist so desires.

2   Related Work

The present work builds upon LyricJam [21], which generates a stream of lyrics in real-time based on the music played by an artist. LyricJam was envisioned as a tool for generating lyric lines in response to live music performance, with the goal of inspiring songwriters to write their own lyrics. The main goal of LyricJam Sonic in to help artists re-discover sounds from their catalogue of past recordings that go with their live music, and to inspire them with new and possibly unexpected sequences of sounds brought together by the system. Both of these tools serve the purpose of supporting and enhancing artists' own creativity.

There is a large body of work in the area of generative music [2,10], which predominantly includes approaches to generating original raw audio [16] or music score [1,4]. A number of approaches address generation of specific instrumental accompaniment to existing music, e.g. drum tracks [13]; however, this body of work is largely not relevant to this paper as our goal is not to generate raw audio or music score.

Some of the earliest approaches to generative music that construct a new musical composition by remixing existing sound sources were pioneered in the 60s and 70s by electronic musicians. Notably, Brian Eno [6] created generative music by capturing the output of multiple magnetic tape recorders running at different speeds, each playing a single note. More recent algorithmic approaches to creating generative soundscape compositions through sampling or retrieving clips of existing recordings include [5,18,19], which combine heuristic rules with ranking functions to retrieve the next clip.

The majority of research using deep learning for music generation is aimed at developing systems that would create music autonomously [10]. The intent behind such systems is to create music that is ready to be used by non-musicians, for example, in commercials or documentaries. In contrast, our intent in this work is to design a tool that would be useful to musicians and composers, who create original music based on their own new and past sound recordings.

The main research contribution of this work is a new bi-modal self-perpetuating neural network architecture, where a lyric line is generated in real-time based on the previously played clip, both of which influence the selection of the next clip. The neural networks in our system are trained to predict the next clip based on the intrinsic properties of the raw audio of the previous clip and its accompanying lyric line. Furthermore, the system is trained to generate lyric lines conditioned on the raw audio, which differs from past works on lyric generation primarily conditioned on rules, e.g. melody and style [3,14,15,17,22].

3 Background

3.1 Variational Autoencoder (VAE)

A variational autoencoder (VAE) [12] is a stochastic neural generative model that consists of an encoder-decoder architecture. The key idea is to learn a posterior distribution that closely approximates the true posterior (also called prior). The distance between the approximate posterior and the prior is minimized using Kullback-Leibler divergence, a distance metric in the probabilistic space. VAE's main advantage over a vanilla autoencoder is learning a continuous latent space as opposed to a deterministic input-output mapping. This allows us to sample diverse outputs from the model at inference time, thus generating new original data, such as lyric lines.

3.2 Generative Adversarial Network (GAN)

A generative adversarial network (GAN) [7] is capable of generating new data samples that match the characteristics of a given data distribution. It consists of two main components: a generator G and a discriminator D. GAN's objective function incentivizes the generator to fool the discriminator by creating fake samples (also called adversarial samples) that closely resemble the training data distribution, while the discriminator tries to distinguish between an adversarial sample from the generator and a real data sample from the ground truth data distribution. By jointly training the generator and discriminator, the GAN learns to generate data that comes from the same distribution as the ground truth.

4 Methodology

The developed system consists of four main components:

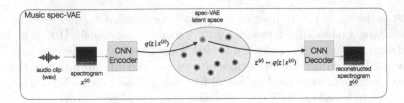

Fig. 1. Training spec-VAE.

1. A Spectrogram Variational Autoencoder (Spec-VAE), which is trained to learn latent representations (latent codes) of the spectrograms corresponding to audio clips.
2. A Text Conditional Variational Autoencoder (Text-CVAE), which is trained to learn latent representations of lyric lines, conditioned on the the latent code of the corresponding audio clips.
3. A Generative Adversarial Network (GAN) that predicts the latent code of the next audio clip given the element-wise addition of the latent codes of the corresponding lyric line and previous audio clip.
4. A Retrieval module, which retrieves the audio clip from the user's collection based on the cosine similarity of the GAN-predicted latent code and the latent codes of the audio clips in the collection.

4.1   Training Spec-VAE

We train the Spec-VAE model to learn the latent representations of audio clips. First, we convert the raw waveform audio files into Mel-spectrogram images using the method proposed in [20]. These spectrograms are then used as input for the spec-VAE.

The encoder transforms the input spectrogram image $x^{(s)}$ into the approximate posterior distribution $q_\phi(z|x^{(s)})$ learned by optimizing parameters ϕ of the encoder[2]. The decoder reconstructs x from the latent variable z, sampled from $q_\phi(z|x^{(s)})$. In our implementation, we use convolutional layers as the encoder, deconvolutional layers as the decoder, and a standard normal distribution as the prior distribution $p(z)$. The VAE loss function combines the reconstruction loss (Binary Cross-Entropy) and KL divergence loss that regularizes the latent space by pulling the posterior distribution to be close to the prior distribution (Figs. 1, 2 and 3).

4.2   Training Text-CVAE

Unlike the vanilla VAE used for encoding spectrograms, we use a conditional VAE (CVAE) for encoding lyrics. The CVAE learns a posterior distribution that is conditioned not only on the input data, but also on a class c: $q_\phi(z|x,c)$.

---

[2] Superscripts $^{(s)}$ and $^{(t)}$ in our notation refer to spectrogram and text, respectively.

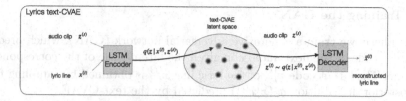

Fig. 2. Training text-CVAE.

Here, we define the class as the audio clip spectrogram corresponding to a given lyric line. Every conditional posterior distribution $q_\phi(z^{(t)}|x^{(t)}, z^{(s)})$ is pulled towards its corresponding prior $p(z^{(t)}|z^{(s)})$. Following prior research on CVAE, all conditional priors are set to the standard normal distribution.

During training, every input data point consists of a lyric line and its corresponding spectrogram. We first pass the spectrogram through the spec-VAE encoder to get the parameters of the posterior distribution (a vector of means and a vector of standard deviations). We then sample from this posterior to get a vector $z^{(s)}$ that is then concatenated with the input of the encoder and the decoder during training. The reason why we used sampling as opposed to the mean $z^{(s)}$ vector is to induce the text-CVAE model to learn conditioning on continuous data, as opposed to discrete classes. This prepares it to better handle conditioning on unseen new spectrograms at inference. Both the encoder and the decoder in the text-CVAE are Long Short Term Memory networks (LSTMs) [9]. The sampled $z^{(s)}$ is concatenated with the word embedding input to every step of the encoder and the decoder. The reconstruction loss is the expected negative log-likelihood (NLL) of data:

$$J_{\text{rec}}(\phi, \theta, z^{(s)}, x^{(t)}) = -\sum_{t=1}^{n} \log p(x_i^{(t)}|z^{(t)}, z^{(s)}, x_1^{(t)} \cdots x_{i-1}^{(t)}) \qquad (1)$$

where ϕ and θ are parameters of the encoder and decoder, respectively. The overall CVAE loss is given by:

$$J = J_{\text{rec}}(\phi, \theta, z^{(s)}, x^{(t)}) + KL(q_\phi(z^{(t)}|x^{(t)}, z^{(s)}|p(z^{(t)}|z^{(s)}))) \qquad (2)$$

where the first term is the reconstruction loss and the second term is the KL-divergence between z's posterior and a prior distribution, which is typically set to standard normal $\mathcal{N}(0, I)$.

Text Generation at Inference Time. We use the trained model to generate novel lyric lines at inference time. The encoder transforms the input sequence of words x into the approximate posterior distribution $q_\phi(z|x)$ learned by optimizing parameters ϕ of the encoder. The decoder reconstructs x from the latent variable z, sampled from $q_\phi(z|x)$. Both encoder and the decoder in our work are recurrent neural networks, specifically, LSTMs.

4.3   Training the GAN

In this phase, we train a generative adversarial network (GAN), which predicts the latent code of the next audio clip given the latent code of the corresponding lyric line. The latent code $z_i^{(t)}$ of the lyric line $x_i^{(t)}$ is obtained by sampling from the posterior distribution $q_\phi(z|x^{(t)})$ predicted by the text-CVAE.

The GAN architecture has a generator G and a discriminator D. For training the GAN, we follow these steps:

1. First, we input the lyric line $x_i^{(t)}$ to the encoder of the text-CVAE in order to obtain the latent code of the lyric line $z_i^{(t)} = \mu_i^{(t)} + \tau(\epsilon \cdot \sigma_i^{(t)})$.
2. After obtaining $z_{i-1}^{(s)}$ and $z_i^{(t)}$, we feed the element-wise addition of these two vectors $[z_{i-1}^{(s)} \oplus z_i^{(t)}]$ to the generator network G, which outputs a predicted latent code of the next audio clip $\hat{z}_i^{(s)}$. As an alternative to element-wise addition, we also evaluated (a) element-wise multiplication (Hadamard product) $[z_{i-1}^{(s)} \circ z_i^{(t)}]$ and (b) weighted element-wise addition $[W_s z_{i-1}^{(s)} \oplus W_t z_i^{(t)}]$, where W_s and W_t are trainable weights matrices learned concurrently with the GAN training routine.
3. The role of the GAN discriminator network D is to tell apart the "real" data from the "generated" data. The element-wise addition of the generator's output $\hat{z}_i^{(s)}$ with the lyric line latent code $z_i^{(t)}$ is the "generated" data $\hat{z} = [z_i^{(t)} \oplus \hat{z}_i^{(s)}]$. Whereas the "real" data $z = [z_i^{(t)} \oplus z_i^{(s)}]$ is the element-wise addition of the lyric line latent code and the actual latent code $z_i^{(s)}$ of the next audio clip $x_i^{(s)}$, which is obtained from the spec-VAE by following the same process as in step 1. The discriminator D tries to distinguish between the two types of inputs ("real" vs. "generated"). This adversarial training regime incentivizes the generator G to match $\hat{z}^{(s)}$ as closely as possible to $z^{(s)}$.

The adversarial loss is formulated as follows:

$$\min_G \max_D V(D, G) = \mathbb{E}_{x \sim \mathcal{D}_{train}}[\log D(z^{(s)}) + \log(1 - D(\hat{z}^{(s)}))] \qquad (3)$$

where $\mathcal{D}_{train}$ is the training data, and each sample $x = \{x_{i-1}^{(s)}, x_i^{(t)}, x_i^{(s)}\}$. We also add an auxiliary MSE loss to the objective function as it is found to stabilize GAN training [11]. The overall loss for the GAN is:

$$J_{GAN} = \min_G \max_D V(D, G) + \lambda_{MSE}||\hat{z}^{(s)} - z^{(s)}|| \qquad (4)$$

4.4   Retrieval Module

The system requires a collection of music audio clips $X_{data}^{(s)}$ that the model would draw from to dynamically create a music or soundscape composition. According to the intention of the application, this dataset could be comprised of the

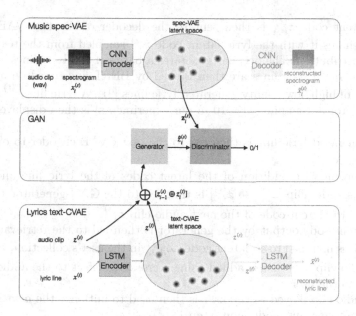

Fig. 3. Training GAN.

recordings from the artist's studio sessions. After the spec-VAE has been trained, we run it in inference mode by feeding each audio clip $x^{(s)} \in X_{data}^{(s)}$ to obtain the latent code $z^{(s)}$ by sampling from the posterior distribution $q_\phi(z|x^{(s)})$. This results in a set $Z^{(s)}$ of all latent codes.

After the GAN generator G outputs the predicted latent code $\hat{z}_i^{(s)}$ of the next audio clip, the latent code is sent to the retrieval module, which uses cosine similarity to rank all latent codes $z^{(s)} \in Z_{data}^{(s)}$ of the audio clips in the music collection $X_{data}^{(s)}$. The latent code $z^{(s)}$ of the next audio clip is selected using either argmax or top-K sampling, where K could be a user-controlled hyperparameter. The audio clip $x_i^{(s)}$ corresponding to the selected latent code is then added using a crossfade effect to the music composition, played to the user of the application.

4.5 Inference

The system can run indefinitely in a fully autonomous mode without user input. In this mode the generated lyric line and currently playing audio clip influence the prediction of the next audio clip, which in turn influences the generation of the next lyric line, and so on. In more detail, the process (Fig. 4) consists of the following steps.

1. The process is initially seeded with a random audio clip $x_{i-i}^{(s)}$ from the user's collection $X_{data}^{(s)}$. Its corresponding latent code $z_{i-i}^{(s)}$ is obtained by sampling from the posterior distribution predicted by spec-CVAE.

2. The latent code $z_{i-i}^{(s)}$ is then sent to the decoder of the text-VAE, which concatenates it with the lyric latent code $z^{(t)}$ sampled from the text-CVAE prior distribution. The decoder generates a batch of 100 lyric lines.
3. The generated lyric lines are then ranked by BERT fine-tuned on a custom dataset of high/low quality generated lyric lines [21]. A lyric line $x_i^{(t)}$ selected through top-K sampling (K=10 in our experiments) is then displayed to the user.
4. The generated lyric line is then fed to the text-CVAE encoder to obtain its latent code $z_i^{(t)}$
5. The element-wise addition of the latent codes of the lyric line and of the previous audio clip $[z_{i-1}^{(s)} \oplus z_i^{(t)}]$ is then fed to the GAN generator G, which predicts the latent code of the next audio clip $\hat{z}_i^{(s)}$.
6. The latent code output by the generator is then fed to the retrieval module where it is matched to all latent codes $Z_{data}^{(s)}$ in the user's collection, returning the audio clip $x_i^{(s)}$ that is added using a crossfade effect to the audio stream played to the user.
7. The audio clip played to the user is then used to initialize the next iteration of the autonomous mode loop (step 1).

At any time, the system can accept user's input in the form of either a live-recorded audio clip or a new lyric line or both (Fig. 5). This is intended as a mechanism for the user to influence the generative process, and steer the system in a certain direction of user's choosing. For example, if the system is playing ambient music, the user can play a heavy drum or bass track to dynamically steer the composition being created into that genre. If the user records an audio clip, it replaces the $x_i^{(s)}$ in Step 7 above, and is used to initialize the subsequent iteration through the system. If the user provides a lyric line, it interrupts the feedback loop in Step 4, replacing the system-generated lyric line. After one iteration, the system returns to using its own generated lyric lines unless the user provides a new line.

5   Data and Evaluation

5.1   Data

Spec-VAE. For training the Spec-VAE, we combine clips from two sources:

1. A lyric-music aligned dataset developed by Vechtomova et al. [20], which contains 22,763 WAV audio clips of original songs by seven music artists in various sub-genres of "Rock", including two artists who create Electronic rock sub-genre with extensive use of synthesizers.
2. Our new dataset of 4814 10-second WAV audio clips (over 13 h) of original electronic instrumental music created by an electronic music artist during studio recording sessions over a period of one year. We will refer to our this dataset as $X_{data}^{(s)}$.

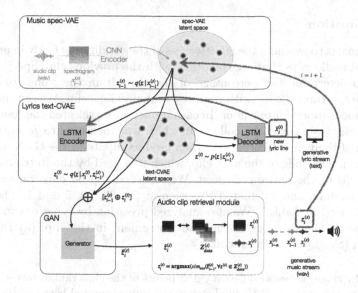

Fig. 4. Inference without user input (autonomous mode).

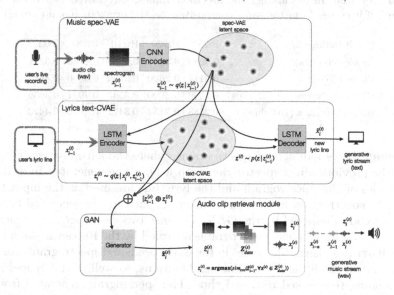

Fig. 5. Inference with optional user input.

Text-CVAE and GAN. To train the Text-CVAE, we use the *aligned* lyric-music dataset [20], which contains 18,000 lyric lines and their corresponding audio clips. To train the GAN, we encode the clips and their corresponding lyric lines using trained Spec-VAE and Text-CVAE, respectively, to obtain a dataset of latent codes. We prepare the latent codes beforehand to reduce GAN's training time.

5.2   Evaluation

Our first goal is to evaluate the effectiveness of the conditional GAN in predicting subsequent audio clips that are congruent with the previous clip. Specifically, we set out to determine the percentage of the clips from the same composition among the top-ranked clips with respect to the previous clip. We ran the system in autonomous mode for one hour. In each iteration, we selected the conditioning clip randomly from the entire collection, which was then used to generate a lyric line. Both of their latent codes were then used as inputs to the GAN module to predict the latent code of the next clip, which was used by the Retrieval module to rank clips in the entire collection. We recorded top 50 clips ranked by the Retrieval module, and calculated Precision at 50, 20, 10, 5 and 1. The results are summarized in Table 1. We also analyzed precision by types of conditioning clips based on the category of the lead instrument in the clip: (a) Drone, (b) Piano and (c) Rhythm/Percussion.

Table 1. Precision values at different cutoff points of the clips ranked by the Retrieval module in response to the GAN-predicted latent code generated based on the previous clip and lyric line. In calculating the precision values, a retrieved clip is considered "relevant" if it comes from the same composition as the previous (conditioning) clip.

Conditioning clips	P@50	P@20	P@10	P@5	P@1
Drone (46 clips)	0.3083	0.3667	0.4087	0.4783	0.5217
Piano (36 clips)	0.2111	0.2600	0.2750	0.3000	0.3889
Rhythm/Percussion (22 clips)	0.2336	0.3000	0.3363	0.4091	0.3636
All instruments (104 clips)	0.2588	0.3157	0.3471	0.4019	0.4423

Figure 6 shows four sets of spectrograms, where the left-most figure in each row is the previous clip's spectrogram with a lyric line generated from it. The latent codes of this spectrogram and the lyric line are used as the input to the GAN. The spectrogram in the second image in each row is generated based on the latent code predicted by the GAN. The last two spectrograms in each row are the top first and second spectrograms returned by the Retrieval module. The Figure illustrates visual differences in the GAN-predicted spectrograms based on different previous clip and lyric line combinations, as well as differences in the corresponding top-ranked retrieved clips. The spectrogram generated from the GAN-predicted latent code bears overall visual resemblance to its conditioning spectrogram, and both have similarities with the top retrieved spectrograms.

The results show that quite a substantial number of top-ranked clips come from the same composition. This is especially the case in the Drone category, where in 52% of cases the top-ranked clip (P@1) comes from the same composition as the conditioning clip. This points to the good ability of the system to predict clips that are musically related to the previous clip, resulting in a coherent music. However, the fact that not all top-ranked clips are from the same composition is essential, as this means that the music can evolve and not be stuck in the same composition.

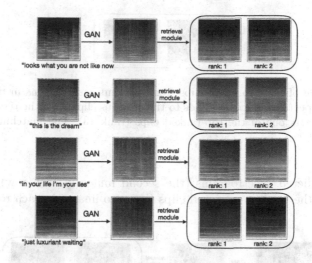

Fig. 6. Conditional GAN spectrogram predictions based on the previous clip and lyric lines. Each row shows the spectrogram of the previous clip with a lyric line generated from it, the generated spectrogram from the GAN-predicted latent code and spectrograms of the top two clips returned by the retrieval module.

One research question we would like to explore is how sensitive the system is in predicting related piano compositions. This is an interesting question, because most piano compositions in our dataset have a motif and are typically played in a specific key, so prediction of the next clip from a composition in a different key may clash with the previous clip and not result in a smooth transition or an aesthetically pleasing music. For the 36 piano clips in Table 1, the total number of predicted clips among the top 10 that come from the same piano composition as the conditioning clip is 99, while the number of clips from a different piano composition is only 10. This suggests that the model is able to predict piano clips that come from the same composition and therefore have similar musical characteristics, such as key. Furthermore, upon listening to the 10 clips that came from a different composition, we observed that they evoke a similar mood to the conditioning clip and are typically played in the same key. Figure 7 demonstrates visual similarities between the spectrograms of two such clips. Specifically, yellow horizontal lines at different heights in the spectrograms correspond to the frequencies of the played notes or chords. This provides further evidence that the GAN is able to generate a latent code representing a clip that is close to the preceding clip at a fine level of granularity, including the notes or chords played.

To analyze the extent that GAN predictions are influenced by lyric lines, we fixed the conditioning spectrogram and varied the lyric lines. Figure 8 shows examples from the system with this setting. The lyric line has an effect on the predicted spectrogram, as can be seen both from the visual differences in

Fig. 7. Example of the two piano clips from different compositions in the same key. The left-hand spectrogram corresponds to the previous clip, while the right-hand spectrogram represents one of the top-ranked clips (rank 13). The matching chords are highlighted.

the GAN-predicted spectrograms (the second image in each row) and in the differences of the top two retrieved clips (last two images in each row).

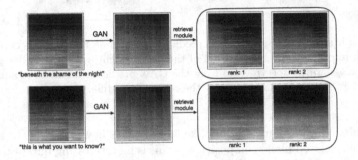

Fig. 8. Conditional GAN spectrogram predictions based on the previous clip and lyric lines. To demonstrate the effect of the lyric line on the prediction of the next clip, the previous audio clip is fixed, while the lyric line is different.

To quantitatively evaluate the effect of the lyric line on the next clip prediction, we randomly selected seven clips in each of the three categories: Drone, Piano, Rhythm. We let the system run for 10 iterations in autonomous mode with a given clip fixed as the conditioning clip. In each iteration, the system randomly selected a lyric line from the top 10 lyric lines generated by text-CVAE conditioned on this clip and ranked by the fine-tuned BERT model. Thus, for each conditioning clip we obtained set N, which is a union of n top-ranked clips across all 10 iterations. We then computed Impact@n as the number of unique clips in set N out of the total number of clips in N. In our experiments, n was set to 10, 5 and 2. Table 2 shows the results of the evaluation per category and for all clips.

The results in the Table show that on average 20.48% of all clips among the top-2 clips predicted for the same previous clip are unique, which indicates that the lyric line has a noticeable effect on the prediction of the next clip. The effect of the previous clip on the prediction is stronger than the effect of the lyric line, which is desirable, as we would like the next clip to be musically consistent with the previous clip. The lyric line therefore acts more as a nudge towards a certain

Table 2. Impact of variable lyric lines on the top-ranked clips, given a fixed conditioning clip. Impact is measured at 10, 5 and 2 top-ranked clips.

Conditioning clips	Impact@10	Impact@5	Impact@2
Drone (7 clips)	0.1443	0.1771	0.1929
Piano (7 clips)	0.1943	0.2171	0.2357
Rhythm/Percussion (7 clips)	0.1471	0.1686	0.1857
All instruments (21 clips)	0.1619	0.1876	0.2048

mood or a musical theme, rather than causing a drastic change in the music composition being created. The variance calculated on the 21 clips is 0.0019, 0.0036 and 0.0045 for the top 10, 5 and 2 sets, respectively. While the variance is low, we observed that some conditioning clips have lower diversity than others. When a clip comes from a composition that is highly dissimilar to the others in the collection, the top-ranked clips are more likely to also come from the same composition. This is because the variations attributed to the lyric line may not be strong enough for the GAN to generate a sufficiently different latent code that would make it close to clips from the other compositions.

6   Listening Tests

We conducted a listening test study to determine whether the system constructs more musically coherent compositions compared to audio segments made of randomly selected audio clips. We recorded 30 one-minute audio segments by running LyricJam Sonic in autonomous mode (test condition). Similarly, we recorded 30 one-minute audio segments by letting the system pick random clips in each iteration (control condition). The segments in control and test conditions were randomly paired, and the order of segments in each pair was randomized.

Two participants familiar with electronic music genre participated in the study. Participant A is a musician and producer, playing guitar and bass. Participant B is a vocalist, having performed as a lead singer in live concerts. Each participant was given the same 30 pairs of audio segments. For each pair, the participant was asked to listen to each audio segment, and select the one which they perceived to be more musically coherent. As a result of the study, 22 and 25 of the test condition clips were determined by Participants A and B, respectively, to be more coherent than the control condition clips. The overall accuracy is 78.3%. This suggests that the compositions created by LyricJam Sonic are perceived to be more musically coherent than segments with randomly selected clips.

7   Conclusions

We presented a real-time system LyricJam Sonic that dynamically creates a continuous stream of music from the clips in the artist's own catalogue of music recordings and generates corresponding lyrics. Using the system, artists can rediscover sounds from their own past recordings and re-contextualize them with the new sounds, creating new and original music compositions or soundscapes.

The self-perpetuating system can work in autonomous mode, whereby a generated lyric line and a previously played audio clip are used to predict the next audio clip, which in turn conditions the generation of the next lyric line. The system can also be used in live performance mode, whereby the musician plays live music, and the system "listens" and creates a continuous stream of music and lyrics in response to live music. The system is available to the public as an interactive web-based application that creates music compositions in real-time based on over 13 h of electronic music recorded by an electronic artist.[3]

The results of a listening study showed that participants perceived the compositions created by LyricJam Sonic to be more musically coherent than audio segments composed of random audio clips. Furthermore, automatic evaluation demonstrated that while the top ranked clips contain many clips from the same composition, they also include clips from other compositions. This suggests that while the created musical compositions have musical coherence, they also have musical diversity, which allow the composition to musically develop over time. Finally, the results of evaluation demonstrated a noticeable impact of generated lyric lines on the selection of the following audio clips.

References

1. Agarwal, S., Saxena, V., Singal, V., Aggarwal, S.: LSTM based music generation with dataset preprocessing and reconstruction techniques. In: 2018 IEEE Symposium Series on Computational Intelligence (SSCI), pp. 455–462 (2018). https://doi.org/10.1109/SSCI.2018.8628712
2. Briot, J., Hadjeres, G., Pachet, F.: Deep learning techniques for music generation. Computational Synthesis and Creative Systems, Springer, Cham (2019). https://doi.org/10.1007/978-3-319-70163-9. https://books.google.ca/books?id=_flrswEACAAJ
3. Chen, Y., Lerch, A.: Melody-conditioned lyrics generation with seqGANs. In: 2020 IEEE International Symposium on Multimedia (ISM), pp. 189–196 (2020). https://doi.org/10.1109/ISM.2020.00040
4. Dua, M., Yadav, R., Mamgai, D., Brodiya, S.: An improved RNN-LSTM based novel approach for sheet music generation. Procedia Comput. Sci. **171**, 465–474 (2020). In: Third International Conference on Computing and Network Communications (CoCoNet'19)
5. Eigenfeldt, A., Pasquier, P.: Negotiated content: generative soundscape composition by autonomous musical agents in coming together: Freesound. In: ICCC, pp. 27–32 (2011)
6. Eno, B.: Generative music. http://www.inmotionmagazine.com/eno1.html (1996). Accessed 17 Dec 2022
7. Goodfellow, I., et al.: Generative adversarial nets. In: Advances in Neural Information Processing Systems 27 (2014)

---

[3] https://lyricjam.ai.

8. Harkins, P.: Digital sampling: the design and use of music technologies. Routledge (2019)

9. Hochreiter, S., Schmidhuber, J.: Long short-term memory. Neural Comput. **9**(8), 1735–1780 (1997)

10. Hunt, S.J., Mitchell, T., Nash, C.: Thoughts on interactive generative music composition (2017)

11. Khan, K., Sahu, G., Balasubramanian, V., Mou, L., Vechtomova, O.: Adversarial learning on the latent space for diverse dialog generation. In: Proceedings of the 28th International Conference on Computational Linguistics, pp. 5026–5034. International Committee on Computational Linguistics, Barcelona, Spain (2020). https://doi.org/10.18653/v1/2020.coling-main.441. https://www.aclweb.org/anthology/2020.coling-main.441

12. Kingma, D.P., Welling, M.: auto-encoding variational Bayes. In: 2nd International Conference on Learning Representations, ICLR 2014, Banff, AB, Canada, 14–16 April 2014, Conference Track Proceedings (2014)

13. Makris, D., Zixun, G., Kaliakatsos-Papakostas, M., Herremans, D.: Conditional drums generation using compound word representations. In: Martins, T., Rodríguez-Fernández, N., Rebelo, S.M. (eds.) EvoMUSART 2022. LNCS, vol. 13221, pp. 179–194. Springer, Cham (2022). https://doi.org/10.1007/978-3-031-03789-4_12

14. Malmi, E., Takala, P., Toivonen, H., Raiko, T., Gionis, A.: DopeLearning: a computational approach to rap lyrics generation. In: Proceedings of the 22nd ACM SIGKDD International Conference on Knowledge Discovery and Data Mining, pp. 195–204 (2016)

15. Oliveira, H.G.: Tra-la-Lyrics 2.0: automatic generation of song lyrics on a semantic domain. J. Artif. General Intell. **6**(1), 87 (2015)

16. van den Oord, A., et al.: WaveNet: a generative model for raw audio. arXiv preprint arXiv:1609.03499 (2016)

17. Potash, P., Romanov, A., Rumshisky, A.: Ghostwriter: using an LSTM for automatic rap lyric generation. In: Proceedings of the 2015 Conference on Empirical Methods in Natural Language Processing, pp. 1919–1924 (2015)

18. Thorogood, M., Pasquier, P., Eigenfeldt, A.: Audio metaphor: audio information retrieval for soundscape composition. In: Proceedings of the Sound and Music Computing Conference (SMC), pp. 277–283 (2012)

19. Turchet, L., Zanetti, A.: Voice-based interface for accessible soundscape composition: composing soundscapes by vocally querying online sounds repositories. In: Proceedings of the 15th International Audio Mostly Conference (2020)

20. Vechtomova, O., Sahu, G., Kumar, D.: Generation of lyrics lines conditioned on music audio clips. In: Proceedings of the 1st Workshop on NLP for Music and Audio (NLP4MusA), pp. 33–37. Association for Computational Linguistics (2020). https://aclanthology.org/2020.nlp4musa-1.7

21. Vechtomova, O., Sahu, G., Kumar, D.: LyricJam: a system for generating lyrics for live instrumental music. In: Proceedings of the 12th Conference on Computational Creativity (2021)

22. Watanabe, K., Matsubayashi, Y., Fukayama, S., Goto, M., Inui, K., Nakano, T.: A melody-conditioned lyrics language model. In: Proceedings of the 2018 Conference of the North American Chapter of the Association for Computational Linguistics: Human Language Technologies, Volume 1 (Long Papers), pp. 163–172 (2018)

Searching for Human Bias Against AI-Composed Music

Dimiter Zlatkov[1](✉), Jeff Ens[2], and Philippe Pasquier[2]

[1] University of British Columbia, Vancouver, BC V6T 1Z1, Canada
dzdimi14@gmail.com
[2] Metacreation Lab, Simon Fraser University, Burnaby V5A 1S6, Canada
https://metacreation.net/human-bias-against-ai-composed-music/

Abstract. With the popularization of musical AI in society comes the question of how it will be received by the public. We conducted an empirical study to investigate the hypotheses that human listeners hold a negative bias against computer-composed music. 163 participants were recruited from Amazon's MTurk to fill out a survey asking participants to rank 5 computer-composed and 5 human-composed musical excerpts based on subjective musical preference. Participants were split into two groups, one informed of correct authorship, the other deceived. The hypothesis, that those in the informed group would rank computer-composed excerpts as lower than human-composed excerpts, was not supported by significant results. We outline potential weaknesses in our design and present possible improvements for future work. A review of related studies on bias against AI-composed music and art is also included.

Keywords: Musical Metacreativity · Computational Creativity · Human-computer Interaction · Bias

1 Introduction

Neural Networks are becoming widely used to automate a number of tasks. They can act as our chauffeurs and personal assistants and they can even outperform humans at skilled tasks such as screening for melanoma (Fogel and Kvedar, 2018). While it is becoming accepted that humans can be outperformed by computers in various logical tasks, the general belief holds that this is not the case when it comes to more creative endeavors. However, AI systems, such as Google's DeepDream, are able to create novel visual artwork using their advanced neural networks and large databases (Fogel and Kvedar 2018; Marzano and Novembre 2017). Beyond visual artwork, computer systems have also created literary works and films (Hong and Curran 2019).

There have been computer-composed musical works as early as 1957 when professors of music, Hiller and Issacson, created the Illiac Suite for strings using the Illiac I computer at the University of Illinois Urbana-Champaign. Today, artificially composed music has come a long way to the point where programs such as audiometaphor.ca (Thorogood et al., 2022) will create a soundtrack

C. Johnson et al. (Eds.): EvoMUSART 2023, LNCS 13988, pp. 308–323, 2023.
https://doi.org/10.1007/978-3-031-29956-8_20

based on only a few user imputed key-words. There are even pop albums being released which use artificially composed music such as Hello World by SKYGGE (helloworldalbum.net 2021). This new field of creative music systems using AI has become known as Musical Metacreativity (https://musicalmetacreation.org). Musical Metacreativity (MuMe) systems range from tools meant to help users create music to more autonomous systems, which can both compose and play their own novel works.

Computational creativity presents a number of philosophical questions, particularly whether an AI system can be creative. Where general intelligence can often be measured by various tests, creative works do not have an optimal solution. The evaluation of creative works is fundamentally a subjective process. Critics argue that because many MuMe systems are trained using large databases of human-composed work, their compositions are not capable of being creative since they are imitations of past work (Jennings 2010). However, it is also true that artistic inspiration often comes from the work of others (Jackson 2017). Even though computer-composed works are indistinguishable from that of humans, there are still many people who argue these works are not "human-like" (McCarthy 2007). The question still remains whether people are willing to accept the creativity shown by MuMe systems. More specifically, do people hold a bias against computer created artwork such as music?

2   Background

2.1   Bias Against Computer Composed Artwork

Rather than looking at a bias against AI-composed artwork in the form of musical compositions, Norton et al. (2013) present a system called DARCI which was designed to render images to match a list of adjectives. Their system is based on Colton's creative tripod (2008) which posits there are three necessary behaviours for a system to be considered creative: skill, imagination, and appreciation. Colton defines skill as the ability to produce functional or quality artifacts which are recognized as members of their intended domain. Colton's definition of imagination adds that these artifacts must be original and meaningful in some way. Finally, Colton's appreciation is the ability of the system to evaluate its own works.

DARCI demonstrates creativity through Colton's three necessary behaviours. It demonstrates skill through creating original images that correlate with the appropriate adjectives it is given, imagination by generating these unpredictable yet non-random images, and appreciation by evaluating how strong of a match each image is to a database of adjective semantics. The images created by DARCI are created through an evolutionary process. In this process, a myriad of possible images is defined through a specific encoding called a genotype, and the rules used to transform the genotype into the desired output is called a phenotype. A population of random genotypes are evaluated based on the qualities of their respective phenotypes. This evaluation then goes through a fitness function which determines which genotypes pass on to future generations of the evolutionary system. DARCI uses a fitness function which uses artificial neural networks to model user aesthetics and image features as the input to the neural network. DARCI is unique in that it focuses on computational creativity rather than pure evolutionary art. Therefore, DARCI's fitness function is made up of many neural networks each corresponding to a specific adjective, this way DARCI's fitness function does not measure a single kind of aesthetic sense but rather an overall sense of what an image means.

In addition to creating DARCI, Norton et al. developed a set of metrics for evaluating their system as well as an online survey to measure the novelty and quality of DARCI's work. Their survey had 6 questions all of which were answered using a five-point likert-type scale. The questions were specially designed to assess the creativity behind DARCI's creations. They asked participants whether they like the image, whether they think the image is novel, whether they would use the image as a desktop wallpaper, whether they had seen the image before, if they thought the image was hard to make, and if it was creative. The images chosen for the survey were created from the same source image and modified by one of ten adjectives: bright, cold, creepy, happy, luminous, peaceful, sad, scary, warm and weird.

The results of their survey suggest that people do in fact consider some of DARCI's creations to be creative. It was found that adjectives which describe emotion (peaceful, scary, happy, sad, and creepy) scored the highest on average in the survey. Adjectives which only described particular attributes (bright, warm and luminous) scored lowest. Interestingly, the highest scoring image was created off the adjective weird, which while not necessarily an emotion, is consistent with certain components of creativity such as novelty.

Similar to Norton et al. (2013), Ragot et al. (2020) conducted an experiment examining whether there is a negative perception towards AI-created paintings. They recruited 565 participants to evaluate paintings, produced by AI or humans, based on four dimensions: liking, perceived beauty, novelty, and meaning. Similar to previous studies, they used a priming effect stating whether each work was created by AI or a human. Participants were split into an AI-condition group and a human-condition group. Those in the AI condition were primed to believe that all the paintings they were presented were by AI, and those in the Human condition were told the paintings presented to them were created by human artists.

Participants in each group were presented with 8 different paintings randomly selected from 40 total paintings: 10 portraits created by AI, 10 landscapes created by AI, 10 landscapes created by humans, and 10 portraits created by humans. Participants in both groups were actually shown a mix of paintings created by AI and human artists. After rating each painting on the four dimensions, participants were asked if they remember whether the paintings they were presented were created by AI or humans as a manipulation check. Afterwards, participants were told the origin of the paintings were manipulated and asked to guess the origin of four randomly selected paintings. This was conducted as a modified Turing-test to avoid any bias in evaluation.

Ragot et al. (2020) found that participants did in fact rate art presented as AI-created to be significantly less liked and perceived as less beautiful, novel, and meaningful than those presented as being created by humans. Their results support those of Moffat and Kelly in the wider question of assessing a bias against computational creativity. They also found there was higher recognition of correct authorship for human paintings than AI paintings in their modified Turing-test.

Ragot et al. present these results to possibly be an effect of an inter-group bias. Inter-group bias generally refers to the tendency to evaluate one's own membership group (in-group) more favourably than a non-membership group (out-group) (Hewstone et al. 2002). In-group members show higher trust, cooperation, empathy, and positive regard to other members of their group, but not to out-group members. This discrimination towards those in the out-group could explain why participants rated human-created art (in-group) higher than AI-created (out-group) paintings. A key moderator of inter-group bias is threat. Threat can be defined in terms of the in-group's social identity, goals, values, position in the hierarchy, or even its existence (Hewstone et al. 2002). Ragot et al. propose that, in terms of their experiment, participants could view the existence of high quality computational creativity as a threat to their in-group (humans). Fear of AI systems often presents itself with the idea of people's jobs being displaced due to AI automation (McClure 2017). Perhaps it is due to a fear of human creativity being replaced that human artists are being rated higher than AI artists.

2.2 Bias and Musical Metacreativity

Multiple experimental studies have been conducted to test how people perceive AI-created artwork. Rather than testing the creativity of AI created art, Moffat and Kelly (2006) looked for a proposed bias against computer-composed music. Their experiment asked twenty participants to discern whether each composition was composed by a human or computer in a Turing test-like fashion. Participants were first tested on their music knowledge/experience and then subsequently divided into 'Musician' and 'Non-Musician' groups. The participants in both groups were given six one-minute long musical excerpts, three of these were human-composed and the other three composed by various computer systems. These pieces came in the form of three different styles: "Bach", "Strings", and "Free-form Jazz". After listening to each composition, participants indicated how much they "liked" a particular piece on a 5-point Likert scale. After this round, the authorship of each piece was revealed and the participants were then asked to rate each piece again, although in a disguised manner asking how willing they would be to buy, download, or recommend each piece to someone.

Moffat and Kelly found that their participants appeared to show prejudice against the AI-composed pieces, however, these findings were not strong enough to be deemed statistically significant due to the small sample size. They did, however, find significant data that participants seemed to consistently prefer human-composed music to computer-composed music. They also found that participants were good at determining which pieces were computer-composed. Surprisingly, non-musicians outperformed musicians at this task.

Pasquier et al. (2016) built on the work of Moffat and Kelly (2006), conducting an empirical study investigating whether listeners hold a bias against computer composed music. Their study sought to improve upon previous studies, which had the issue of participants trying to outsmart the procedure by trying to pick up on "clues" to determine song authorship. By removing the Turing test-like condition from their study, Pasquier et al. tried to remove this difficulty from their study. They also attempted to remove any practice effects that may have happened from the non-randomized order of stimuli in previous studies.

Unlike Moffat and Kelly's (2006) study, Pasquier et al. (2016) divided their participants into three groups: Informed, Naive, and Revealed. In the Naive condition, participants were unaware of song authorship whereas in the informed condition the participants were explicitly told the author for each piece of music. In the Revealed group, participants first heard each piece without knowing the song authorship. Then, in the second round of listening, the authorship of each piece was explicitly told for each song. This condition of the experiment allowed them to check if there would be any "reaction" effect where the newfound knowledge of authorship could cause a drastic change in how people rate the music.

Because they did not include a Turing test-like section to their study, Pasquier et al. did not divide their participants into musician and non-musician groups. However, they did include a demographics questionnaire which included questions about age, gender, university major, country of birth and number of years living in Canada. They also asked each participant how much experience they had with computer programming languages as a measure for computer literacy. This measure was asked after the experiment to increase deception.

Where Moffat and Kelly decided that their stimuli should be of three different styles, Pasquier et al. chose to limit their musical selection to only one style. They created three unique computer-composed pieces in the style of "contemporary string quartet" and paired them with three other human-composed pieces that were of similar structural characteristics, taking into account tempo, polyphony, rhythm, and dynamics for pairing each piece. They believed that this pairing technique would offer better results than Moffat and Kelly who only paired pieces by style. All of the pieces used in Pasquier et al. were performed by the same string quartet who were given an equal amount of time to practice and perform each piece. The performers were unaware of which pieces were composed by humans and which ones were artificial. All of the artificial works were created by the software of Arne Eigenfeldt, a Canadian composer and long-term collaborator of Dr. Philippe Pasquier.

Participants were given a URL to an online survey where they were presented with video recordings of each piece one at a time. Between each musical piece, participants were shown a "palate-cleanse" and each participant was shown a random order of pieces in order to reduce practice effects. After listening to each piece, participants were then asked to rate each piece on four different attributes using a 50-point bipolar scale. The four dimensions measured were: 'Good-Bad', 'Like-Dislike', 'Emotional-Unemotional', and 'Natural-Artificial'. They chose this rating method as a more sensitive measure to identify bias, which could be obscured in a more simple rating, such as simply "liking," which was used in previous experiments such as Moffat and Kelly.

Similar to Moffat and Kelly's study, Pasquier et al. did not find any significant results that show there is a bias against musical metacreativity, however, their results suggest that such a bias may exist. Their data showed that listeners were fairly uncertain in their ratings regardless of their knowledge of song authorship. There was a slight skew towards the "bad" and "artificial" dimensions for participants' ratings in the Revealed condition, but these were not significant enough to show any meaningful bias. The authors suggest that replication should be done with a larger sample size and recognize that their study is limited to only certain musical conditions, particularly style.

2.3 Influence of Context and Expectation When Searching for Bias Against AI-composed Music

Hong et al. (2020) conducted a study that examined the influence of people's met or unmet expectations about AI and their assessments of AI-composed music. Their experiment incorporated Expectancy Violation Theory (EVT) which explained individuals' reactions to met or unmet expectations in communication settings (Burgoon et al. 2016). This theory argues that when one's expectations are exceeded, they perceive the outcome as more favourable than if they made no expectations at all. The same effect happens in the negative direction, where if one's expectations were not met, the outcome is perceived as less favourable than if there were no expectations made. This holds that people's evaluation of artwork is biased by their belief of whether or not AI can be creative rather than the artwork itself. Hong et al. hypothesized that those participants who thought AI-composed music was better than expected will give higher ratings than people who think the music meets their expectations and vice versa.

They recruited 299 participants to complete their online study using Amazon's Mechanical Turk (MTurk). Four AI-composed musical songs were used for the study, two of which were in the electronic dance music (EDM) genre, the other two were of the classical genre. A pilot study was first conducted to make sure each piece would be rated similarly. After confirming each song would be rated equal in quality, each participant was presented with a randomly selected song out of the four and asked to listen to it. After listening to their given song, participants were told to report their evaluation of the musical piece using a 9-item scale based off of the "Rubric for assessing general criteria in a musical composition" (Hickey, 1999). This 9-item scale measured the aesthetic appeal, creativity, and craftsmanship of each song. After rating for musical quality, participants were asked how much the music's quality deviated from their expectations. This was measured using a 7-point Likert-type scale based off one used in a previous study by Burgoon et al. (2016). Finally, Participants filled out another 7-point Likert type scale measuring participants' understanding of AI's creativity.

Their results showed a positive relationship between the perception of creative AI and the evaluation of AI-composed music. Their work shows an interesting implication when working with creative AI, namely that people's attitudes towards AI has a large impact on their attitudes towards AI-made creative works. Those who are able to be persuaded that AI can be creatively autonomous are more likely to rate MuMe works as having creative value. This holds in the other direction as well. Those with negative preconceptions of MuMe systems are shown to devalue MuMe compositions significantly.

Hong et al. (2020) found that one's prior expectations and attitudes towards AI have an impact on how they rate AI-composed music, showing that outside factors may play a part in how we perceive AI-composed music. Another important factor which may influence people's attitudes towards AI-composed music may be the cultural context of the music itself. Deguernel et al. (2022) looked to see what effects the culture and context of the music where AI is being applied has on their perception of a piece. They chose a genre where computer authorship is considered to be generally opposed to the music creating process, Irish Traditional Music (ITM). ITM has a strong emphasis on authenticity and etiquette (Hillhouse 2005), there are strong opinions on how tunes should be played and taught and which instruments should be used. Due to the values of ITM, there is a fear of a loss of authenticity of the genre due to commercialization and more modern music production techniques (Xuan & Ying 2022). Given the context of ITM is rooted heavily in human-centered tradition, Deguernel et al. hypothesize that ITM practitioners will show a bias against liking music they believe to be composed by an AI.

To test their hypothesis, Deguernel et al. recruited participants from traditional music programs at the University of Limerick, Ireland and had them complete an experiment where they would listen to 6 pieces of AI-composed Irish Traditional Music. Each piece was hand selected by a professional ITM musician, Padraig O'Connor. After selecting his 6 favorite pieces from a large corpus of 58,105 tunes, O'Connor recorded himself playing each composition on solo accordion, adding stylistic ornamentation and variations as he saw fit to ensure each recording sounded authentic. These 6 pieces were presented to participants during their completion of two tasks, the "Liking task" and the "Authorship task".

In the "Liking Task", participants were presented with each piece of music and asked "How much do you like the tune?" and recorded their answers on a 5-point Likert scale labeled: "Don't like it at all", "Don't like it", "Neutral", "Like it", and "Like it a lot". The scale was not shown until after the song finished playing and participants were encouraged to use the full range of the scale. After rating how much they liked each piece, participants completed the "Authorship Task". In The "Authorship task", participants were asked "How likely do you believe that the tune is composed by a computer?" and given another 5-point Likert scale. The two tasks were completed in this specific order to ensure that ratings in the

"Liking task" are not affected by a prior mention of AI. After completing both tasks, participants filled out a short questionnaire about demographics, musical practice, and familiarity with ITM.

Deguernel found a plausible bias amongst ITM practitioners for these six AI-composed pieces. Participants tended to like the tunes that they deemed likely to be composed by a human, and disliked pieces they believed to be composed by an AI. The difference in results with that of Moffat and Kelly (2006) and Pasqier et al. (2016) help validate the hypothesis of Deguernel et al. (2022), that the context in terms of musical culture and participants plays a role in observing a bias against AI-composed music. However, Deguernel et al. consider their work a pilot study and say that power analysis of their experiment must be conducted to determine the likelihood of a Type-I error.

When searching for a proposed bias against AI-composed music, the work of Moffat and Kelly (2006) and Pasquier et al. (2016) failed to find any significant bias. However when looking at whether people's prior conceptions about AI and musical AI compositions, Hong et al. (2020) and Deguernel et al. (2022) seem to find some effect. It seems that one's prior expectations as well as the cultural context of the music itself, comes to play when seeking for a negative human bias against AI musical compositions.

3 Methods

The present study aims to build off of the work conducted by Moffat and Kelly (2006), as well as Pasquier et al. (2016) in the aim to search for a general listener bias against computational creativity. By also asking participants their opinions on creative AI systems, we also build off the work of Hong et al. (2020) and investigate how people's predispositions towards computational creativity may affect their ratings towards musical metacreativity. We created a ranking survey based on the suggestions of Pasquier et al. (2016), this survey also included demographic questions including polls on musical and technological ability. According to, Yannakakis and Martinez (2015), rank-based questionnaires can help eliminate some of the problems associated with ratings-based questionnaires when evaluating subjective, psychological factors like emotional response, preference, or opinion.

3.1 Participants

Participants for both the pilot and main experiments were recruited using Amazon's Mechanical Turk (MTurk) service. Participants were compensated with a small monetary reward after completing the survey. Those who did not pass attention checks were excluded from the study. There were a total of 163 participants in the main experiment, two of which were excluded for not completing the attention check. The majority of participants identified as male (72%), with only one participant identifying as non-binary.

3.2   Musical Excerpts

The artificially composed pieces were created using the Multi-track Music Machine (Ens and Pasquier 2020). This system uses the transformer architecture to generate multi-track music by providing users with a fine degree of control over iterative re-sampling. The system is based on an auto-regressive model which is capable of generating novel music from scratch in a wide variety of genres and styles by using a multitude of preset instruments. Tracks can also be generated using MIDI track input as 'inspiration,' re-sampling the piece into further layers of musical composition.

Each computer-composed musical excerpt was paired with a similar human-composed work of the same style. These pairings were made based on rhythm, tempo, and tonality. These pieces were then later rated in a pilot study to ensure that each pair of human and computer-composed music would be rated equally when participants were blind to song authorship. All of the pieces used in the final experiment were of the same genre: contemporary pop.

3.3   Procedure

To test that the musical quality of the computer-generated pieces are on par with human-composed pieces of the same style, a short pilot study was conducted on Amazon MTurk. Participants were told to rate around 30 pieces to test for "equality". These 30 pieces were a mix of computer-composed and human-composed. For this survey, participants were not told about the authorship of each piece, they simply rated each song in terms of musical quality on a 7-point Likert scale and given no further information than what they could hear. This was done to ensure that computer composed pieces would be ranked equally as human-composed pieces. There was no significant difference in ratings between human-composed and computer-composed pieces during the pilot study.

In the primary study, participants were randomly divided into two groups: Deceived and Informed. After providing consent, participants completed a brief attention check. As seen in Fig. 1, participants in both groups were presented with ten different musical excerpts (around 10 s each) of mixed authorship for the first period of the experiment. In this period, participants were blind to song authorship and given the opportunity to play and pause each clip as they pleased. Playtime was tracked to ensure that each participant listened to each clip for an adequate amount of time. One of the audio clips presented to the participants was an audio message telling each participant to rate that clip at the bottom of the list, this was used to ensure participants were paying attention to the experiment. Participants were then told to rank each clip against one another by dragging their favourite to the top of the list and their least favourite to the bottom. After completing this first round of ranking, both groups entered the second period of the experiment where they would listen to the four musical excerpts again, however, this time they were told the authorship of each musical excerpt. The Informed group was told the actual composer for each piece, whether it was composed by a human or computer, whereas the Deceived group was told the wrong composer for each song. In other words, the pieces that were composed by a human were presented as being composed by a computer system and vice versa in the Deceived group. For this second period, participants were asked how likely they would be to listen to each song on their own as a measure of how much each participant liked each piece. This measure was used to ensure that participants would not catch on to the purpose of the study.

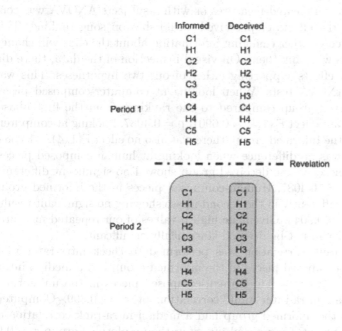

Fig. 1. Experiment design showing the two conditions, Informed and Naive. The yellow shaded area represents the period where authorship is told for each song. The blue shaded represents the songs where authorship was told deceptively. C1, C2, C3, C4, C5 refer to computer composed pieces where H1, H2, H3, H4, H5 refer to human-composed pieces respectively, the order of all pieces was randomized for each participant. (Color figure online)

After completing the second period of listening and ranking each musical piece, participants were asked to provide some general demographic information such as their age, gender, and country of residence. The questionnaire also included a number of questions asking participants how many years they have spent studying or playing music, as well as other questions to assess their level of computer literacy. After completing the questionnaire, each participant was debriefed on the purpose of the study if they chose to do so and were provided with contact information in case they had any further questions regarding the study.

We hypothesize that those in the Informed group will rank computer-composed music as lower than human-composed music. Secondly, we predict that there will be a positive correlation between people's technological literacy and attitudes towards computational creativity and their overall rankings of computer-composed music.

4   Results

To analyze our data, a repeated measures or within-subjects ANOVA was conducted to compare the effects of perceived authorship on song ranking. This measure looked to see whether changing information about the clips will change the way participants will rank them. On visual inspection of the data, there did not seem to be any effects supporting either of our two hypotheses. This was confirmed by the ANOVA tests. When looking at computer-composed pieces ranked in the deceived group compared to the rankings from the first phase, there was no significant effect $F_{(1,4)} = 0.6901$, $p = 0.4087$. Looking at computer-composed pieces in the Informed group, there was also no effect $F_{(1, 4)} = 1.0684$, $p = 0.3044$. There was no difference when looking at human-composed pieces. Human-composed pieces in the deceived group showed no significant difference $F_{(1, 4)} = 0.6901$, $p = 0.4087$. Human-composed pieces in the informed group were also not rated differently in the second phase showing no significant results $F_{(1, 4)} = 1.0684$, $p = 0.3044$. Given the high p-values of our repeated measures ANOVA, our results can not be deemed statistically significant.

A Kendall-Tau rank correlation was performed to check intra-rater rank agreement. Human-composed pieces in the informed group had a median intra-rater rank correlation of $\tau = 0.6$. Human-composed pieces in the uninformed group had a median intra-rater rank correlation of $\tau = 0.399$. Computer-composed pieces in the informed group had a median intra-rank correlation of $\tau = 0.6$ as well as computer-composed pieces in the uninformed group $\tau = 0.6$. According to Akoglu (2018), a τ of 0.6 would be considered either moderate or strong whereas a τ of 0.39 would be on the boundary of weak and moderate. Given our results are in this range we can conclude that intra-rater ranking is fairly consistent in our experiment.

5   Discussion

Along with the past work of Moffat and Kelly (2006), as well as Paquier et al. (2016), the present study sought to find out if humans hold a negative bias against computer-created music. As with past studies, it seems like the anticipated bias seems to be mostly anecdotal and not as strong as we may have presupposed. Although there seems to be a slight skew towards ranking human-composed clips as higher, this effect is not strong enough to be deemed statistically significant. The first hypothesis tested, that those in the informed group would rate human-composed music as better than computer-composed music, was not supported by the repeated measures ANOVA analysis and subsequent T-tests.

The second hypothesis this experiment tested sought to examine whether people's attitudes towards technology and artificial intelligence, in general, would affect their ratings of music that were thought to be computer-composed. Unlike the previous findings of Hong et al. (2020), we did not find any significant correlation between participants' comfort level when using the computer and their rankings of computer-composed music. It must be mentioned, however, that Hong et al. (2020) were searching for an effect in people's expectation violation, not their direct ratings. In addition to not finding any significant evidence for the second hypothesis, there did not seem to be any effect between people's musical capability and their rankings of computer-composed music. This lack of an effect goes against the previous findings of Moffat and Kelly (2006), which found musicians rated computer-composed pieces as lower than non-musicians.

A contributing factor to the lack of support for the second hypothesis may be the population used for the study. While MTurk provides an easily accessible population of participants from all over the world, a large portion of this population is highly comfortable with using technology given the online nature of the platform, 65 out of 161 participants indicated they are "very proficient with computers" and 15 participants indicated they were "experienced programmers". This led the population in our current study to not be necessarily representative of the world population. The majority of participants in the current study reported their ages as under 40, with only 2 participants identifying as 65+, leaving particularly older age groups largely unrepresented. Attitude towards new technologies is often stable within generations, which may make it difficult to convince some, especially older, age groups that AI music is equal to human-composed music (Chung et al. 2010; Niehaves and Platfout 2014). As shown by Deguernel et al. (2022), the context and culture surrounding the music can have an impact on the way it is perceived. Future work could potentially look to see if those of a higher age group (50+) would have any differences on their outlook to computational creativity, as people in this age range are typically less comfortable with new technologies.

Although sample sizes for the current study were adequate, the high variability in the data, as well as the method of ranking clips, suggests that the design should be simplified. Perhaps a more direct comparison, using fewer clips that sound more distinct from one another, could provide different results. Another factor was that some participants reported that having to drag the 10 clips in their desired order was cumbersome and not intuitive. This could have caused listeners to become confused about which clip they were ranking. It may be possible the length of the audio clips were quite short, which in conjunction with the similarity in style between clips, may have led to the high variability in rankings among pieces regardless of their authorship. Finally, although there was no evidence in the data to support this, there is a slight chance that human and computer-composed pieces were distinguishable to listeners even while blind to authorship. There was a subtle characteristic in the computer-composed tracks that set the note velocity to a single value, which made these tracks sound slightly louder. Although this was later corrected by remastering the tracks, there is a very slight chance that astute listeners would have been able to pick up on this minute characteristic.

5.1   Future Work

Although the present study failed to find any significant bias against computer-composed music, this does not mean that such a bias does not exist. Our results are in line with that of Pasquier et al. (2016), however, they go against the findings of Moffat and Kelly (2006), Hong et al. (2020), and Ragot et al. (2020). There does not seem to be a clear consensus on whether there is a general bias against computationally creative artwork and more work must be done on the topic in order to answer our hypothesis.

Future studies could improve on the current design by creating a more direct comparison of musical clips. Rather than presenting 10 different clips to the participant at once, clips could be presented in pairs to keep the design simple. Given that many participants were overwhelmed when trying to rank 10 clips at once, this could prove to be a much simpler and easier to understand design. Further work in this field, one that does not choose to use a ranking-based approach, for example, could perhaps implement a more well-established music rating system such as the "Rubric for assessing general criteria in a composition assignment" (Hickey 1999), which was used by Hong et al. (2020). Their scale measured multiple factors such as aesthetic appeal, creativity, and craftsmanship. Using such a scale could provide more information into what exactly people perceive differently between human-composed and computer-composed works or whether there is any difference at all.

Perhaps instead of looking for a general perceived bias, future work should focus on finding a specific inter-group bias when evaluating AI-created art or music, as suggested by Ragot et al. (2020). Because of the widespread fear of losing one's employment due to AI automated systems (McClure 2017), it may be the case that this plays a significant role on how people perceive music created by such systems. Perhaps people are scared that their own creativity, something perceived to be innately human, could eventually be replaced by a non-human entity.

6   Conclusion

This study sought to search for a negative bias against music created by artificial intelligence systems. Similar to the previous study by Pasquier et al. (2016), we did not succeed in achieving our goal of finding such a bias. We outlined the reasoning behind our lack of significant results and presented a number of possible improvements for future research in this field of study.

Research like the present is part of a wider exploration of human-computer interaction. As artificial intelligence technologies continue to permeate through society, more studies that examine the interactions between humans and these computer systems should be conducted. The general consensus for a bias against computers and AI, especially in the field of creativity, is widely held anecdotally, yet there is still not much work done proving the evidence of such a bias.

Furthermore, music has been shown to provide many health benefits and is used for multiple clinical purposes (Alty et al. 1997; Cross 2014; Hargreaves et al. 2005). By starting to gain a more in-depth understanding of the way humans perceive and interact with metacreative systems, perhaps it may one day be possible for AI to generate creative works for a specific purpose, such as music therapy. In the same way that Open-AI's DALL-E 2 is meant to help artists with their creative process, MuMe systems can aid in the creative process and even help form new genres of music. In the same way that digital audio workstations (DAW), such as Logic and FL studio, have revolutionized music production, MuMe systems provide an excellent tool for artists to come up with new ideas and put them into practice.

This paper discusses the current state of knowledge in the search for a bias against computationally creative systems and paves the way for future work to increase the growing scope of knowledge on the subject. Although we did not find any significant results in our experiment, there does still seem to exist such a bias against computational creativity as presented by Moffat and Kelly (2006) and Ragot et al. (2020), and Deguernel et al. (2022). More work is needed to come to a consensus on whether or not there exists a bias against computer composed music and computational creativity as a whole.

References

Akoglu, H.: User's guide to correlation coefficients. Turkish J. Emerg. Med. **18**(3), 91–93 (2018)

Alty, J.L., Rigas, D., Vickers, P.: Using music as a communication medium. In: CHI'97 extended abstracts on human factors in computing systems, Atlanta, GA, 22–27 March, pp. 30–31. ACM, New York (1997)

Burgoon, J.K., Bonito, J., Lowry, P., et al.: Application of Expectancy Violations Theory to communication with and judgments about embodied agents during a decision-making task. Int. J. Hum. Comput Stud. **91**, 24–36 (2016)

Brewer, W., Treyens, J.: Role of schemata in memory for places. Cogn. Psychol. **13**(2), 207–230 (1981)

Chung, J.E., Park, N., Wang, H., et al.: Age differences in perceptions of online community participation among non-users: an extension of the Technology Acceptance Model. Comput. Hum. Behav. **26**(6), 1674–1684 (2010)

Nass, C., Moon, Y.: Machines and mindlessness: Social responses to computers. J. Soc. Issues **56**(1), 81–103 (2000)

Cope, D.: Virtual music: computer synthesis of musical style. MIT Press (2004)

Cross, I.: Music and communication in music psychology. Psychol. Music **42**(6), 809–819 (2014)

Déguernel, K. Maruri-Aguilar, H., & ; Sturm, B. L. T. (2022). Ken Déguernel, Bob L. T. Sturm, Hugo Maruri-Aguilar. Investigating the relationship between liking and belief in AI authorship in the context of Irish traditional music. CREAI 2022 Workshop on Artificial Intelligence and Creativity

Miranda, E.R.: Artificial intelligence and music: an artificial intelligence approach to sound design. Comput. Music J. **19**(2), 59 (1995)

Ens, J., Pasquier, P.: MMM: exploring conditional multi-track music generation with the transformer. arXiv preprint arXiv:2008.06048 (2020)

Fogel, A.L., Kvedar, J.C.: Artificial intelligence powers digital medicine. Npj Digital Med. 1(1) (2018)

Goldman, S.: Will openai's dall-e 2 kill creative careers? VentureBeat (2022, July 26). Retrieved October 2, 2022. https://venturebeat.com/ai/openai-will-dall-e-2-kill-creative-careers/

Hargreaves, D.J., MacDonald, R., Miell, D.: How do people communicate using music. In: Miell, D., MacDonald, R., Hargreaves, D.J. (eds.) Musical Communication, pp. 1–26. Oxford University Press, New York (2005)

Hillhouse, A.N.: Tradition and innovation in Irish instrumental folk music (T). University of British Columbia (2005). Retrieved from http://open.library.ubc.ca/collections/ubctheses/831/items/1.0092099

Hello World album the first album composed with an artificial intelligence. SKYGGE. (n.d.). https://www.helloworldalbum.net/

Hewstone, M., Rubin, M., Willis, H.: Intergroup bias. Annu. Rev. Psychol. 53(1), 575–604 (2002)

Hong, J., Curran, N.M.: Artificial intelligence, artists, and art: attitudes toward artwork produced by humans vs. artificial intelligence. ACM Trans. (2019)

Hong, J.W., Peng, Q., Williams, D.: Are you ready for artificial Mozart and Skrillex? An experiment testing expectancy violation theory and AI music. new media & society (2020)

Istok, E., Brattico, E., Jacobsen, T., et al.: "I love Rocken' Roll"-music genre preference modulates brain responses to music. Biological Psychol. 92(2), 142–151 (2013)

Jackson, T.: Imitative identity, imitative art, and AI: artificial intelligence. Mosaic: Interdisciplinary Critical J. 50(2), 47–63 (2017)

Jennings, K.E.: Developing creativity: artificial barriers in artificial intelligence. Mind. Mach. 20(4), 489–501 (2010)

Norton, D., Heath, D., ; Ventura, D.: Finding creativity in an artificial artist. J. Creative Behav. 47(2), 106–124 (2013)

Marzano, G., Novembre, A.: Machines that dream: a new challenge in behavioral-basic robotics. Procedia Comput. Sci. 104, 146–151 (2017)

McCarthy, J.: From here to human-level AI. Artif. Intell. 171(18), 1174–1182 (2007)

McClure, P.K.: "You're fired," says the Robot. Soc. Sci. Comput. Rev. 36(2), 139–156 (2017)

Moffat, D.C., Kelly, M.: An investigation into people's bias against computational creativity in music composition (2006)

Niehaves, B., Plattfaut, R.: Internet adoption by the elderly: employing IS technology acceptance theories for understanding the age-related digital divide. Eur. J. Inf. Syst. 23(6), 708–726 (2014)

Pasquier, P., Burnett, A., Maxwell, J.: Investigating listener bias against musical metacreativity. In: Proceedings of the Seventh International Conference on Computational Creativity, pp. 42–51 (2016, June)

Ragot, M., Martin, N., Cojean, S.: Ai-generated vs. human artworks. A perception bias towards artificial intelligence? Extended Abstracts of the 2020 CHI Conference on Human Factors in Computing Systems (2020)

Thorogood, M., Pasquier, P., Eigenfeldt, A.: Audio Metaphor: Audio Information Retrieval for Soundscape Composition. In: Proceedings of the 9th Sound and Music Computing Conference, Copenhagen (2012)

An ai-generated artwork won first place at a state fair fine arts competition, and artists are pissed. VICE. (2022, August 31). Retrieved January 30, 2023, from https://www.vice.com/en/article/bvmvqm/an-ai-generated-artwork-won-first-place-at-a-state-fair-fine-arts-competition-and-artists-are-pissed

Yannakakis, G.N., Martınez, H.P.: Ratings are overrated! Frontiers in ICT **2**, 13 (2015)

Xuan, W.Z., Ying, L.F.: The development of Celtic Music Identity: Globalisation and media influences. Media Watch **13**(1), 34–48 (2022)

Short Talks

Fabric Sketch Augmentation & Styling via Deep Learning & Image Synthesis

Omema Ahmed(ID), Muhammad Salman Abid(✉)(ID), Aiman Junaid(ID), and Syeda Saleha Raza(ID)

Habib University, Karachi, Pakistan
{oa04320,ma05045}@alumni.habib.edu.pk, aj05161@st.habib.edu.pk,
saleha.raza@sse.habib.edu.pk

Abstract. This paper introduces a two-fold methodology of creating fabric designs and patterns, using both traditional object detection and Deep Learning methodologies. The proposed methodology first augments a given partial sketch, which is taken as an input from the user. This sketch augmentation is performed through a combination of object detection, canvas quilting, and seamless tiling, to achieve a repeatable block of a pattern. This augmented pattern is then carried forward as an input to our variation of the pix2pix GAN, which outputs a styled and colored pattern using the sketch as a baseline. This design pipeline is an overall overhaul of the creative process of a textile designer, and is intended to provide assistance in the design of modern textiles in the industry by reducing the time from going to a sketch to a pattern in under a minute.

Keywords: deep learning · GAN · image synthesis · quilting · textile pattern · fabric pattern · object detection

1 Introduction

The textile industry has been the rampart of Pakistan's economy. It contributes more than 60% to the total export earnings of the country, accounts for 46% of the total manufacturing and provides employment to 38% of the manufacturing labor force and 9% of GDP [1]. However, fashion design in Pakistan is more or less an entirely manual process, with little to no automation involved. It requires hours of manual labor to design a prototype pattern, and it goes through multiple iterations and feedback loops to get it finalized. The pattern itself also has to be designed from scratch, and its quality is entirely contingent upon the human designer.

This paper addresses both aspects of the problem in fashion designing in Pakistan; speeding up the design process and bringing these sketches to life. This is similar to how Microsoft developed an AI to reduce the time taken to design Indian jewelry designs [10], and this paper applies a similar approach on the domain of fashion design.

© The Author(s), under exclusive license to Springer Nature Switzerland AG 2023
C. Johnson et al. (Eds.): EvoMUSART 2023, LNCS 13988, pp. 327–340, 2023.
https://doi.org/10.1007/978-3-031-29956-8_21

In this paper, image quilting techniques are applied to generate an augmented sketch from a given input sketch, in the form of a single flower or a collection of multiple flowers. For the second part of the problem, Conditional Generative Adversarial Networks (CGAN) are used to make these augmented sketches come to life by reproducing a photo-realistic rendition [3] in the form of an actual fabric pattern.

The rest of the paper is organized as follows: Sect. 2 explores the existing work done in this domain. Section 3 presents the preparation of the dataset used, and Sect. 4 details the methodologies employed in this paper. Section 5 gives a summary of our results, showcasing how the sketch has evolved from a given input. Finally, Sect. 6 concludes the paper, and Sect. 7 mentions the possible future work in this domain.

2 Related Work

The task of generating a realistic image from a sketch is a very challenging prospect, and has been the focus of research of various researchers in academia. In their paper [9], Gao et al. proposed a new neural architecture called EdgeGAN, which is used for automatic image generation from scene-level freehand sketches. A paper by Tian et al. [20] focuses on the colorization of images of logo sketches, to automate the process of manual logo and icon design. This paper builds upon the work done by Pix2Pix [12], with their contributions present in the architecture to adapt to the task of logo colorization. One similar project on Image-to-Image translation has also been done by Fan et al. [7] using Conditional GANs on single objects. Another paper by Hu, M., Guo, J [11] implements a convolutional generative network with encoder-decoder architecture and facial attributes classifier, that colorizes sketches of humans faces.

All the works currently done in the domain of sketch-to-image translation are restricted to either single objects and scenes, or facial features, both of which have a structural similarity in the objects to be colorized. This paper applies the Pix2Pix model [12] in the domain of fabric patterns.

According to the previous literature, a lot of work has been done on Image-to-Image translation using GANs. Different neural nets with conventional and hybrid models are used to generate photo-realistic Images. The implementations become challenging if user sketch input is taken into account to synthesize augmented textures. Some work on texture synthesis is presented below.

Alexei A. Efros and William T. Freeman [6] proposed in their paper 'Image Quilting for Texture Synthesis' a simple yet effective image-based method for generating novel textures. The quilting algorithm takes an input image and synthesizes a new, visually appealing, texture by stitching together small patches of an existing image. First, the image to be synthesized undergoes a raster scan to find a block that satisfies the overlap constraint. Once a block is found, the surface error between the new block and the old blocks is computed. The boundary of the new block is then determined by calculating the minimum cost path along the surface, and finally the chosen block is pasted onto the final texture.

These steps are repeated for every block until a high-quality image is obtained. In addition to this, an extended algorithm for texture transfer is also presented in the paper which involves rendering an image in the style of a different image.

Another similar paper on texture synthesis has been published by Alexei A. Efros and Thomas K. Leung [5], where they propose a non-parametric method. This method synthesizes a new texture from an initial seed, one pixel at a time. Their algorithm uses the Markov random field model to find similar neighborhoods in the input image to the pixels neighborhood. Then, it randomly selects one such neighborhood and takes its center to be the newly synthesized pixel. This method generates synthetic textures by preserving as much local structure as possible.

An interesting approach taken by Lin Liang, Ce Liu, Ying-Qing Xu, Baining Guo, and Heung-Yeung Shum in their paper [15] is to use a real-time patching algorithm for synthesizing high-quality textures. The model first computes the local conditional probability density function (PDF) for each patch from the source image. The pixels are then synthesized incrementally on to the output texture. Lastly, feathering is used to blend the patch boundaries and synthesize a high quality texture.

The proposed pipeline for this paper is novel in its approach by being one of the first attempts to utilize a variety of objects for Image-to-Image translation, particularly in images dealing with sketches. Existing solutions are mostly focused on a specific category or object; a drawing of a piece of furniture, or a sketch of a fruit, whereas this work combines multiple categories of sketches into its supported function. Users may draw a sketch of a flower, a leaf, a butterfly, or all three objects combined, and it would still be able to generate a viable design.

For sketch augmentation, there are little to no existing works related to completing, or rather augmenting, user sketches. This part of the pipeline had to be built from scratch; the idea of using objects from a partial pattern, and then supplementing those with similar objects on a new canvas, was left unexplored in current literature. There have been some attempts at auto-completion of a given incomplete object [13,21], however very limited work is done to generate, or extend, existing images/objects in this domain. This paper proposes an algorithm that can leverage these image processing techniques, augmenting a partial pattern input by the user into a structure of a repeatable sketch.

3   Dataset

The dataset has been sourced from various open-source collections, and has been processed to convert it into a usable form for our CGAN model. Our dataset comprises approximately 1500 images, with the final dataset including ~ 500 images from all three categories.

3.1   Data Collection

The dataset was scraped from Patternbank, The Sketchy Database and Kaggle Dataset for butterfly, floral and leaf categories. The collected dataset was then manually filtered to discard any images present which did not belong to the afore-mentioned categories. The filtered dataset then undergoes image resizing to be converted into images of dimensions 256 × 256, to be passed as training input to our CGAN model. A few examples of the collected dataset are shown in Fig. 1.

Fig. 1. Examples of the collected dataset, which includes images of three categories: leaf, butterfly and floral

3.2   Data Processing

In order to train the CGAN image-to-image translation model, images from the collected dataset are converted to their corresponding sketches through various image processing techniques.

For this purpose, Canny Edge Detection [2] is applied to all the images after they are converted to gray-scale. The resulting images are then combined together, including both their original pattern and the corresponding sketches. An example from the resultant combined dataset is shown in Fig. 2.

4   Methodology

This section provides an overview of the methodology to generate photo-realistic images from an input sketch. The process comprises the following two steps: user sketch to sketch augmentation, and augmented sketch to Photo-Realistic conversion.

Fig. 2. Example of application of Edge Detection on source (left) with its result (right)

4.1   Sketch Augmentation

The first step in augmentation is further divided into the following steps:

- Object Detection
- Content-Based Image Retrieval
- Texture Synthesis

The following sub-sections will describe, in detail, what process each step follows.

Object Detection. This technique is used to extract objects from the input sketch and store them as separate images. The following steps are done to perform object detection:

- Load the source image and apply background removal to extract useful pixels from the image.
  Background removal removes any white/non-transparent pixel(s) from the sketch that are not part of the drawing.
- Convert image to gray-scale, and apply a filter for Gaussian blurring
- Perform Canny Edge Detection [2] and dilate [4] image to increase the size of the foreground object.
- Find contours and iterate through the loop to extract region of interest (ROI) i.e. objects in the source image.
- Save each extracted object in their respective file format. i.e. (.png or .jpg)

Content-Based Image Retrieval. CBIR is a technique which searches for similar images in our local dataset based on a given source image. The process begins with extracting features from both the input image and the dataset using a feature extraction algorithm. Then, it calculates the similarity distance between images present in the dataset and the query image, finally retrieving all similar images with the lowest distance from the dataset.

Our problem uses CBIR to retrieve similar objects from the dataset to the ones present in our input image. It uses a pre-trained, VGG-16 CNN model [19] to extract features from the image dataset and stores them in a feature vector. The previously separated objects are then imported as input images in the retrieval algorithm. After performing feature extraction, and comparing them with the previously saved feature vectors, we compute the distance between the query image and the images in the dataset using the L2 Norm [16]. The shorter the distance between images, the more similar they are to each other. Lastly, two images with the lowest distance are then retrieved from the dataset for further use in augmentation.

The image retrieval algorithm proceeds as follows:

- Apply feature extraction using VGG-16 on local dataset
- Apply feature extraction on input image
- Calculate the similarity between the preceeding two vectors using L2 Norm

- Search for the most similar results (lowest distance ranked highest) and return top two

Additionally, even if object retrieval is not used, the algorithm is designed to add a random shape object with random size from our fixed set of shapes – currently limited to stars and circles – to the design. This makes the generated design more stochastic, and results in more organic patterns. Some examples of randomly added shapes can been seen in Figs. 3 and 7. This is an optional feature, and can be toggled according to user preference.

Texture Synthesis. A final synthesized texture with 256×256 dimensions is augmented from the processed images using the following techniques:

Grid Texture Synthesis. Generate arbitrarily large textures by stitching together patches of the same size in a grid-like form. The process involves two user controlled parameters: source image and size, where source image is the user drawn sketch, and size is the total number of patches on the final image in each row/column. Once user inputs are defined, the next step is to calculate the block size by dividing the dimensions of the final image with the input size. Here, the block size B determines the size for each patch to be placed on the augmented image.

From the feature arrays obtained through feature extraction on both the local dataset and the input image, pick a random object and apply image resizing to obtain dimensions of $B \times B$. This ensures that every patch is of same size and a uniform grid can be obtained. The patches are placed side-by-side on the final image, forming a structured grid of objects as the final output.

To the final image, edge enhancement filters are applied to enhance contour fidelity and overall quality. To cater to the transparent background of each patch image while pasting, a bitmap is generated from the alpha channel of the image, where *true* indicates all non-transparent pixels and *false* otherwise. Lastly, using the bitmap, all *true* pixels of the array are set to black and all others to white, outputting a black-and-white augmented sketch.

Random Texture Synthesis. Generate patterns by randomly stitching together different patches of varying sizes. Using the objects from Object Detection and Content-Based Image Retrieval, a random object is chosen to be placed on the canvas. Different transformation techniques are applied to the selected object, which include:

- Rotation: The patch undergoes any one of the four randomly selected rotation methods; horizontal flip, vertical flip, mirroring and counterclockwise rotation.
- Resize: The patch dimension is randomly resized in the range of $[200 - 600]$.

Additionally, the algorithm also checks for an existing patch while pasting a new object on the canvas, to ensure that there are no visually-jarring overlaps in the final sketch. To do this, once a random position on canvas is selected to paste the new object, the image pixels of the new location on canvas is checked for any existing black image pixels, and if there are none present, the new object is pasted. Otherwise, another location is randomly selected for the object. This process keeps repeating until an empty space is found. However, in case the locations are chosen more than 10 times for one patch, the process is halted, assuming that there are already enough objects present on the canvas. An example of the augmented pattern produced from given sketch is shown below.

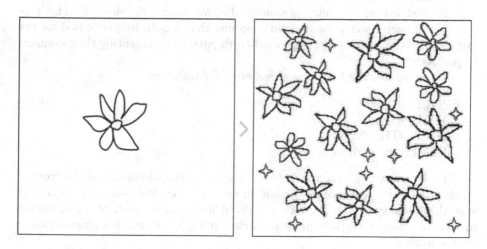

Fig. 3. Sketch Augmentation using Random Texture Synthesis on a sample sketch (left) with its result (right)

4.2   Sketch to Photo-realistic Image

This paper proposes a design generation technique based on Generative Adversarial Networks (GANs), which consist of two neural net models; a generator and a discriminator. It is a slight modification from the original pix2pix structure. The model architecture is detailed below:

Generator. The generator in the model is based on a modified U-Net architecture, adopted from Ronneberger et al. [18], comprising of an encoder and a decoder. Each block in the encoder includes a convolution layer, followed by batch normalization and activation function, Leaky ReLU. Each block in the decoder includes a transposed convolution layer, followed by batch normalization, dropout layer, and an activation function, ReLU. The loss function for the generator is binary cross-entropy loss and MAE (mean absolute error).

Discriminator. The discriminator in our CGAN is a convolutional PatchGAN classifier, where each block consists of a convolution layer, followed by batch normalization and an activation function, Leaky ReLU. The loss function for the discriminator is the binary cross-entropy loss.

We also make use of the Adam optimizer [14], with the learning rate set for both the generator and discriminator at 2e−4.

The images used for our training dataset go through several pre-processing techniques, as described in Sect. 3. These include random jittering and mirroring of the images to introduce stochastic behavior, and normalizing the images in the range of $[-1, 1]$ for them to be compatible with the neural-net layers.

For the purposes of training the GAN, we combine each pattern in the dataset with its sketch output, and concatenate the two images side-by-side. This represents, for each image, the domain mapping that has to be performed for the sample with the sketch and the ground-truth pattern representing their domains respectively.

The model has the following parameters for training:

- EPOCHS: 10,000
- BATCH SIZE: 4
- IMG WIDTH: 256px
- IMG HEIGHT: 256px

These parameters were tuned to get the best approximations of the ground-truth images as a result of prediction on user sketches. Post-training, the model was able to successfully generate a stylized image as a result of a user sketch input. An example of the generated styled image after sketch augmentation is shown in Fig. 4.

Fig. 4. Augmented image (left) converted into a photo-realistic image (right) via pix2pix

4.3   Tiling

The process of tiling makes use of an open-source library by Artëm G., named img2texture [8]. It creates a single tile-able block from a source image using image overlapping, to hide the seams when the image is tiled using multiple copies concatenated with each other to form a larger, repeating pattern. The parameters of the tiling process include the tile size (an N×N image tile), and the overlap percentage that controls the degree of overlap in the source image. Results from this tiling process are shown in Sect. 5.

Although similar to the augmentation process in Sect. 4, this process provides a birds-eye view on the final colourised image rather than the original sketch. Fabric patterns are better evaluated when they can be viewed on a larger-scale, hence the option to be able to visualize this on a dimension of the user's choosing.

5   Experiments and Results

This section contains experimentation results with different parameters, which includes sketches created using Grid Style (deterministic block placement) and Random (random block placement) Augmentation.

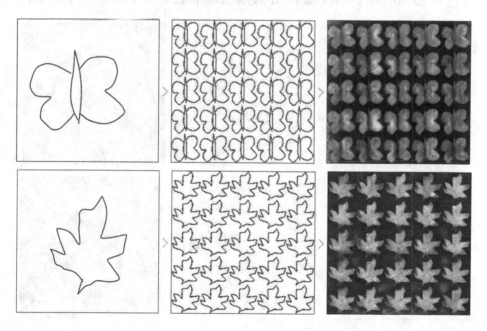

Fig. 5. End-to-end process via **Grid Style Augmentation** without object retrieval

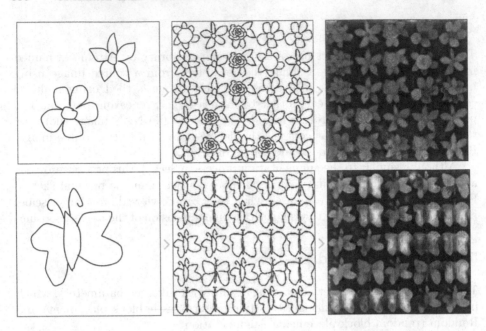

Fig. 6. End-to-end process via **Grid Style Augmentation** with object retrieval

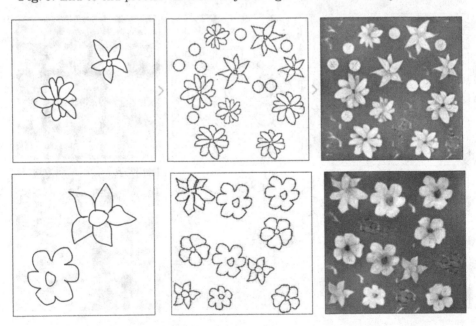

Fig. 7. End-to-end process via **Random Augmentation** without object retrieval

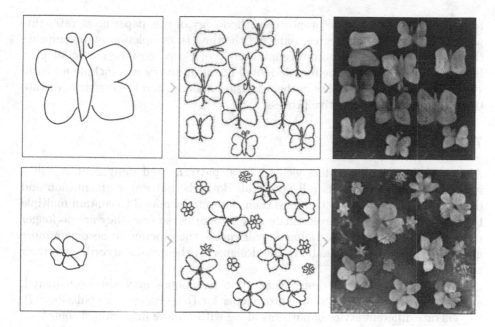

Fig. 8. End-to-end process via **Random Augmentation** with object retrieval

5.1   Discussion on Results

Figures 5 and 7 are examples of results obtained through both augmentation techniques, albeit without using Object Detection. Figures 6 and 8 are results that incorporate Object Detection in the process (see Sect. 4.1).

Given the empirical evidence of the performance of both augmentation techniques that are used to generate patterns, the Grid Texture Synthesis technique performs an order of magnitude faster than the Random Texture Synthesis. The average time for an end-to-end process for Grid-based augmentation was less than 10 s, whereas the latter averaged around 60–75 s based on the number of iterations for checking patch overlaps. The test runs were carried out on a compute-optimized Microsoft Azure F4s v2 instance (4 vCPUs, 8 GiB RAM). We identify areas of performance in the slower technique, and discuss these in Sect. 7, along with discussion on how the actual outputs may be evaluated for quality.

Given the potential application of this pipeline for designers, we intend this to act as a productivity tool that they can experiment their sketches on. A new design always goes through multiple iterations before it is perfected, and we intend this tool to provide the initial source of inspiration.

6   Conclusion

The majority of the textile design process is contingent upon the designer going through the various iterations of their designs, before choosing a viable candidate to move forward with. The proposed methodology of this paper accelerates this process by a significant degree, reducing the time to completion by preempting the design of the final pattern through the use of Deep Learning methodologies. By making use of this pipeline, designers in the industry can include an early visualization of their sketches in their decision-making, and leverage it to dictate the direction of their creative process.

7   Future Work

Although this pipeline caters well to floral patterns and designs, it is still a reduction in scope from its full potential. Both the pattern augmentation and design generation modules can have their datasets increased to contain multiple types of objects, perhaps even entire new categories, so that they are no longer restricted to floral designs. Doing so will enable the pipeline to become a more generalized tool, further enabling the designer in their tasks according to their needs.

Additionally, different design generation techniques may also be explored, such as the use of the Swapping Autoencoder for Deep Image Manipulation [17], to generate different styles of patterns along with a more fine-grained control on the texture of the image. This may prove to be a replacement for our CGAN, as the Autoencoder outputs boundaries and textures that are quite an improvement over those present in textures generated with the CGAN (albeit with a much lower degree of control what gets colored).

As for our performance concerns referred to in Sect. 5.1, we found our Random Texture Synthesis augmentation technique to be lacking in terms of compute efficiency. The technique iterates with the time complexity of $O(N^2)$ where N is the number of blocks in the augmented image, causing the time taken for a single block to be an order of magnitude higher than Grid Texture Synthesis. In future works, we can explore an alternative, more optimized algorithm that can perform the same with a reduced time complexity, as we have observed that the Random Texture Synthesis generates patterns which have more utility or scope of application – Grid textures may be a very niche application of this tool. Exploiting inherent parallelism in the algorithm is also an option, in case the algorithmic complexity does not change in the foreseeable future.

A notable part of the results is the subjectivity in human preference for art, due to which we are yet to propose a formalized evaluation metric. This is an avenue for further research, where user surveys can be conducted to learn what aspects of patterns are most important – turning those aspects into a potentially automated evaluation system.

References

1. Ahmad, Z.: Research report on Pakistan's textile industry analysis. SSRN Electron. J. (2009). https://doi.org/10.2139/ssrn.1651789
2. Canny, J.: A computational approach to edge detection. IEEE Trans. Pattern Anal. Mach. Intell. **PAMI-8**(6), 679–698 (1986). https://doi.org/10.1109/tpami.1986.4767851

3. Chen, W., Hays, J.: SketchyGAN: towards diverse and realistic sketch to image synthesis. In: 2018 IEEE/CVF Conference on Computer Vision and Pattern Recognition, pp. 9416–9425. Salt Lake City, UT, USA (2018). https://doi.org/10.1109/CVPR.2018.00981

4. Dougherty, E.R.: An introduction to morphological image processing. SPIE-Int. Soc. Opt. Eng. (1992)

5. Efros, A., Leung, T.: Texture synthesis by non-parametric sampling. In: Proceedings of the Seventh IEEE International Conference on Computer Vision, vol. 2, pp. 1033–1038. IEEE, Kerkyra (1999). https://doi.org/10.1109/ICCV.1999.790383

6. Efros, A.A., Freeman, W.T.: Image quilting for texture synthesis and transfer. In: Proceedings of the 28th Annual Conference on Computer Graphics and Interactive Techniques. SIGGRAPH 2001, pp. 341–346. Association for Computing Machinery, New York (2001). https://doi.org/10.1145/383259.383296

7. Fan, L., Krone, J., Woolf, S.: Sketch to image translation using GANs - LISA.FAN. https://lisa.fan/Resources/SketchGAN/sketch-image-translation.pdf. Accessed 25 Jan 2023

8. github-actions: rtmigo/img2texture: Cli for converting images to seamless tiles (2021). https://github.com/rtmigo/img2texture_py. Accessed 22 May 2022

9. Gao, C., Liu, Q., Xu, Q., Wang, L., Liu, J., Zou, C.: SketchyCOCO: image generation from freehand scene sketches. In: 2020 IEEE/CVF Conference on Computer Vision and Pattern Recognition (CVPR), pp. 5173–5182. IEEE, Seattle (2020). https://doi.org/10.1109/CVPR42600.2020.00522

10. Gupta, K., Damani, S., Narahari, K.N.: Using AI to design stone jewelry, November 2018. https://doi.org/10.48550/arXiv.1801.00723. Accessed 3 Feb 2023

11. Hu, M., Guo, J.: Facial attribute-controlled sketch-to-image translation with generative adversarial networks. EURASIP J. Image Video Process. **2020**(1), 1–13 (2020)

12. Isola, P., Zhu, J.Y., Zhou, T., Efros, A.A.: Image-to-image translation with conditional adversarial networks. In: 2017 IEEE Conference on Computer Vision and Pattern Recognition (CVPR), pp. 5967–5976. IEEE, Honolulu (2017). https://doi.org/10.1109/CVPR.2017.632

13. Karimi, P., Davis, N., Grace, K., Maher, M.L.: Deep learning for identifying potential conceptual shifts for co-creative drawing (2018). https://doi.org/10.48550/ARXIV.1801.00723. Accessed 27 Jan 2023

14. Kingma, D.P., Ba, J.: Adam: a method for stochastic optimization (2014). https://doi.org/10.48550/ARXIV.1412.6980. Accessed 3 Feb 2023

15. Liang, L., Liu, C., Xu, Y.Q., Guo, B., Shum, H.Y.: Real-time texture synthesis by patch-based sampling. ACM Trans. Graph. **20**(3), 127–150 (2001). https://doi.org/10.1145/501786.501787

16. Moravec, J.: A comparative study: L1-norm vs. l2-norm; point-to-point vs. point-to-line metric; evolutionary computation vs. gradient search. Appl. Artif. Intell. **29**(2), 164–210 (2015). https://doi.org/10.1080/08839514.2015.993560

17. Park, T., Zhu, J.Y., Wang, O., Lu, J., Shechtman, E., Efros, A.A., Zhang, R.: Swapping autoencoder for deep image manipulation. In: Advances in Neural Information Processing Systems. Curran Associates Inc., Red Hook (2020). https://doi.org/10.48550/arXiv.2007.00653

18. Ronneberger, O., Fischer, P., Brox, T.: U-net: convolutional networks for biomedical image segmentation. In: Navab, N., Hornegger, J., Wells, W.M., Frangi, A.F. (eds.) MICCAI 2015. LNCS, vol. 9351, pp. 234–241. Springer, Cham (2015). https://doi.org/10.1007/978-3-319-24574-4_28

19. Simonyan, K., Zisserman, A.: Very deep convolutional networks for large-scale image recognition. In: 3rd International Conference on Learning Representations (ICLR 2015), pp. 1–14. Computational and Biological Learning Society, San Diego (2015). https://doi.org/10.3390/electronics10040497
20. Tian, N., Liu, Y., Wu, B., Li, X.: Colorization of logo sketch based on conditional generative adversarial networks. Electronics **10**(4) (2021). https://doi.org/10.3390/electronics10040497
21. Xing, J., Wei, L.Y., Shiratori, T., Yatani, K.: Autocomplete hand-drawn animations. ACM Trans. Graph. **34**(6) (2015). https://doi.org/10.1145/2816795.2818079

Transposition of Simple Waveforms from Raw Audio with Deep Learning

Patrick J. Donnelly$^{(\boxtimes)}$ (ID) and Parker Carlson (ID)

Oregon State University, Corvallis, OR 97331, USA
{donnellp,carlspar}@oregonstate.edu

Abstract. A system that is able to automatically transpose an audio recording would have many potential applications, from music production to hearing aid design. We present a deep learning approach to transpose an audio recording directly from the raw time domain signal. We train recurrent neural networks with raw audio samples of simple waveforms (sine, square, triangle, sawtooth) covering the linear range of possible frequencies. We examine our generated transpositions for each musical semitone step size up to the octave and compare our results against two popular pitch shifting algorithms. Although our approach is able to accurately transpose the frequencies in a signal, these signals suffer from a significant amount of added noise. This work represents exploratory steps towards the development of a general deep transposition model able to quickly transpose to any desired spectral mapping.

Keywords: Deep learning · FFT · Music · Timbre · Transposition

1 Introduction

Transposing a recorded audio file to a different set of frequencies is a difficult research task. Automatic transposition requires scaling frequency by a fixed multiplicative constant while preserving the time scale. A system that is able to transcribe audio in near real-time would have a variety potential uses ranging from automatic pitch correction during live vocal performances to intelligent filtering and transformations to improve hearing aid-mediated perception.

Changing the pitch is often the desired goal of audio manipulation, such as when making adjustments to the intonation of vocal lines or for the transposition of an audio recording to another musical key. In other circumstances, a change in pitch is an undesired side effect, often resulting from resampling or time-stretching manipulations. Whichever the goal, such audio manipulations must balance a trade-off between temporal and spectral fidelity.

In this paper, we investigate the automatic transposition of audio directly from raw audio signals. We present a novel approach to train deep recurrent neural networks to learn new frequency mappings in order to perform transposition directly on the time-domain signal itself. Currently, other deep learning approaches rely on Fourier analysis, feature extraction, spectral manipulation,

C. Johnson et al. (Eds.): EvoMUSART 2023, LNCS 13988, pp. 341–356, 2023.
https://doi.org/10.1007/978-3-031-29956-8_22

and resampling or resynthesis, which all require time to compute. To compensate, near real time systems must use small windows for spectral analysis, which reduces the fidelity of the frequency analysis by limiting the number of samples available. Our proposed approach tasks the neural network to approximate the spectral transformation, without explicit use of the Fast Fourier Transform (FFT) algorithm. To our knowledge, this is the first attempt to train deep neural networks for music transposition directly from the raw audio signal.

2   Related Work

Following the success of deep learning approaches for speech synthesis (see review [10]), these techniques were soon adapted for other audio learning tasks. Researchers have applied deep learning in a number of musical tasks including signal processing [20], content description [19], information retrieval [3], recommendation systems [23], and composition [8].

Although many recent approaches and models for audio deep learning have focused primarily upon speech synthesis, some models also support music synthesis. One such example is Wavenet [18], a deep generative model for generating speech and music that uses a fully connected convolutional network with dilation factors to generate output audio one sample at a time. Another recent example is Differential Digital Signal Processing [6], which provides a differentiable framework for traditional signal-processing techniques to enable audio generation with smaller models. Despite these recent advances in deep learning of music, many challenges and unsolved research tasks persist [2].

In the remainder of this section, we briefly review techniques, algorithms, and studies related to automatic music transposition. First, we review signal processing approaches for time-scale modification of pitch. Next, we discuss the fast Fourier transform and the phase vocoder. Lastly, we review the recent approaches for automatic pitch correction and intonation adjustment.

2.1   Time-Scale Approaches

Before the availability of deep-learning approaches, researchers employed signal processing approaches to attempt to manipulate audio frequency. When an audio signal is resampled and played back at it original sampling rate, both the pitch and duration of the signal will change. To decouple this dependency, researchers developed numerous resampling and time-scale modification approaches. These algorithms facilitate applying changes of pitch with retaining the original duration, or conversely, they allow changes of duration without changing the pitch.

Noteworthy examples of such time-domain algorithms include the Synchronised Overlap-Add (SOLA) [22], Time-Domain Pitch Synchronous Overlap-Add (TD-PSOLA) [16], Waveform Similarity Overlap-Add (WSOLA) [24], and Adaptive Overlap-add (AOLA) [11] algorithms. More recent approaches, such as Canonical Time Warping (CTW) [26] have also been employed to aid in aligning vocal corrections with the accompaniment mixture [13].

2.2   Phase Vocoder

The fast Fourier transform (FFT) is an efficient algorithm to compute the discrete Fourier transform (DFT) in $\mathcal{O}(n \log n)$ multiplications. In the context of deep-learning, Bengio and LeCun describe the FFT as a deep representation of the Fourier transform [1]. And researchers have found that performing convolutions as products in the Fourier domain can accelerate training and inference of convolutional neural networks [15]. FFT-based deep representations have also been shown to help reduce complexity while performing inference on embedded platforms [12]. However, one recent study failed to learn the FFT representation within $\mathcal{O}(n \log n)$ scaling using a deep linear neural network [17].

A phase vocoder (see tutorial [5]) is technique for analyzing and synthesising an audio signal. Originally developed for synthesizing the human voice, the phase vocoder has been a popular technique in electroacoustic music since the 1970's. A phase vocoder uses the FFT to decompose a signal into narrow frequency bands, which are used to tune a series of bandpass filters. A carrier signal, such as a noise source, is then sent through these filters to synthesis the original sound.

In addition to its other creative uses, a phase vocoder can be used to time stretch a signal without changing the pitch and to manipulate pitch without changing the signal's duration. One study proposed a modulation vocoder for selective frequency pitch transposition [4]. The authors compared their approach against a commercial algorithm in a small subject study. The participants judged the vocoded transposition satisfactory with respect to pitch and harmony but preferred the propriety approach for preservation of timbre. Another study explored adaptive pitch shifting with a non-linear phase vocoder to perform global and local intonation adjustments in recordings of a cappella singing [21].

2.3   Pitch Correction

Pitch correction is an important tool in audio engineering and music production. This is the task of altering the pitches in an audio stream, usually with the intent to tune the notes to a specific track or reference frequency. Autotune[1], VariAudio[2], and Melodyne[3] are commercial tools that enable manipulation of pitch, primarily intended for corrections of intonation in vocal tracks.

Current approaches to pitch correction typically require a pipeline that includes a reference track as input, a DFT, spectral feature or filter manipulation, and post-processing to reconstruct the adjusted audio as the output. These approaches are usually limited by how much transposition they are able to perform. Many auto-tuner systems work best for transposition of small musical intervals but may sound distorted when transposing larger intervals, such as those greater than an octave. Furthermore, these approaches are usually not well suited to audio streams with accompaniment or other background vocals.

---

[1] http://www.antarestech.com/products/.

[2] https://www.steinberg.net/cubase/new-features/.

[3] https://www.celemony.com/en/melodyne/what-is-melodyne.

Other approaches have attempted data-driven approaches for automatic pitch correction. In one such example, the authors trained deep neural networks using Constant Q-Transform features on a dataset of short excerpts of amateur karaoke performances [25]. In subjective listening tests, study participants preferred the automatically auto-tuned examples compared to the original 49% of time.

3   Dataset

We begin our exploration of audio transposition directly from raw audio by investigating the four basic waveform shapes used in music synthesis. We consider the pure sine tone, as well as the non-sinusoidal square, triangle, and sawtooth waveforms. In Fig. 1, we illustrate the waveform and spectra for each waveform. We synthetically-generate the waveforms for the dataset of our experiment, rather than using an existing dataset of sounds or music so that we can precisely control the frequencies of the sounds used in training and testing.

In this section we provide an overview of our process to synthetically generate our dataset, our mapping between our input and our desired output based on transposition step size, and our partition between training and test examples.

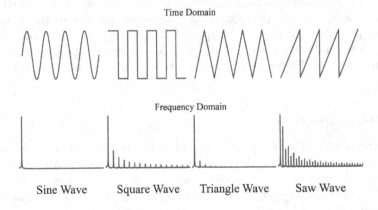

Fig. 1. Time domain (top) and frequency domain (bottom) for each waveform.

3.1   Waveform Generation

For each waveform shape, we generate one second examples with fundamental frequencies linearly spaced along the frequency range using a spacing of 0.5 Hz. We generate all signals with amplitudes of 0.1 to avoid potential clipping in our output. These waveforms are saved as a single channel WAV file with a sampling rate of 16 kHz. Given our sampling rate and the Nyquist limit, we explore frequencies up to 8 kHz.

3.2   Input and Output

In order to evaluate different transposition step sizes, ranging from one semitone (minor second) to 12 semitones (octave), we generate 12 separate datasets each with a different mapping from input frequencies to corresponding output frequencies. Because we evaluate the octave as our largest possible transposition, we consider frequencies in the range [20, 4000] for inclusion as input examples for training our neural network.

For each waveform, we consider examples in our input range and identify the corresponding output resulting in the transposition from semitone s in the range [1, 12]. Using intervals from equal temperament, we calculate $d = 2^{\frac{s}{12}}$ as a multiplier to find the frequency of a tone raised by s semitones. For each semitone up to an octave, we map each input frequency in range [20, 4000] Hz to the corresponding transposed frequency in range $[20 \times d, 4000 \times d]$ Hz.

3.3   Train and Test Split

In this work we seek to train a deep neural network to transpose audio directly from raw audio. In other words, we desire the trained network to approximate the Fourier transform over the range of possible input frequencies. For this reason, our approaches does not lend itself to a traditional partition of examples into train examples and held-out test examples. Instead we seek to learn to approximate a general transposition function for any input frequency, using a model trained only on a discrete and finite sampling of frequencies in the input range.

Specifically, we train the neural network with a set of fundamental frequencies covering the entire range linearly spaced by 1.0 Hz. We then test our models using frequencies spaced between the training examples. This ensures that test examples are evenly spread throughout the training domain. For example, examples with fundamental frequency F_0 of 440, 441, 442, etc. are added to the training partition, while examples with F_0 of 440.5, 441.5, and 442.5, etc. are held out for testing. This spacing of 0.5 was chosen to ensure an equal distance in frequency between any test example and its neighboring train examples.

In total, we have 12 datasets, one for each semitone [1, 12], each including all four wavesforms: sine, square, triangle, and sawtooth. Each dataset contains 15,920 examples for training and 15,920 examples for testing.

4   Experimental Design

In this section we describe our choices of deep learning architecture, model parameters, and training paradigm. We also discuss the two pitch shifting algorithms against which we compare our results as well as the algorithm we use to measure the noise of our generated transpositions.

4.1 Model

Based on preliminary experiments, we select a recurrent network topology consisting of three stacked layers of Bidirectional LSTMs (BiLSTMs) each with 1024 hidden units, with a recurrent dropout of 0.4, and batch normalization performed between the BiLSTM layers. We train using `RMSProp` [27] with a learning rate of 0.001 and a batch size of 64. We initialize our model weights and parameters based on a random seed, which we hold constant across our experiments.

When training the model, we split each input audio stream into a sequence of chunks of 1024 samples. This ensures that each input chunk contains a full cycle 20 Hz, the lowest frequency in our input set. The last input chunk of the audio stream is zero-padded to a multiple of 1024. For each experiment, we train our models for 250 epochs. We use a fixed stopping criteria to better support comparison between the results of our different experiments.

4.2 Evaluation

We compare our results against two open-source pitch shifting algorithms. First, we consider the pitch-shifting algorithm in Librosa v.0.8.1[4], a Python package for music and audio analysis. This naïve implementation stretches the signal in time, then resamples to ensure the output's duration is the same of the original. Second, we consider the pitch shifting algorithm implemented by Rubber Band[5]. This algorithm is known for its robust qualities, such as handling audio transients and its ability to pitch shift up to three octaves.

We also calculate the signal-to-noise ratio (SNR) to evaluate the noisiness of our generated output. We use the implementation in Matlab[6] to measure the noise in decibels relative to the carrier (dBc). This implementation expects a sinusoidal input and uses a modified periodogram with a Kaiser window of $\beta = 38$ and explicitly excludes the first six harmonics as not noise.

5   Results

For each of the 12 semitone steps, we train a separate recurrent neural network to transpose audio. The models are trained on the raw audio signal with examples from all four waveforms, sampled in the range [20, 4000] Hz. These models generate a transposed signal that approximates the desired transposition step size. The output of the model does not require any post-processing and can be saved directly as a WAV file.

We compare our transpositions against those generated by the Librosa and Rubberband baseline algorithms, described in Sect. 4.2. We examine the quality of our generated output in three ways. First, we examine how well our generated signal matches the expected signal in the time domain. Second, we examine the accuracy of generated signal in the frequency domain. Lastly, we measure the amount of noise generated in our signal, compared to the expected signal.

---

[4] https://librosa.org/doc/main/generated/librosa.effects.pitch_shift.html.
[5] https://breakfastquay.com/rubberband/.
[6] https://www.mathworks.com/help/signal/ref/snr.html.

5.1 Error in the Time Domain

We begin by comparing our generated signal against the expected transposed audio. We calculate the mean squared error (MSE) between the two signals in the time-domain. This approach, used by [9], measures how well our system is able to precisely match the expected output signal, sample by sample. We compare our result against the Librosa and Rubberband algorithms in Table 1.

Across the different transposition sizes, our approach achieves a slightly lower MSE error than the other two algorithms. However, we do not find this difference statistically significant. Moreover, from the manual examination of individual examples, we observe that the lower performance of Librosa and Rubberband relative to our approach is likely due to small differences in phase. In other words, our approach was better able to match the exact phase of the expected transposed signal. The other approaches, which rely on time-stretching and resampling, were slightly out of phase with the expected output. This contributed to a higher overall error using this comparison, and does not necessarily reflect the quality or accuracy of the transposed frequencies.

Table 1. Mean Squared Error (MSE) in the time domain for each model across the different transposition sizes.

Model	Sine	Square	Triangle	Saw	Overall
Our Approach	0.0089	0.0113	0.0071	0.0070	**0.0086**
Librosa	0.0094	0.0129	0.0078	0.0077	**0.0094**
Rubberband	0.0100	0.0138	0.0081	0.0080	**0.0100**

In left column of Fig. 2 we breakdown this performance by transposition step-size for each waveform. Overall we found similar trends between our algorithm and the two baseline algorithms for many of the transpositions. Unexpectedly, we found that all approaches had more difficulty with some transpositions for certain waveforms. For example, all three algorithms demonstrated increased error for step size $s = 3$ for sine waves, $s = 6$ for triangle waves, and $s = 7$ and $s = 9$ for sawtooth waves. Although we expected a more even performance across the different transposition sizes, we instead found a variance between step sizes and that this variance was fairly consistent regardless of the algorithm.

To facilitate direct comparison between our different experiments, we used a consistent model architecture and a fixed number of training epochs, as described in Sect. 4.1. To further reduce error, we might benefit from exploring different network topologies or varying model training time for each transposition size and waveform. However, the similarities of error between the different algorithms imply that certain transposition sizes are more difficult than others. More work is needed to understand if this is a side-effect of our choices in experiment design. As future work, we will investigate the effects of the example length, sample rate, and the frequency resolution of training set.

5.2   Error in the Frequency Domain

We also compare the frequencies in our generated signal against those in the expected transposed audio. We calculated the MSE between the two signals in the frequency-domain, after performing a FFT over each one second excerpt. This measures how well our system was able to precisely match the frequency content of the expected transposition. As before, we compare our result against the Librosa and Rubberband algorithms in Table 2.

Table 2. Mean Squared Error (MSE) in the frequency domain of each model across the different transposition sizes.

Model	Sine	Square	Triangle	Saw	Overall
Our Approach	0.00000209	0.00000265	0.00000168	0.00000163	**0.00000201**
Librosa	0.00000163	0.00000188	0.00000209	0.00000150	**0.00000177**
Rubberband	0.00000247	0.00000338	0.00000199	0.00000195	**0.00000245**

In the right column of Fig. 2 we show the error in frequency domain by transposition step-size and waveform type. Overall we found similar performance between our approach and the two baseline algorithms. Just as we observed with MSE of the time-domain signal, we find that our approach often had increased error in frequency-domain for certain step sizes and that those steps were similarly difficult for our baseline algorithms.

5.3   Signal to Noise

When listening to our generated transpositions, it is clear that our approach generates significantly more noise than the two baseline algorithms. To assess the noisiness of our generated signal, we calculate the signal to noise ratio (SNR), using the approach described in Sect. 4.2. We compare the average noise of our transpositions against the baseline algorithms in Table 3. Here we confirm that our approach generates significantly more noise than the other approaches.

For all waveform types we narrowly achieve a positive SNR dBc value, indicating there is more signal than noise in our generated transpositions. We also observe that we performed marginally better with the sine and triangle waveforms compared to the square or sawtooth waveforms. This implies, unsurprisingly, that our approach works better with simpler timbres and struggles more with complex timbres with a high number of frequency components. The baseline

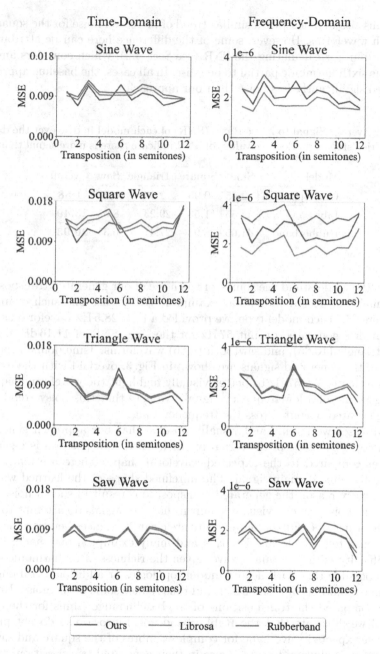

Fig. 2. Error by transposition size in semitones and for each waveform type. The left column shows MSE of the time domain signal and the right column shows MSE of the signal in the frequency domain.

algorithms also demonstrate a similar trend of increased noise for the square and sawtooth waveforms. However, some of the difference here can be attributed to our algorithm for calculating the SNR (see Sect. 4.2), which considers anything above the sixth harmonic partial to be noise. In all cases, the baseline approaches generated significantly less noise than our approach.

Table 3. Average Signal to Noise Ratio (SNR) of each model in dBc over the different transposition step sizes. Positive values of SNR indicate there is more signal than noise.

Model	Sine	Square	Triangle	Saw	Overall
Our Approach	2.65	0.95	2.49	0.23	**1.58**
Librosa	44.55	11.53	29.23	8.65	**23.49**
Rubberband	41.05	8.93	27.41	6.31	**20.93**

To better understand the source of the noise of our generated transpositions, we manually inspected individual examples. We discuss one such example in detail here. For each model type, we provided a $F_0 = 28.5$ Hz waveform as input. These models generated output 57 Hz for the sine (SNR of 11.16 dBc), square (7.50), triangle (10.65), and sawtooth (6.86) waveforms. Time domain representations of the generated signals are shown in Fig. 3, overlaid with the expected waveform shape. We 57 Hz to better visually highlight the waveform shape over the long period of a low frequency signal. We note that this noisy trend occurs for all generated output across the frequency range.

In all cases we readily and visually observe that our approach generates a lot of noise. The shape of the signal of the learned transposition is jagged and erratic as compared to the expected waveform shape. There are many errant peaks in the time domain signal. The unrefined shape of the learned waveform is the primary reason the generated transpositions result in such a noise signal. Despite the noise, we can visually confirm that our signals do attempt to match the waveform shape. Furthermore, there are clear differences between the shapes of our generated waveforms. For this and many examples, our model had the most difficulty with the square wave, given the richness of its harmonic series.

It is possible that with deeper model topologies or additional training time, our signals would become smoother and thus reduce some of the noise. However, we also compared the transpositions of the baseline algorithms for these tasks. In Fig. 4 we illustrate that the Rubberband and Librosa also do not precisely match the expected waveshape for complex timbres of the square and sawtooth waveform, contributing to some noise in their generated transpositions.

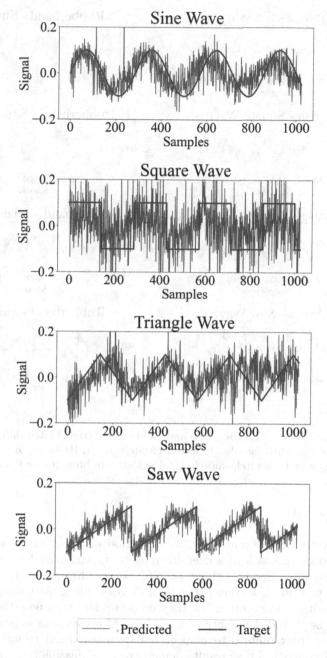

Fig. 3. Example of one octave ($s = 12$) transposition 57 Hz, comparing the generated signal (blue) to the true signal (orange). (Color figure online)

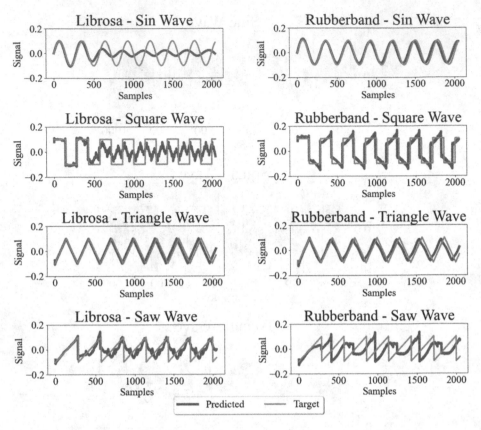

Fig. 4. In some cases, the Librosa (left) and Rubberband (right) pitch shifting methods also struggle to capture complex timbre accurately for 57 Hz example. The target is shown in orange and the pitch-shifted signal is shown in blue. (Color figure online)

6 Discussion

In this work we explore automatic transposition directly from raw audio signals. This novel approach uses deep-learning to approximate Fourier processes to transpose audio directly without pre-processing, filtering, explicit use of the fast Fourier transform, feature extraction, or resampling. Although our generated transpositions are quite noisy, they do accurately transpose the frequency content by the intended number of semitones. Moreover, our approach readily transposes complex timbres to map each harmonic partial to its appropriate transposed frequency. These results demonstrate the feasibility of an approach to transpose audio directly from the raw time domain signal.

6.1   Limitations and Future Work

Despite these encouraging initial results, there remain many research questions, challenges, and opportunities to improve our transpositions. Here we discuss the limitations of our approach in the context of our plans for future work.

One of the primary limitations of our approach is the amount noise found in the output signal, as compared to other approaches. Much of this noise results from the overall jagged shape of our generated signal, which are imperfect approximations of the true waveforms. To improve our approximation, we will explore other network architectures, including deeper recurrent networks, convolutional networks, and differential DSP models [6]. Furthermore, we will explore dilated convolution, as used by Wavenet [18], to improve the shape of our generated output. Additionally, we will investigate post-processing approaches to reduce noise, such as the use of machine learning models to filter noise as well as signal processing techniques to smooth our generated waveform. The noise in our signal appears to be Gaussian distributed and would be a good candidate for existing denoising algorithms. Lastly, we will investigate how the signal to noise ratio is affected by the transposition size.

In this initial exploration, we limited ourselves to four basic waveform shapes. In future work, we will investigate this approach on more complex examples, including both speech and polyphonic music. We will consider large established datasets like Nsynth [7], which contains MIDI pitch references from which we can estimate fundamental frequencies [14]. Towards our goal of creating a generalizable transposition model, we will also investigate transposing real world datasets of sounds and music. For these, we will not have an example of the expected transposition and thus must consider other evaluation strategies.

A persistent challenge in generative audio tasks using deep learning is measuring the efficacy of the approach and the quality of the generated audio. In the task of vocal correction, one study found a lack of generalizability of existing metrics used between studies [14]. In this study, we synthetically generated a dataset of audio signals such that we would have the input and transposed output signals. This enabled us to directly compare the time-domain signal of our true and generated outputs using mean-squared error. As future work, we will consider other evaluation approaches and metrics to enable examination of the efficacy of our system in which the correct transposed audio is not available.

First, we will evaluate the correctness of the frequencies of the transposed output. In the present study, we were unable to achieve sufficient resolution in the spectral domain given our design choice of one-second examples and a 16 kHz sampling rate. As future work, we will train our models with longer 10 s audio examples and with a higher sampling rate to ensure we have the fidelity to identify the fundamental frequency with a resolution of 0.1 Hz. However, this will significantly increase model training time. Second, we will design a subjective listening test to evaluate the efficacy of our approach to transpose complex pieces of music in which we do not have ground-truth comparisons.

6.2   Novelty and Applications

There are many useful real-world applications for an intelligent system that can perform transposition in near real-time. If a robust solution was available, it could perform pitch corrections or intonation adjustments during live concert performances. Automatic music transposition would enable near real-time audio transpositions from one musical key to another. This could serve as useful tool for musicologists, composers, arrangers, and those learning new pieces of music. Additionally, such a system would be of great use to recording engineering and music producers who could quickly transpose audio samples to other keys.

Unlike other approaches our approach could readily be adapted to apply non-linear frequency transpositions. Models could be trained to map from any individual frequency to any desired frequency. This could be used to transpose between different tuning systems, such as transforming equal temperament to just temperament. Moreover, such a system enables the possibility to remap timbre directly, perhaps allowing the timbral transformation from one instrument to another, or the conversion of one spoken voice to sound like another.

We motivate our work by the desire to perform non-linear frequency transpositions to improve the intelligibility of music for those that listen through cochlear implants. If there existed a mapping between true frequencies and those pitches the patient perceives, collected through patient-specific tests by an audiologist, researchers could investigate timbral filtering of sounds to improve perception of complex sounds. For example, such a system could simplify complex timbres by attenuating certain and very specific frequencies. In such a way, researchers could begin adapting sound directly to the patient's perception, which results from the surgeon's placement of electrodes along the cochlear during implantation.

6.3   Conclusion

In this study, we demonstrate the feasibility of a deep learning approach to transpose raw audio signals without explicit spectral analysis or manipulation. By training neural networks with generated waveforms covering the range of possible frequencies, we are able to automatically transpose signals with fundamental frequencies not explicitly included in the training set. For a model trained for a specific transposition step size and waveform shape, we successful transpose the signal to the desired frequencies and timbre, while approximating the original waveform shape. Although our current approach produces considerably more noise compared to established algorithms, we identify numerous avenues to improve the audio quality of our generated transpositions. This study represents the first investigation of a general transposition engine trained on simple waveforms and able to transpose to any non-linear spectral mapping.

References

1. Bengio, Y., LeCun, Y., et al.: Scaling learning algorithms towards AI. Large-Scale Kernel Mach. **34**(5), 1–41 (2007). https://doi.org/10.7551/mitpress/7496.003.0016
2. Briot, J.-P., Pachet, F.: Deep learning for music generation: challenges and directions. Neural Comput. Appl. **32**(4), 981–993 (2018). https://doi.org/10.1007/s00521-018-3813-6
3. Choi, K., Fazekas, G., Cho, K., Sandler, M.: A tutorial on deep learning for music information retrieval. arXiv preprint arXiv:1709.04396 (2017). https://doi.org/10.48550/arXiv.1709.04396
4. Disch, S., Edler, B.: Frequency selective pitch transposition of audio signals. In: 2011 IEEE International Conference on Acoustics, Speech and Signal Processing (ICASSP), pp. 29–32. IEEE (2011). https://doi.org/10.1109/ICASSP.2011.5946320
5. Dolson, M.: The phase vocoder: a tutorial. Comput. Music. J. **10**(4), 14–27 (1986). https://doi.org/10.2307/3680093
6. Engel, J., Hantrakul, L.H., Gu, C., Roberts, A.: DDSP: differentiable digital signal processing. In: International Conference on Learning Representations (2020). https://openreview.net/forum?id=B1x1ma4tDr
7. Engel, J., et al.: Neural audio synthesis of musical notes with wavenet autoencoders. In: International Conference on Machine Learning, pp. 1068–1077. PMLR (2017). https://doi.org/10.48550/arXiv.1704.01279
8. Hernandez-Olivan, C., Beltran, J.R.: Music composition with deep learning: a review. arXiv preprint arXiv:2108.12290 (2021). https://doi.org/10.48550/arXiv:2108.12290
9. Jawahir, A., Haviluddin, H.: An audio encryption using transposition method. Int. J. Adv. Intell. Inform. **1**(2), 98–106 (2015). https://doi.org/10.26555/ijain.v1i2.24
10. Khalil, R.A., Jones, E., Babar, M.I., Jan, T., Zafar, M.H., Alhussain, T.: Speech emotion recognition using deep learning techniques: a review. IEEE Access **7**, 117327–117345 (2019). https://doi.org/10.3390/app9194050
11. Lawlor, B., Fagan, A.D.: A novel efficient algorithm for music transposition. Organ. Sound **4**(3), 161–167 (2000). https://doi.org/10.1017/S135577180000306X
12. Lin, S., Liu, N., Nazemi, M., Li, H., Ding, C., Wang, Y., Pedram, M.: FFT-based deep learning deployment in embedded systems. In: 2018 Design, Automation and Test in Europe Conference and Exhibition (DATE), pp. 1045–1050. IEEE (2018). https://doi.org/10.23919/DATE.2018.8342166
13. Luo, Y.J., Chen, M.T., Chi, T.S., Su, L.: Singing voice correction using canonical time warping. In: 2018 IEEE International Conference on Acoustics, Speech and Signal Processing (ICASSP), pp. 156–160. IEEE (2018). https://doi.org/10.1109/ICASSP.2018.8461280
14. Luo, Y.J., Lin, Y.J., Su, L.: Toward expressive singing voice correction: On perceptual validity of evaluation metrics for vocal melody extraction. arXiv preprint arXiv:2010.12196 (2020). https://doi.org/10.48550/arXiv.2010.12196
15. Mathieu, M., Henaff, M., LeCun, Y.: Fast training of convolutional networks through FFTs. arXiv preprint arXiv:1312.5851 (2013). https://doi.org/10.48550/arXiv.1312.5851
16. Moulines, E., Charpentier, F.: Pitch-synchronous waveform processing techniques for text-to-speech synthesis using diphones. Speech Commun. **9**(5–6), 453–467 (1990). https://doi.org/10.1016/0167-6393(90)90021-Z

17. Nye, M., Saxe, A.: Are efficient deep representations learnable? arXiv preprint arXiv:1807.06399 (2018). https://doi.org/10.48550/arXiv.1807.06399

18. van den Oord, A., et al.: Wavenet: a generative model for raw audio. arXiv preprint arXiv:1609.03499 (2016). https://doi.org/10.48550/arXiv.1609.03499

19. Peeters, G., Richard, G.: Deep learning for audio and music. In: Benois-Pineau, J., Zemmari, A. (eds.) Multi-faceted Deep Learning, pp. 231–266. Springer, Cham (2021). https://doi.org/10.1007/978-3-030-74478-6_10

20. Purwins, H., Li, B., Virtanen, T., Schlüter, J., Chang, S.Y., Sainath, T.: Deep learning for audio signal processing. IEEE J. Sel. Top. Sig. Process. $13(2)$, 206–219 (2019). https://doi.org/10.1109/JSTSP.2019.2908700

21. Rosenzweig, S., Schwär, S., Driedger, J., Müller, M.: Adaptive pitch-shifting with applications to intonation adjustment in a cappella recordings. In: 2021 24th International Conference on Digital Audio Effects (DAFx), pp. 121–128. IEEE (2021). https://doi.org/10.23919/DAFx51585.2021.9768268

22. Roucos, S., Wilgus, A.: High quality time-scale modification for speech. In: ICASSP'85. IEEE International Conference on Acoustics, Speech, and Signal Processing, vol. 10, pp. 493–496. IEEE (1985). https://doi.org/10.1109/ICASSP.1985.1168381

23. Schedl, M.: Deep learning in music recommendation systems. Front. Appl. Math. Stat. 44 (2019). https://doi.org/10.3389/fams.2019.00044

24. Verhelst, W., Roelands, M.: An overlap-add technique based on waveform similarity (WSOLA) for high quality time-scale modification of speech. In: 1993 IEEE International Conference on Acoustics, Speech, and Signal Processing, vol. 2, pp. 554–557. IEEE (1993). https://doi.org/10.1109/ICASSP.1993.319366

25. Wager, S., Tzanetakis, G., Wang, C.i., Kim, M.: Deep autotuner: a pitch correcting network for singing performances. In: ICASSP 2020–2020 IEEE International Conference on Acoustics, Speech and Signal Processing (ICASSP), pp. 246–250. IEEE (2020). https://doi.org/10.1109/ICASSP40776.2020.9054308

26. Zhou, F., Torre, F.d.l.: Canonical time warping for alignment of human behavior. In: Proceedings of the 22nd International Conference on Neural Information Processing Systems. NIPS 2009, pp. 2286–2294. Curran Associates Inc., Red Hook (2009). https://doi.org/10.5555/2984093.2984349

27. Zou, F., Shen, L., Jie, Z., Zhang, W., Liu, W.: A sufficient condition for convergences of Adam and RMSPROP. In: Proceedings of the IEEE/CVF Conference on Computer Vision and Pattern Recognition, pp. 11127–11135 (2019). https://doi.org/10.48550/arXiv.1811.09358

AI-Aided Ceramic Sculptures: Bridging Deep Learning with Materiality

Varvara Guljajeva[1,2]([✉])[ID] and Mar Canet Sola[3][ID]

[1] CMA, Hong Kong University of Science and Technology (Guangzhou),
Guangzhou, China
varvarag@ust.hk

[2] ISD, Hong Kong University of Science and Technology, Hong Kong, China

[3] Baltic Film, Media and Arts School, Tallinn University, Tallinn, Estonia
mar.canet@tlu.ee

Abstract. With the advent of neural networks as powerful tools for generating various forms of media, so-called 'Deep Learning' (DL) has entered the sphere of art production. The concept of creative artificial intelligence (AI) has become part of popular discourse around 2D digital image-making, but can AI exceed the limitations of 2D media and be applied creatively in more tactile 3D media such as sculpture? In this paper, we describe what happens when AI is applied in a real-life production line, from concept to physical object. The article presents a case study that explore DL's potential for creating a tactile sculpture guided only by text prompt and a 3D model. In the production process, we mix several methods, including neural, digital, and traditional, to achieve the final results. In terms of methodology, this is an artistic study that explores existing DL tools for 3D object generation and later manufacturing in 3D printed ceramics. In the study, we use practice-based research methods to explore what happens when modern technology meets traditional ways of production, such as pottery. Further, we discuss reference art projects that have utilised AI, lessons learned, and the potential use of DL tools in art production. The aim of the paper is to explore new meanings and to open new avenues for investigation that emerge by bringing together creative AI with materiality.

Keywords: Ceramics · AI sculpture · 3D AI · 3D printing · interdisciplinary · physical AI · deep learning · AI art · creative AI · hybrid process · practice-based research · artistic research

1 Introduction

Artificial Intelligence, and especially Deep Learning, has triggered enormous interest and critique within the art world just as it has in other fields. Through their practice, artists are exploring the new meanings and possibilities that this technology offers. Many image-, sound-, and text-based works with AI models,

Both authors contributed equally.

Fig. 1. Final result of ceramic sculpture. Different views of *Mermaid in green jelly and pink feather* (Nymph) from the sculpture series *Psychedelic Forms* by Varvara & Mar. (Color figure online)

such as GAN and GPT, have been created. Terms like 'AI art' and 'GAN art' are now on the research agenda for many scholars, including Aaron Hertzmann [9], Lev Manovich [16] and Joanna Zylinska [29], who discuss AI as introducing a new paradigm and potentially transforming the established role of the artist. Well-known examples of GAN art include *Mosaic Virus*[1] (2019) by Anna Ridler and *Memories of Passersby I*[2] (2018) by Mario Klingemann.

At the same time, new tools are being released daily. Today, the discussion has shifted from GAN models to Stable Diffusion, Midjourney, DALLE and similar CLIP-based models, which introduce a paradigm shift into the main workflow and characteristics of AI art. The accelerated pace of the field of AI and DL (Krenn et al., 2022) is a source of excitement for researchers in creative AI, but also presents key challenges.

Although AI is often described as a very powerful tool [14,24,28], but when applied to art production in installation or 3D formats difficulties are soon encountered. As we have described in our previous work, the development of an autonomous interactive robotic art installation has been challenging as neither CLIP models nor robot control software are designed to talk to each other [6]. In this article, we focus on discovering ways to generate 3D objects with neural networks and introduce DL technology into traditional art production methods, such as pottery. More precisely, the art project entitled *Psychedelic Forms*[3] by artist duo Varvara & Mar (also the authors of the article) is a series of ceramic sculptures (see Fig. 1) that serve as a case study in this paper. The project investigates the current potential of DL for creating a physical form guided by text

---

[1] http://annaridler.com/mosaic-virus.

[2] https://underdestruction.com/2018/12/29/memories-of-passersby-i/.

[3] https://var-mar.info/psychedelic-forms/.

prompt and 3D model. By mixing modern and traditional production methods, we explore the potential and limitations of neural nets in the context of artistic practice. Psychedelic Forms reinterprets and translates ancient sculptures into contemporary forms. The project considers co-creative AI and the shared agency between various production methods-modern, traditional, autonomous, or manual. Thus, it offers a novel approach to bridging DL technology with materiality.

The practice-based research methodology was used to answer research questions: What happens when neural networks meet materiality? And who has the agency in this production process? Our methodology is explained in a section below. Before analysing the available DL models for 3D object generation from a practical point of view, we would like to discuss several reference art projects that aim to extend 2D Generated Adversarial Network (GAN) outputs to the sculpture format.

1.1   Reference Works

The artwork *Dio*[4] (2018) by Ben Snell aims to extend GAN's 2D output to 3D physical space by producing sculptures from the shredded material of the computer used to produce the GAN 2D images of that same exact sculpture. Unfortunately, the author does not elaborate further on the production processes involved. A similar realization strategy-from GAN-generated images to 3D models by way of manual conversion-can be seen in Egor Kraft's *Content Aware Studies*[5] (2018). His work is a relevant reference in terms of re-visiting, or rather restoring, ancient culture through the vista of machine learning. Kraft sees the emerging objects as "[...] synthetic documents of emerging machine-rendered history" [12].

The artwork *Metamultimouse*[6] (2022) by Matthew Plummer-Fernández is a digital and physical figurine that used a GAN to generate the texture applied to the 3D-modelled shape [20]. The artwork, which is sold as a 100-edition NFT, has two materialities, one as a video animation of the figurine and a second as a digital colour-printed sculpture that appears different in each edition as each uses a single different frame of the video. The pieces are printed in Nylon using HP's Multi Jet Fusion (MJF) technology, which supports very complex geometries and full colour [22]. The double materiality of the piece appears to be a common feature of 3D projects that flirt with AI technology.

In contrast to the generative approach with GANs, another art project from the same year as *Psychedelic Forms, AI Sculpting*[7] (2022) by the group Onformative, demonstrates 3D object generation by applying reinforcement learning. In their technical paper, *Reinforcement Learning applied to sculpting: A technical story of AI craftsmanship*, the group described how an AI model was trained

---

[4] http://bensnell.io/dio/.

[5] https://egorkraft.com/.

[6] https://feralfile.com/artworks/metamultimouse-wzq?fromExhibition=
doppelganger-jgz.

[7] https://onformative.com/work/ai-sculpting.

to sculpt in a 3D environment [23]. Starting with a cube and 3D object as an input, the model begins to subtract mass from the shape step by step until the final form is reached. Various different digital substractive tool heads determine the final aesthetics of the surface. Although this method for using a model to produce the final object is reminiscent of the traditional way of sculpting from a block of stone, the project remains in the digital realm. This again illustrates the gap between the DL models' output and that of physical manufacturing, which this paper explores.

Currently, there are very few artworks that would generate a 3D object directly from a neural net. In *Psychedelic Forms* the generated output was an '.obj' file that was later crafted into the ceramic sculptures and colouring also had an important place in the process and final aesthetics of the sculptures. Because of this, and the way project brings together modern and traditional ways of production, it introduces a novel method to art creation.

2   Case Study: Psychedelic Forms - Text-to-Ceramics

Fig. 2. Final result of ceramic sculptures. *Snake and Angel on Blue Moon* (Dionysus) from the sculpture series *Psychedelic Forms* by Varvara & Mar. (Color figure online)

The project described in this article is inspired by classical sculptures, an ancient way of producing ceramics, and the limitations of DL models when it comes to tactile object production. Seeing that the GAN models are limited to 2D, we began to experiment with options how to achieve a tactile object from the neural net. By using a 3D object and text prompt, we guided the AI model that stylized the mesh. After numerous iterations of object generation using different input models and text prompts as inputs, based on the criteria the most fitting models were chosen (among the AI art community this process is known as 'curating' AI output). Later, the model was manually altered and prepared for 3D printing in clay. Glazing happened by hand, often inspired by the AI-generated object's vertex colours. The artwork demonstrates the embodied experience and transformation of a DL model through artistic practice.

The project name *Psychedelic Forms* came during the process. Psychedelic refers to unexpected or unexplored imagination that a human mind has not seen before. Although the original input was well-known ancient sculptures, like Venus, the AI model was capable of stylizing the mesh with the inputted text prompt in such a way that the new form was hardly recognizable (see interaction process in Fig. 3). The complex form emerged because of the co-creation between human and AI, and interplay with material and production processes (see Figs. 1, 2, 8, 9). The hybrid processes that combine digital, physical, chemical and ancient craft worlds are embedded into this project and described below.

Our choice to work with clay was connected with the ancient origin of the sculptures that were used as conceptual ready-mades. According to Meredith and Barnett, "a conceptual readymade is a contemporary artist's use of a classical work selected as a key point of reference taken out of time" [17]. Our idea was to re-interpret the classics through neural net algorithms and again return to the antique through matter.

The art project brings together ancient culture, craft, poetry and emerging AI tools. This practice-based research revisits known forms and processes from a novel angle and sets out new possibles.

2.1 Methodology

In this exploratory research project, practice-based research methodology was applied to analyse the available DL models for generating 3D models and to investigate how this technology could become part of an artistic production process in ceramics. The study combines DL with CAM technology and traditional craftsmanship. According to Candy et al., since the emphasis is on the creative process and the artefacts that contribute to new knowledge, the current research can be classified as practice-based (Candy et al. 2006). Although at first glance a practice-based approach may seem subjective, expert knowledge that stems from a long experience in art practice and applied research has significant value that should not be underestimated. The idea of 'art-as-research' is that art creates new theories in the act of making. According to Busch: "Art and theory, in effect, are nothing more than two different forms of practice interrelated through a system of interaction and transferences." (Busch 2011). Hence, the practice can provide valuable insights that can be helpful for understanding how innovative technology can be applied in art making and what kind of shift in the relationships of these processes it introduces. When it comes to the procedure of applied methodology, the project had four stages. Each demonstrates a translation process:

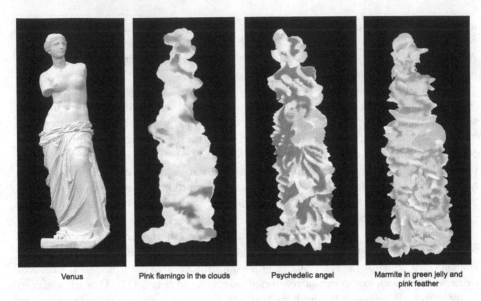

Venus   Pink flamingo in the clouds   Psychedelic angel   Marmite in green jelly and pink feather

Fig. 3. Iteration process: Text2Mesh outputs using the same input 3D object and different text prompts

1. **From exploration to concept: research.** Field research into existing DL models for 3D shape generation was conducted. When a possible tool was identified, the artistic concept and hypothesis of the project were developed. The hypothesis was that DL can be used for the translation of different processes and semiotic spaces, thereby contributing to creativity and value in the context of art and creative AI.
2. **From neural to digital: curating.** The selection and creation of inputs (3D object and text), iteration and curating of generated results belong to this next phase. The criteria for selection were set by the experts (the artists). There were two main criteria: a model had to be reparable, possible to produce, and it should have an interesting form. The first two criteria are more of a technical decision, while the last is purely artistic. At this point, the artist decides if the generated result is artistically valuable or not. This process is known as 'curating' in the AI art community.
3. **From digital to physical: fabrication.** This is the 3D-printing process in clay. This phase demonstrates the interplay between digital fabrication technology, material, and traditional pottery.
4. **From physical to chemical: coloring and firing.** This is the final phase where drying, multiple firing and glazing processes took place.

2.2   Translation Processes

The artistic project started with a simple question back in 2021: Is it possible to generate a 3D object with the available DL models? Later, when the practice-based research project evolved, we began to explore the relationships between different agencies and the meanings behind them. Although in recent years, there has been lots of research in the field of 3D with deep learning [7, 21, 26, 27], we found it difficult to find any model that could be applied in our project. We conducted our own field research on various different DL models. Such algorithms as 3D GAN, pytorch3D, Texture synthesis to 3D, and Text2Mesh, were tested. Based on the selection criteria, the freshly released Text2Mesh model was identified as the most suitable for the project. The model deforms a 3D object's mesh guided by text input [11, 18]. Text2Mesh is an AI model that transforms object mesh based on text input using CLIP, which is very similar to what CLIP can do in image models but in 3D [18]. Simply speaking, we could create, or more precisely alter, the input 3D model with our words. Here it is important to note that the model does not create a new object but modifies the existing one. In the developers' words: "[...] we consider content as the global structure prescribed by a 3D mesh, which defines the overall shape surface and topology. We consider style as the object's particular appearance or affect, as determined by its colour and fine-grained (local) geometric details" [18].

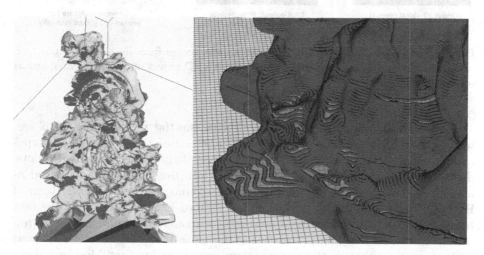

Fig. 4. An overhang problem with a Text2Mesh generated model. The image on the left indicates problematic areas in red. The image on the right shows problem with steep angle. (Color figure online)

The particular DL model used in this project and others that deploy CLIP or GPT algorithms work at the level of the concept and not the meaning. Thus, the generated mesh reflects upon the decoded concepts recalled by the Dadaistic short poems submitted by the artists as text prompts for the algorithm. This illustrates a new way of navigation, detached from meaning in the 'concept-plus-image' neural space and then superimposed onto the surface of the 3D model.

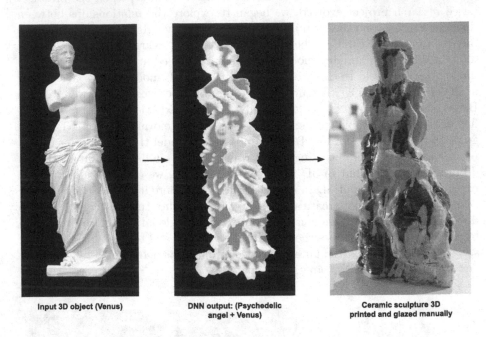

Input 3D object (Venus)   DNN output: (Psychedelic   Ceramic sculpture 3D
                                     angel + Venus)   printed and glazed manually

Fig. 5. The visual change during the translation process from an original input to a ceramic sculpture. Case study with Venus as input 3D object and *psychedelic angel* as text prompt.

Following the 'exploration to concept' phase was the 'neural to digital' phase, which involved the selection and creation of inputs, iteration, curation, and reparation processes. After experimenting with many digital objects and text inputs, digital copies of antique sculptures were fed into the algorithm and served as a point of departure for the creative process-Venus, Nymph, Double Herm of Herodotus and Thucydides, Dionysis, and Milford Lane-together with the text prompts created by the artists to style or re-mould the input form. Figure 3 demonstrates the iteration process using the same 3D model but with different text prompts. Although the vertex colour generated by Text2Mesh was later sometimes used as an inspiration for glazing, it did not play a role in the process of selecting an object for the next phase. The selection decision was based on the criteria set out at the beginning.

The last step in this phase was reparation and remodelling. In addition to a broken mesh, there was a big issue with the overhanging parts that was an obstacle for 3D printing in ceramics (see Fig. 4). In order to solve this problem, the generated 3D models were manually altered and overhangs were minimised.

In addition, aesthetic variations of the model were made. Figures 5 and 6 depict the translation process that the original model underwent.

The 'digital-to-physical' phase demonstrated several challenges too. First, printing large formats and complex shapes in clay is problematic because, in contrast to plastic, clay does not dry in minutes but takes days or even weeks if the environment is humid. Several attempts to 3D print failed by sculptures collapsing or falling in the middle of the printing. For that reason, instead of an autonomous process, printing was constantly supervised by human. Figure 7 illustrates the production pipeline of ceramics work: preparing clay and inserting it into the 3D printer's tank, supervising the 3D printing process, the first firing (after drying), and glazing. All sculptures in ceramic have 25–35% of infill. Only one sculpture is hollow inside, which was printed in recycled plastic. The translation process to a physical object revealed its tactile shape, but the material and production method also intervened in the visual aesthetics of the artefacts. Clay added extra risk and tension to the process as it required a lot of care and introduced its own limitations and aesthetics. The gravity and density of the clay influence the form, as do the drying and firing processes.

The last phase, 'physical to chemical', involved glazing and firing in the kiln and so colour was applied. In ceramics, glazing is a chemical process where glazing powder mixed with water is applied by hand to pre-fired pottery (see Fig. 7). Pouring technique was used for colouring the sculptures. This was an artistic decision that resonated with the process of AI, where control is heavily disturbed by chance. This method served as a metaphor for the thermodynamics of meaning spaces in latent spaces.

The selected colour pallet was sometimes inspired by the generated objects, but not always, as Fig. 6 clearly demonstrates. The colour selection was an artistic decision, too, based on the aesthetic taste of the artists as is common in most art practices. Nevertheless, the final outcome could not be entirely controlled because the technique of glazing technique is a little like painting blind as the colours appear only after the chemical reaction at high temperature during the firing process. Before firing, one does not see the colour or the composition and so the final result is always a surprise when it emerges from the kiln (see Figs. 1, 2, 8, 9). The colour aspect of this project is, therefore, quite unique as it has rarely, if ever been explored by artists working with AI tools.

3 Discussion on Agency and Creativity Within AI-Aided Processes

Although DL technology offers many novel formats, they all tend to remain within the digital realm. As the experiences of this study and the demonstrations of the reference projects show, practitioners quickly encounter the limitations of AI models when moving into the physical space. This project shows vividly that the output of the AI model was not ready for being reproduced in physical material form. Many laborious hours were spent fixing and preparing the 3D models for print. In addition to gravity, clay sets its own limits to what shapes

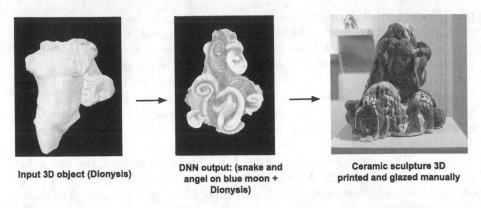

Input 3D object (Dionysis)   **DNN output: (snake and angel on blue moon + Dionysis)**   **Ceramic sculpture 3D printed and glazed manually**

Fig. 6. The visual change during the translation process from an original input to a ceramic sculpture. Case study with Dionysis as input 3D object and *snake and angel on blue moon* as text prompt. (Color figure online)

Fig. 7. Working with ceramics (from left to right): 1. mixing clay with water and filling a 3D printer tank with it. 2. supervising 3D printing process 3. glazing process.

Fig. 8. Final result of ceramic sculpture. *Snake and Angel on Blue Moon* (Dionysus) from the sculpture series *Psychedelic Forms* by Varvara & Mar. (Color figure online)

can be produced finally. Having said that, we should mention that all partners with whom we have worked in ceramics and industrial 3D printing said that our project involved some of the most complex shapes they had ever printed. Thus, the project pushed not only the limits of AI but also of CAM technology. However, as the number of scientific publications and related applications on DL continues to grow quickly and processes develop [13] so these limitations may soon be overcome as new and better options for shape generation appear.

By introducing clay as the material of the process, the destruction and alternation of the form continued to deform and evolve. The interplay of neural networks, digital, physical and chemical processes created new meanings and experiences. We see this as a meaningful and tangible way to navigate latent space. Such transversal and translation processes are related to the ideas of New Materialism, a cultural theory that de-territorializes disciplines by cutting across them rather than acting in opposition, and that sees agency as a two-way relationship instead of an act of possession [3,5,25]. The research project has demonstrated the four phases of creation that AI-aided ceramics sculptures underwent, and how each tool, technique, and material carried its own agency. This interplay of agencies was made explicit through the applied methodology.

More precisely, the research design was based on the four translation phases: exploration to concept, neural to digital, digital to physical, and finally, physical to chemical. It can be summarized as being a text-to-ceramics approach that led to unexpected results from the initial inquiry of our research questions. Charles Peirce, who is the father of formal logic and semiotics, describes the principles of inference as being based on induction that classifies, deduction that reduces choices, and abduction which explains [19]. This creates a combined approach that leads to new ideas as a source of creativity [1]. According to Yuri Lotman, art is the ultimate, most effective process of meaning-making [15]. Thus, this research project creates unexpected connections that provide for the creation of new meanings that emerge from foundational models like CLIP, and the allows the imprecision of translations to provide for a space of potentiality.

Semiotician Yuri Lotman describes the translations of various semiotic spaces as a strategy for augmenting creativity [15], and this can now be explored by utilising deep learning models such as CLIP to act as foundational models. Lotman investigated how new information be generated with translations into different mediums, such as image-to-text or text-to-image. In simple terms, by translating semiospheres, new sources of creativity emerge [8]. The authors of On the Digital Semiosphere describe how different semiotic spaces are incommensurable, and so efforts to translate between them constitute the very essence of creativity [8].

Translation systems can prompt the emergence of mutations that artists can use creatively in combination with generative deep learning models. The translation of media through such models translates the text to image and transfers the outcome to a distinct semiotic space which, according to Yuri Lotman, may be different from the input prompt and that the AI model unaware of. This phenomenon is related to Bender and Gebru's findings of language models [4] and is associated with the complexity of dealing with dimensional spaces of meaning. Accordingly, artists can benefit from the potential of AI navigations with text prompts as a creative inspiration and for generating accidental creative outcomes.

According to Yuk Hui's concept of 'cosmotechniques' there is no singular cosmos but instead there are multiple cosmologies that "[...] imply ways of knowing and being that cannot be simply rejected because they don't comply with modern scientific theories" [10]. In craft, multiple cosmologies can also be found, where knowledge manifests through manual work and materiality. Therefore it is vital that the universal cosmos of high-tech (for example, the foundational models used in DL are an example of a singular cosmos) does not erase or constrain our cultural knowledge, but instead is revisited and constantly reinterpreted. Ultimately, it is vital that there be sufficient opportunities for emerging relationships between the agencies instead of opposition.

This paper presents the potential of DL technology for craftwork when working with material. As Sofian Audrey emphasises, it is critically important that artists engage with AI technology in their practice because that is where ethics meet aesthetics. In his words: "[...] artists reveal both the inadequacies of AI and its untapped potential by imagining, through the materiality of AI itself [...]" [2]

Fig. 9. Exhibition view. Final result of ceramic sculpture. *Clouds in my head* (Double Herm of Herodotus and Thucydides) from the sculpture series *Psychedelic Forms* by Varvara & Mar.

4 Conclusions

The paper introduces the novel production method that is text-to-ceramics, which embeds AI into ceramics. The practice-based research methodology was applied to explore creative dialogue and potential between DL technology and sculpture-making in ceramics. The artistic project discussed here juxtaposes archaic craft method and ancient culture with innovative AI technology. The production of AI-assisted ceramic sculptures had four translation phases: exploration to concept, DL to digital, digital to physical, and physical to chemical. This introduces a new translation pipeline that contributes to the field of art and creative AI.

The project's conceptual and experimental properties helped connect the dots between new technologies and ancient production methods. The project demonstrates the artistic exploration of latent space and its translation to material, colour and shape. As a result, recent DL technology mixed with artisan techniques and processes offered irregular transformations that contribute to creativity and augment imagination. In other words, irregular mutations can lead to new creations that would not happen otherwise.

Finally, the study shows that creative AI can be applied to 3D form generation, and most importantly that DL technology can be used in sculpture making. However, this is currently possible only in combination with other digital and CAM tools, manual labour, and craftsmanship. This means that AI is often merely a departure point in artists' creative practice. Thus, the fear that AI will replace artists remains speculation. Nonetheless, in this project DL technology has enabled the artists to achieve unexpected results that would not have been possible without it. As demonstrated in *Psychedelic Forms*, AI can assist in the creative realisation of fictional ideas, thereby pushing the boundaries of technology and acting as a catalyst for the further development of creative AI.

Acknowledgement. *Psychedelic Forms* is an art project by Varvara & Mar realised partly in a residency at Espacio Open in Bilbao (Spain), and partly during Art Mesh project curated by Torun Ekstrand. The project was exhibited as a solo show entitled *Psychedelic Trips* at Alalimón Gallery (Barcelona, Spain) as from 10.06.22 until 18.10.22., which was part of ISEA2022 side program. We would like to express our gratitude for their help with production to Godot Studio and the Ceramics department at the Estonian Academy. MSC is supported as a CUDAN research fellow and ERA Chair for Cultural Data Analytics, funded through the European Union's Horizon 2020 research and innovation program (Grant No. 810961). VG is supported by HKUST(GZ).

References

1. Anderson, D.R.: Creativity and the Philosophy of C.S. Peirce, vol. 27. Springer, Dordrecht (2013). https://doi.org/10.1007/978-94-015-7760-1
2. Audry, S.: AI for good: why artists are key to improving machine learning technologies (2022). https://tiltwest.medium.com/ai-for-good-why-artists-are-key-to-improving-machine-learning-technologies-417f64923a6f. Accessed 30 Nov 2022

3. Barad, K.: Meeting the Universe Halfway: Quantum Physics and the Entanglement of Matter and Meaning. Duke University Press, Durham (2007)
4. Bender, E.M., Gebru, T., McMillan-Major, A., Shmitchell, S.: On the dangers of stochastic parrots: can language models be too big? In: Proceedings of the 2021 ACM Conference on Fairness, Accountability, and Transparency, pp. 610–623 (2021)
5. Braidotti, R.: Metamorphoses: Towards a Materialist Theory of Becoming. Wiley, Hoboken (2013)
6. Mar Canet Sola and Varvara Guljajeva: Dream painter: exploring creative possibilities of AI-aided speech-to-image synthesis in the interactive art context. Proc. ACM Comput. Graph. Interact. Tech. **5**(4), 1–11 (2022)
7. Gao, J., et al.: Get3D: a generative model of high quality 3D textured shapes learned from images. In: Advances in Neural Information Processing Systems (2022)
8. Hartley, J., Ibrus, I., Ojamaa, M.: On the Digital Semiosphere: Culture, Media and Science for the Anthropocene. Bloomsbury Publishing, New York (2020)
9. Hertzmann, A.: Can computers create art? In: Arts, vol. 7, p. 18. MDPI (2018)
10. Hui, Y.: Art and Cosmotechnics. U of Minnesota Press,Minneapolis (2021)
11. Jetchev, N.: Clipmatrix: text-controlled creation of 3D textured meshes. arXiv preprint arXiv:2109.12922 (2021)
12. Kraft, E., Kormilitsyna, E.: On content aware and other case-studies: historical investigations at blazing ultra resolution. In: Proceedings of Art Machines 2: International Symposium on Machine Learning and Art, pp. 88–93 (2021)
13. Krenn, M., et al.: Predicting the future of AI with AI: high-quality link prediction in an exponentially growing knowledge network. arXiv preprint arXiv:2210.00881 (2022)
14. Lee, R.S.T.: Quantum finance. In: Iyengar, S.S., Mastriani, M., Kumar, K.L. (eds.) Quantum Computing Environments. Springer, Cham (2020). https://doi.org/10.1007/978-3-030-89746-8_5
15. Lotman, J.M.: Universe of the Mind: A Semiotic Theory of Culture. Indiana University Press, Bloomington (1990)
16. Manovich, L.: AI and myths of creativity. Archit. Des. **92**(3), 60–65 (2022)
17. Meredith, H., Barnett, S.: 'Contemporary classicism' copy of a copy: appropriating classical statues as conceptual readymades. CLARA **8** (2021)
18. Michel, O., Bar-On, R., Liu, R., Benaim, S., Hanocka, R.: Text2mesh: text-driven neural stylization for meshes. arXiv preprint arXiv:2112.03221 (2021)
19. Peirce, C.S.: Collected papers of charles sanders peirce, vol. 5. Harvard University Press (1974)
20. Plummer-Fernández,   M.:   Metamultimouse   (2022).   https://www.plummerfernandez.com/works/metamultimouse/. Accessed 20 Oct 2022
21. Poole, B., Jain, A., Barron, J.T., Mildenhall, B.: Dreamfusion: text-to-3D using 2D diffusion. arXiv preprint arXiv:2209.14988 (2022)
22. Shapeways: Full color nylon 12 (MJF). https://www.shapeways.com/materials/nylon-12-full-color-3d-printing-with-multi-jet-fusion. Accessed 30 Oct 2022
23. Tensen, M., Hahn, A., Kiefer, C.: Reinforcement learning applied to sculpting: a technical story of AI craftsmanship (2022). https://docs.google.com/document/d/1Ug88-bEwDw1oJA_BRlZwM9Aj7RM-HfrUBiLd8mKVPoQ/edit#heading=h.9uxkp9yefic. Accessed 30 Nov 2022
24. Thomas, M.: The future of AI: how artificial intelligence will change the world (2022). https://builtin.com/artificial-intelligence/artificial-intelligence-future. Accessed 30 Oct 2022

25. van der Tuin, I., Dolphijn, R.: New Materialism: Interviews & Cartographies. Open Humanities Press, London (2012)
26. Wang, N., Zhang, Y., Li, Z., Fu, Y., Liu, W., Jiang, Y.-G.: Pixel2Mesh: generating 3D mesh models from single RGB images. In: Ferrari, V., Hebert, M., Sminchisescu, C., Weiss, Y. (eds.) ECCV 2018. LNCS, vol. 11215, pp. 55–71. Springer, Cham (2018). https://doi.org/10.1007/978-3-030-01252-6_4
27. Wu, S., Jakab, T., Rupprecht, C., Vedaldi, A.: Dove: learning deformable 3D objects by watching videos. arXiv preprint arXiv:2107.10844 (2021)
28. Yongjun, X., et al.: Artificial intelligence: a powerful paradigm for scientific research. Innovation **2**(4), 100179 (2021)
29. Zylinska, J.: AI Art: Machine Visions and Warped Dreams. Open Humanities Press, London (2020)

OSC-Qasm: Interfacing Music Software with Quantum Computing

Omar Costa Hamido[1(✉)] and Paulo Vitor Itaboraí[2]

[1] Coimbra, Portugal
[2] ICCMR, University of Plymouth, Plymouth, UK
paulo.itaborai@plymouth.ac.uk
https://omarcostahamido.com/about

Abstract. *OSC-Qasm* is a cross-platform, Python-based, OSC interface for executing Qasm code. It serves as a simple way to connect creative programming environments like Max (with *The QAC Toolkit*) and Pure Data with real quantum hardware, using the Open Sound Control protocol. In this paper, the authors introduce the context and meaning of developing a tool like this, and what it can offer to creative artists.

Keywords: Quantum Computing · Music · QAC · networking · creative programming

1 Context

The use of Quantum Computing (QC) without a Computer Science degree is no longer a far-fetched idea. Recent literature shows that in the Arts, Humanities, and Music, there are creative researchers exploring both theoretical and practical applications of QC in their respective fields [14,15].

The development of QC, in itself, is an international and multi-layered effort to articulate both hardware and software in a synergetic dance that allows valid computations to be performed. Very recently, IBM released their expanded roadmap to develop QC. In it, they explain with great detail and enthusiasm the (literal) shape of future quantum processors and when they plan to release them. This description is accompanied with a graph showing the focus areas and layers of this roadmap, including: systems, kernel, algorithms, and models [3].

While there seems to be very distinguishable paths into developing quantum hardware (which is enough to tell most QC hardware companies apart), as in the different approaches of a quantum computer (superconducting, photonics, annealing, etc.), the quantum software development is not in the same position yet [5]. The current main focus on chemistry simulation, machine learning, and noise mitigation algorithms is far from representing the large breadth of QC applications that are currently being explored [16,22]. After all, the purpose of pursuing a universal quantum computer is precisely to be able to compute any type of problem, and especially very complex ones [1].

Without any other incentive to do it otherwise, most quantum software development so far has been pursued independently, and closely connected to different

QC programming frameworks. In the majority of cases, it means that researchers are writing Python scripts - even though this may not always be preferable. In the end, they all are dealing with quantum circuits that share a very consensual textbook definition across all layers and fields of development. At this point it becomes relevant to refer to OpenQASM (Open Quantum Assembly Language) as an important effort to unify different QC frameworks in their discrete-variable quantum circuit definition [2].

In the field of Computer Music, in particular, the first author has explored the practical implications of QC for creative practice, focusing on Quantum-computing Aided Composition (QAC) [7,9]. This work has enabled the development of *The QAC Toolkit*, a software toolkit for musicians and artists to build, run, and simulate quantum circuits, using the Max visual programming environment [8,13].

The QAC Toolkit represents a shift in the focus of the quantum software development from the QC frameworks and into the Computer Music ecosystem. At the same time, it doesn't diverge entirely - it still allows users to export Qiskit code as well as Qasm [4]. One of the main reasons for this, is to still be able to connect with the real quantum hardware that only the original frameworks provide. In [7] this was originally proposed with an object called [och.qisjob], inspired by the *QisJob* project by Jack Woehr [23].

In late 2021, with the advent of the QuTune Project and the 1st International Symposium on Quantum Computing and Musical Creativity (ISQCMC), we witnessed an increased interest in enabling musical applications to execute quantum circuits [12,19]. The *OSC-Qasm* project was thus born, as a direct descendant from [och.qisjob], and an attempt to abstract the required process for executing quantum circuits, in the form of Qasm code, just by simply exchanging OSC messages.[1]

1.1   OSC Protocol

The Computer music community has long explored different network communication implementations for exchanging, mapping and controlling musical parameters within and between computers, digital instruments and software. One of the most prominent protocols for this purpose is Open Sound Control (OSC).

OSC is a protocol based on UDP networks and consists of a strictly formatted binary message. As a consequence, this protocol comprises an inherent Client-Server communication logic. In other words, there is a Server listening to incoming UDP messages (as long as they are formatted according to the OSC convention) and one (or multiple) Clients connecting to it, sending information.

Most importantly, this message is structured in two parts. There is a *path* and a *message* (that can also be a list). The *path* is an arbitrarily chosen word that declares the subject or place to which this message is addressed to - similar to URLs pointing to webpages, or pointing to a file on disk using a directory

---

[1] *OSC-Qasm* software and source code is available for download [10].

path. The Server is then capable of parsing and routing the received information according to their respective paths.

2 OSC-Qasm Server

The *OSC-Qasm* project includes both a server application and client patches. The *OSC-Qasm Server* is a cross-platform, Python-based, application that runs quantum circuits on quantum backends (both simulators and real hardware) using *Qiskit* [4,18]. It consists of an OSC server that listens to incoming messages with Qasm scripts from any OSC client. The structure of these messages uses the OSC path /QuTune and a list of 1 to 3 values that include, in order: the Qasm code, the number of shots, and the backend name (see Fig. 3).

Once *OSC-Qasm Server* receives a valid message, it will automatically transform the Qasm script into a qiskit.QuantumCircuit() object. The circuit is then either simulated in one of the Qiskit simulator backends or executed as an IBMQ job on an available backend [11].[2] Finally, the resulting aggregated counts of the computation are sent back to the client as an OSC message with the path /counts.[3]

Originally a set of Python scripts, *OSC-Qasm Server* is, as of version 2.0.0, distributed as a standalone application for Linux, MacOS, and Windows. By default, it runs with a graphical user interface (GUI), but it can also be executed as a Command-Line Interface (CLI), provided the necessary arguments are included. Sections 2.1 and 2.2 will explain in more detail their affordances and differences.

2.1 GUI

Intended to be accessible to users who don't code in Python, the Graphic User Interface of *OSC-Qasm server* displays all the customizable parameters on a small window, as depicted in Fig. 1.

On the left side, there is a network setup form with a Start/Stop button at the bottom. These fields are relevant to the configuration of the OSC server. By default, the server is set to run locally (meaning that both server and client will be running on the same device), listening on UDP port 1416 and sending messages back to another application on the same machine (IP 127.0.0.1), listening on port 1417. Optionally, if enabled, the Remote field allows the server to listen to incoming messages from other machines in a network (see Sect. 2.3).

---

[2] Running jobs on real IBMQ hardware requires user credentials. The number of available backends depends on these credentials. At the time of this writing, several backends are publicly available with a free IBMQ account. Additionally, IBM provides cloud-based backend simulators for larger jobs.

[3] Additional information is also sent back under the /info path, and errors are flagged with the /error path. It is important to note that the receiving client is, in fact, also running an OSC server.

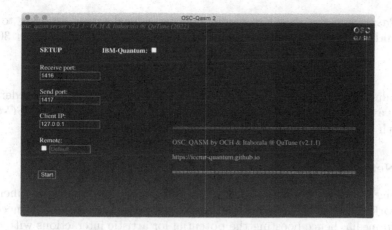

Fig. 1. OSC-Qasm Graphic User Interface

Without further customization, *OSC-Qasm* will use a Qiskit simulator named *qasm_simulator* to run the quantum circuits. As explained before, a client can request a specific backend by sending its name with the OSC message (see more in Sect. 3). However, in order for *OSC-Qasm Server* to be able to access real IBM Quantum hardware in the cloud, it is necessary to enable the IBM-Quantum checkbox, that reveals a second column of additional fields. These allow the user to fill the necessary account token, and specify the Hub, Group, and Project details to use.

The right side reveals a console-like monitoring area of up-scrolling text where it displays information about the running processes. After changing all the necessary options, the user only needs to click the Start button to boot the OSC server. If the server boots successfully, two lines will appear in the monitoring section, displaying the server arguments and declaring itself ready to exchange OSC messages.

2.2 CLI

In alternative to running the GUI, the *OSC-Qasm Server* can also be executed as a *Command-Line Interface* program from a Terminal or a Command Prompt, using the --headless flag. The default network configurations are identical to the GUI version. The main difference of the CLI mode is that the server boots automatically, as soon as the program is executed, greeting the user with an identical stream of messages as in the GUI monitor section (see Fig. 8).

OSC network configurations can be changed with positional arguments on the command-line execution. For instance, the following command launches

OSC-Qasm Server in headless mode, booting a local server listening to port 3000, and sending the results to a machine with IP 192.168.0.1, on port 3005.[4]

```
$ ./OSC_Qasm_2 3000 3005 192.168.0.1 --headless
```

Similarly, additional options such as `Remote`, and loading IBMQ credentials, are also available using other flag arguments. Additionally, launching *OSC-Qasm* just with the `--help` flag will list all of the available CLI options.

2.3   Network Distribution

A strong advantage of using the OSC protocol is the fact that it is inherently a network protocol. This facilitates more complex mappings and connectivity between media, hence boosting the potential for artistic interactions with quantum computers. The current Client-Server logic allows for a more distributed network setup.

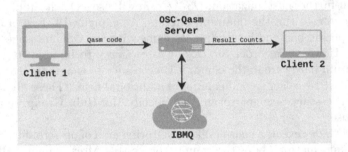

Fig. 2. A general network distributed *OSC-Qasm*

As indicated in Fig. 2, the *OSC-Qasm Server* can be hosted on a separate machine, as long as it is accessible through a local network or the Web. A dedicated machine for *OSC-Qasm Server* could be useful not just for complex networking but also for offloading the cost of running heavier quantum simulations.

In a possible networked scenario, one client could connect to the server to send Qasm code. *OSC-Qasm* then parses the quantum circuit and runs it on the requested backend. The retrieved computation results could then be sent to a third machine that would be, for example, responsible for mapping these results with an ongoing real-time synthesis patch in Max/MSP or SuperCollider (see Sect. 3).

To allow *OSC-Qasm Server* to be remotely accessible, the `Remote` option must be enabled. This can be achieved with the checkbox on the GUI, or with

---

[4] The user needs to either navigate to the directory where the application is, or drag and drop it onto a terminal window. The Windows version would start with `OSC_Qasm_2.exe`, and the MacOS version with `OSC_Qasm_2.app/Contents/MacOS/OSC_Qasm_2`.

the `--remote` flag in CLI mode. When this option is selected, *OSC-Qasm* will automatically look for the device's main network adapter IP address to host the server. If the server machine is connected to multiple networks (Ethernet, Wifi, VPN, Cellular), the machine's IP address on the desired network can be added, in the field box on the GUI, or as an argument to the flag on the CLI.

3 OSC-Qasm Clients

As explained in Sect. 1.1, the advantage of using an OSC server for receiving Qasm code is that it can come, technically, from *any* OSC-enabled client. More specifically, the authors have explored three Computer Music environments as clients. The only necessary requirement is a correctly formatted OSC message including some valid Qasm code defining a quantum circuit with measurement gates.

The following sections explain each implementation with more detail. The included toy example patches all share a sample 2-qubit circuit (known as Bell State, see Fig. 4). The complete OSC message sent by each client can be seen in Fig. 3.

```
/QuTune "OPENQASM 2.0;
include \"qelib1.inc\";
qreg q[2];
creg c[2];
h q[0];
cx q[0],q[1];
measure q[0] -> c[0];
measure q[1] -> c[1];
" 1024 qasm_simulator
```

Fig. 3. An example *OSC Qasm Client* message

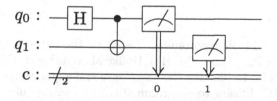

Fig. 4. First Bell state circuit

3.1 Max

As explained in Sec. 1, *OSC-Qasm* is a direct descendant from the [och.qisjob] Max object. Consequently, Max was the most natural first choice for an *OSC-Qasm Client* [13]. In Max, the user can programmatically create a quantum circuit and retrieve the respective Qasm code definition using *The QAC Toolkit* package [8]. Then, a [udpsend] object can be used to send an OSC message to the *OSC-Qasm Server*. Finally, a [udpreceive] object is required in order to receive the computed results back from the server (see Fig. 5).

For convenience, and better integration with Max and *The QAC Toolkit* workflow, all the objects related to the message formatting and OSC network exchange were encapsulated in a Max abstraction named [osc_qasm] (see Fig. 9). Both a help patch and a complete reference page are included with this abstraction.

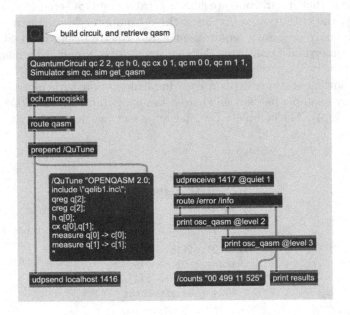

Fig. 5. *OSC-Qasm* Client in Max

3.2 Pure Data

Another Computer Music environment tested by the authors, with comparable features to Max, is Pure Data (Pd) [17]. Unlike Max, however, Pd doesn't include by default objects for formatting OSC messages and a quantum-computing aided composition (QAC) library to programmatically generate quantum circuits.

To address the OSC message formatting we make use of the osc-v0.2 external that can be installed from the included externals library navigator. As for the Qasm code, we decided to include a fixed Qasm string definition in a message object. This Qasm code can be obtained or generated elsewhere. The user must only pay some attention to the Pd message syntax when copying it over. Figure 6 depicts the Pd example patch that is capable of reproducing the *OSC-Qasm Client* message.

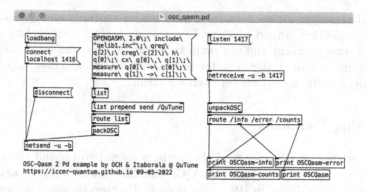

Fig. 6. *OSC-Qasm Client* in Pd

3.3   SuperCollider

```
// ============ OSC Callback function / Message Processing ============
(
~callback = { |msg, time, addr|

    "%: %\n".postf(msg[0].asString.basename, msg[1]);
};

// Setting the callback function to different paths
OSCdef(\info, ~callback, '/info');
OSCdef(\error, ~callback, '/error');
OSCdef(\counts, ~callback, '/counts');
);

// ===== Open send/receive UDP IP & ports for OSC communnication =====
(
thisProcess.openUDPPort(1417); // Receive Port
~sendAddr = NetAddr.new("127.0.0.1", 1416); // Send IP & Port
~shots = 1024; // The number of times the circuit will be run
~backend = ~backendsSimulators[0]; // qasm_simulator
);

// =================== Send Qasm Code to OSC-Qasm! ===================
(
~sendAddr.sendMsg("/QuTune", "OPENQASM 2.0;
include \"qelib1.inc\";
qreg q[2];
creg c[2];
h q[0];
cx q[0],q[1];
measure q[0] → c[0];
measure q[1] → c[1];
", ~shots, ~backend);
'Sending script to OSC-Qasm'.postln;
);
```

Fig. 7. OSC-Qasm Client in SuperCollider

By the same token, the authors have also explored SuperCollider as an *OSC-Qasm Client* [21]. Similar to Pd, it does not have a library for *QAC*, however, its text-based paradigm makes it more convenient for writing OpenQASM scripts by hand. In addition, SuperCollider relies strongly on OSC protocol for communications between `sclang` and `scsynth`. It has useful functions for defining OSC messaging and parsing structures, for instance, `NetAddr.sendMsg` and `OSCdef`. Figure 7 depicts the SuperCollider example client.

4   Closing Remarks

With the release of *OSC-Qasm 2* the authors intend to expand the range of users who are currently working with QC. The *OSC-Qasm* software and source code is publicly available in [10], with compiled server applications for Linux, MacOS, and Windows, as well as client example patches for Max, Pure Data, and SuperCollider.

This project is motivated by the need to directly connect music software with real quantum hardware, and to establish a framework and a platform-agnostic implementation for doing so. The authors decided to use the OSC protocol, which is often found in creative programming environments and applications. This allows *OSC-Qasm* to be used easily with platforms other than those explored in this paper.

This paper intends to be a contribution to the overall discussion about the use of QC in Music with a focus on a proposed framework and implementation. However, identifying the compositional methods that are unnatainable with non-quantum technology is outside the scope of this paper.

For performance purposes, *OSC-Qasm* enables a real-time connection with quantum hardware available in the cloud. However, it is important to note that the current workflow in state-of-the-art machines can lead to much larger latencies than in typical networked OSC use case scenarios. These are caused by the *execution* and *queuing* times. The execution time depends on the system, type of circuit, and amount of measurements, but typically takes ~ 0.1 min to process [20]. More importantly, quantum cloud services contain queuing times that can vary greatly (from seconds to hours), depending on the access and priority levels that the user has with the provider.

The attention given to networking capabilities also implies that this software can be used in one or multiple machines, on the LAN or across the internet. In that regard, the authors have already successfully experimented connecting two machines across countries. In this experiment, a Max client was used running on a MacOS laptop, located in Plymouth, UK. The server was running on a Linux (Pop!OS) machine in São Paulo, Brazil (see Figs. 8 and 9). The computers were connected through a Hamachi VPN network [6].

In the future the authors plan to add the possibility to include user/community developed plugins that will expand how *OSC-Qasm* parses Qasm scripts. For example, allowing it to perform some pre-processing. Moreover, plugins could enable automatic jobs containing multiple quantum circuit iterations with a changing variable, and other higher level quantum algorithms.

Lastly, this tool opens possibilities for multi-user collaborations. It is the authors' hope that musicians and artists will explore, improve, and make the best out of this tool, and feel inspired to apply QC into their artistic workflows.

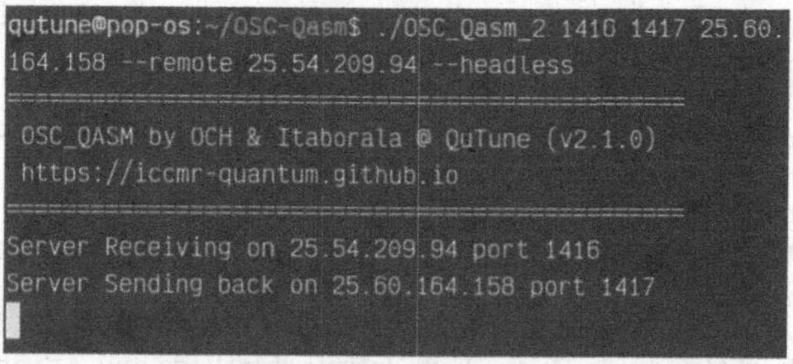

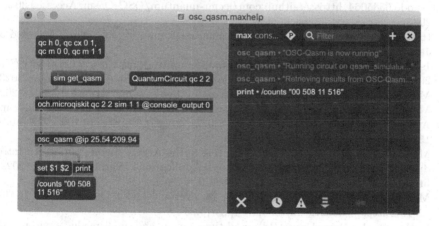

```
qutune@pop-os:~/OSC-Qasm$ ./OSC_Qasm_2 1416 1417 25.60.
164.158 --remote 25.54.209.94 --headless
======================================================
OSC_QASM by OCH & Itaborala @ QuTune (v2.1.0)
https://iccmr-quantum.github.io
======================================================
Server Receiving on 25.54.209.94 port 1416
Server Sending back on 25.60.164.158 port 1417
```

Fig. 8. *OSC-Qasm Server* used on the distributed network experiment

Fig. 9. Patch used for the distributed network experiment

Acknowledgment. The authors would like to thank other members of the QuTune Project [19], as well as all the interested participants at the 1st ISQCMC. This work was done thanks to the support from EPSRC QCS Hub.

References

1. Bernstein, E., Vazirani, U.: Quantum complexity theory. In: Proceedings of the Twenty-Fifth Annual ACM Symposium on Theory of Computing. pp. 11–20. STOC 1993, Association for Computing Machinery, New York, NY, USA, Jun 1993. https://doi.org/10.1145/167088.167097
2. Cross, A.W., Bishop, L.S., Smolin, J.A., Gambetta, J.M.: Open Quantum Assembly Language. arXiv (Quantum Physics), Jul 2017. https://arxiv.org/abs/1707.03429. Accessed 30 Jan 2023

3. Gambetta, J.: Expanding the IBM Quantum Roadmap to anticipate the future of quantum-centric supercomputing. IBM Research Blog, May 2022. https://research.ibm.com/blog/ibm-quantum-roadmap-2025. Accessed 30 Jan 2023

4. Gambetta, J., et al.: Qiskit/qiskit: Qiskit 0.23.6 (Feb 2021), https://zenodo.org/record/4549740. Accessed 30 Jan 2023

5. Gill, S.S., et al.: Quantum computing: a taxonomy, systematic review and future directions. Technical report, arXiv (Emerging Technologies), September 2021. https://arxiv.org/abs/2010.15559. Accessed 30 Jan 2023

6. Hamachi by LogMeIn. https://www.vpn.net/. Accessed 30 Jan 2023

7. Hamido, O.C.: Adventures in Quantumland. Ph.D. thesis, UC Irvine (2021), https://escholarship.org/uc/item/93c1t8vx. Accessed 30 Jan 2023

8. Hamido, O.C.: The QAC Toolkit (2021). https://www.quantumland.art/qac. Accessed 30 Jan 2023

9. Hamido, O.C.: QAC: quantum-computing aided composition. In: Miranda, E.R. (ed.) Quantum Computer Music: Foundations, Methods and Advanced Concepts, pp. 159–195. Springer, Cham (2022). https://doi.org/10.1007/978-3-031-13909-3_8

10. Hamido, O.C., Itaboraí, P.V.: OSC-Qasm, May 2022. https://doi.org/10.5281/zenodo.6585064, https://github.com/iccmr-quantum/OSC-Qasm. Accessed 30 Jan 2023

11. IBM-Quantum online portal. https://quantum-computing.ibm.com/. Accessed 30 Jan 2023

12. 1st International Symposium on Quantum Computing and Musical Creativity (Nov 2021), https://iccmr-quantum.github.io/1st_isqcmc/. Last Accessed 30 Jan 2023

13. Max MSP, Cycling '74 (Feb 2021), https://cycling74.com/. Accessed 30 Jan 2023

14. Miranda, E.R. (ed.): Quantum Computer Music: Foundations, Methods and Advanced Concepts. Springer, Cham (2022). https://doi.org/10.1007/978-3-031-13909-3

15. Miranda, E.R. (ed.): Quantum Computing in the Arts and Humanities: An Introduction to Core Concepts, Theory and Applications. Springer, Cham (2022), https://doi.org/10.1007/978-3-030-95538-0

16. Montanaro, A.: Quantum algorithms: an overview. NPJ Quant. Inf. **2**(1), 1–8 (2016). https://doi.org/10.1038/npjqi.2015.23

17. Puckette, M.: Pure Data (Pd): real-time music and multimedia environment. https://msp.ucsd.edu/software.html. Accessed 30 Jan 2023

18. Qiskit webpage. https://qiskit.org/. Accessed 30 Jan 2023

19. QuTune Project webpage. https://iccmr-quantum.github.io/. Accessed 30 Jan 2023

20. Ravi, G.S., Smith, K.N., Gokhale, P., Chong, F.T.: Quantum computing in the cloud: Analyzing job and machine characteristics. arXiv (Quantum Physics), Mar 2022. https://arxiv.org/abs/2203.13121. Accessed 30 Jan 2023

21. SuperCollider webpage. https://supercollider.github.io/. Accessed 30 Jan 2023

22. Upama, P.B., et al.: Evolution of quantum computing: a systematic survey on the use of quantum computing tools. Technical report, arXiv (Quantum Physics), Apr 2022. https://arxiv.org/abs/2204.01856. Accessed 30 Jan 2023

23. Woehr, J., Hamido, O.C.: QisJob, February 2021. https://doi.org/10.5281/zenodo.4554481. Accessed 30 Jan 2023

EvoDesigner: Aiding the Exploration of Innovative Graphic Design Solutions

Daniel Lopes[1,2]([✉]) [iD], João Correia[1,2] [iD], and Penousal Machado[1,2] [iD]

[1] University of Coimbra, CISUC, DEI, Coimbra, Portugal
{dfl,jncor,machado}@dei.uc.pt
[2] LASI, University of Minho, Braga, Portugal

Abstract. Graphic Design (GD) artefacts, like posters on the streets or book covers on store shelves, often compete with each other to be seen, catch attention and communicate effectively. Nevertheless, due to the democratisation of GD and because finding disruptive aesthetics might be time-consuming, graphic designers often follow existing trends, lacking disruptive and catchy visual features. *EvoDesigner* aims to assist the exploration of distinctive GD aesthetics by employing a genetic algorithm to evolve content within two-dimensional pages. The system takes the form of an extension for *Adobe InDesign*, so both human designers and the machine can alternately collaborate in the creation process. In this paper, we propose a method to automatically evaluate the generated posters by assessing their dissimilarity to the output of an auto-encoder that was trained with a set of posters posted at typographicposters.com by graphic designers worldwide. The results suggest the viability of the evaluation method to recall large sets of images and therefore be used to compute an image dissimilarity degree. Furthermore, the proposed method could be used for evolving GD posters that can be deemed as new when compared to the training set.

Keywords: Graphic Design · Evolutionary · Automatic · Innovation · Novelty · Poster · Layout

1 Introduction

The goal of Graphic Design (GD) artefacts (e.g posters or book covers) might be either to communicate information objectively or to be susceptible to personal interpretation. Either way, these must first of all capture the public's attention. Thus, for standing out over other GD artefacts, such as other posters on the streets or other book covers on store shelves, designers commonly work towards enhancing aesthetics. Nevertheless, as GD is increasingly democratised

This research was funded by national funds through the FCT—Foundation for Science and Technology, I.P., within the scope of the project CISUC—UID/CEC/00326/2020 and by the European Social Fund, through the Regional Operational Program Centro 2020 and is partially supported by *Fundação para a Ciência e Tecnologia*, under the grant SFRH/BD/143553/2019.

C. Johnson et al. (Eds.): EvoMUSART 2023, LNCS 13988, pp. 383–398, 2023.
https://doi.org/10.1007/978-3-031-29956-8_25

and shared thought out social media, television, stores, and on the streets, designs tend to often converge into trendy aesthetics, lacking disruptive eye-catchy features.

In some contexts, besides aesthetics, other techniques might have the potential to catch the public's attention, such as the employment of dynamic or interactive features, e.g. in the creation of moving posters [17]. Nonetheless, such approaches might not be practicable in every context, not only due to technicalities but also because of time and budget restrictions. Therefore, finding innovative visual solutions that arouse the public's curiosity might always be a fundamental assignment in the GD area, either for creating static, moving, or interactive artefacts.

The present work aims to hasten the exploration of innovative GD solutions by employing a conventional Genetic Algorithm (GA) to automatically evolve a given number of items, such as text-boxes, images, and geometric shapes, within two-dimensional pages.

In existing work [10], a similar GA was tested by evolving pages towards given target posters in the PNG format. This system was developed as an extension for *Adobe inDesign*, a widely used GD desktop-publishing software, so both human designers and the system may alternately collaborate in the creation process by editing and evolving designs, using the same software. Fitness is assessed by calculating the negated Mean Squared Error (MSE), i.e. the similarity, between the generated posters (also in PNG) and a given target poster. The bigger the similarity, the better. Thus, this existing system is not optimised for exploring innovative visual solutions, but for approximating the layouts of given target posters (e.g. approximating the layout of a given poster using a different set of page items).

In this paper, we implement the aforementioned system and we present and test a more robust fitness assignment method, aiming to assess the degree of dissimilarity of the generated pages compared to a given set of existing posters.

The present method is based on an auto-encoder trained with a data set of 4620 GD posters, posted at the website typographicposters.com by graphic designers worldwide. The goal is for the model to memorise common features of the posters present in the training data set, avoiding the need to compare n to n images, i.e. avoiding the need to compare a generated poster to the whole data set of existing posters and therefore compute a minimum similarity value, which would be time and resource-consuming.

In theory, the poorer the ability of the auto-encoder to reconstruct (recall) a given poster (and other types of images or even noise), the more distinct it might be from the posters in the data set, mainly, the most ordinary ones. Thus, contrary to conventional auto-encoders, our model should slightly overfit the training data, i.e. the most dissimilar an input is, the worst must be its reconstruction.

In the remainder of this paper, existing computational systems and methods for seeking the generation of innovative designs are reviewed. Thereafter, an overview of the implemented system is made. As the evolutionary abilities of the presented GA were already tested in existing work [10], the approach in this paper focuses on the detailed description of the dissimilarity fitness assignment method and the respective testing experiments.

The results suggest the viability of using an auto-encoder to assess whether or not a PNG poster resembles existing ones on a given training set or whether its aesthetics differ enough to be deemed as more or less innovative, compared to the same data set. Also, the results revealed that the reconstruction loss returned by the auto-encoder can be successfully used for fitness assignment in evolutionary systems, aiding the exploration of GD solutions that differ from a given data set.

Nevertheless, human designers might still be crucial in the process to identify results with potential and post-edit them to create final GD artefacts. Furthermore, to achieve a more robust system, we suggest the presented dissimilarity metric be complemented by other aesthetic metrics, e.g. for assessing page balance, legibility, or whether the results resemble a given GD style.

2   Related Work

The present paper aims to contribute by introducing and testing a practical method for evaluating images regarding their degree of dissimilarity compared to a set of existing images, particularly, ones in the domain of GD, such as posters or book covers. In that sense, this section reviews existing computational systems and methods for aiding the exploration of innovative images.

To overcome the difficulties of objectifying and automating aesthetics, interactive evolutionary approaches have been often endorsed [4,6,7,16,18]. Such techniques can allow users to drive evolutionary systems towards results they believe are more innovative. Nevertheless, the manual evaluation of outputs can often lead to time-consuming creation processes.

Besides, there has also been work towards the automation of creative processes. In art and design, authors have been endorsing a high variety of approaches for generating pleasing results, such as assigning fitness by measuring how long viewers look at the artefacts [17], measuring image complexity [14], measuring balance through weight maps [9] or even training neural networks to learn the opinion of users about whether an image regards low or high aesthetics [20]. However, none of the latter methods might be insightful about the innovation degree of the generated artefacts. In future developments, we aim that our system can explore both pleasing and innovative solutions. However, in this paper, only the innovation issue is addressed by assessing a dissimilarity degree compared to existing artefacts.

A reasonable existing approach to promote innovation is the use of Machine Learning (ML) to explore the latent space between two or more images [8,15], i.e. exploiting interceptions. However, such approaches can often lead to variations of existing styles [19], perhaps because the search space tends to be too near to the training examples. Another shortcoming of many ML projects is the generation of bitmap images rather than editable ones, limiting the usability of the results in real GD applications. Even so, there is also some work exploring the latent space of vectorial representations [1,12] and combining ML and evolutionary approaches to explore results that deviate from the training images by a given degree [11]. Furthermore, approaches combining interactive and automatic evaluation methods might be noteworthy for integration in GD processes [11,13].

3   Approach

For aiding the creation of innovative GD posters, we propose the deconstruction of the problem into three main tasks: (i) the creation of a system for automatically generating variations of posters, (ii) finding a method for directing the generation process towards dissimilar solutions compared to a set of existing posters, and (iii) letting human designers fine-tune the results or get inspired by these to design final artefacts.

3.1   Evolutionary Engine

As mentioned in previous sections, we solved the generation issue by implementing an evolutionary engine, based on a conventional GA, able to control *Adobe inDesign* to edit given page items within two-dimensional pages [10].

The system's basic initialisation setup must start with the manual creation of a new *inDesign* document followed by the insertion of items into pages, i.e. text-boxes, geometric shapes, or images. Thereafter, using a developed interface, the user must select which of the created pages must be considered in the initial population and set the desired population size. Optionally, one might select items that must be present in every individual (page) of the population, e.g. a given text box containing the poster's title. Such items are referred to as mandatory. If the number of selected pages is bigger than the set population size, the latter is automatically updated to match. Instead, if the number of selected pages is smaller than the population size, new pages are automatically created by crossing over and mutating the selected ones, until the size of the initial population matches the defined population size. Also during the initialization process, missing mandatory items are added to any pages where these are missing. For further details, refer to the engine our system was based on [10].

After initialisation, the evolutionary engine evaluates individuals and checks for termination criteria. i.e. (i) whether a fitted individual was found (not considered in current experiments), (ii) whether a given number of runs was completed, or (iii) whether the user ordered the evolutionary process to stop, by pressing a "Stop Generation" button on the interface. After one of the aforementioned criteria was matched, the process terminates and the user can manually edit the results using *inDesign* and/or start evolving again some intended pages, using the same or different parameters.

Otherwise, if no termination criteria were matched, the individuals will be selected using a tournament method of size 2 and an elite of 1 (such values might be editable by the user in further iterations). Thereafter, a new population is created by crossover and mutation and the system proceeds to the evaluation state again, repeating the process in a loop.

```
Page {
    PageItems: [
        Items: {
            shape_of_surrounding_box: one_of_available_constants,
            size: array_of_numbers,
            position: array_of_numbers,
            z_position: integer,
            flipping_mode: one_of_available_constants,
            blending_mode: one_of_available_constants,
            opacity: number,
            background_colour: one_of_available_constants,
            background_tint: number,
            stroke_colour: one_of_available_constants,
            stroke_tint: number,
            stroke_weight: number,
            rotation: number,
            shearing: number,
            text_size: number,
            typeface: one_of_available_constants,
            justification: one_of_available_constants,
            vertical_alignment: one_of_available_constants,
            letter_spacing: number,
            line_height: number,
            ... (other not-used properties of Item)
        }
        ... (other Items)
    ]
    ... (other properties of Page)
}
```

Fig. 1. Schematic representation of a genotype (only for the sake of the example; property names and value types might not be fully accurate).

Representation. In our system, a phenotype consists of the native rendering of *inDesign* pages. These may contain text boxes, geometric shapes, or images, defined by several positioning, geometry, and style properties, natively stored by *inDesign* in the JSON format. Thus, genotypes consist of JSON objects containing all the properties of a page and the items contained in it (refer to Fig. 1 for a schematic representation of the genotype). Although several other properties are available, only the following have been considered so far: the shape of the surrounding box of the items, their size, position and order (z-position), flipping and blending mode, opacity, background colour/gradient, background tint, stroke colour/gradient, stroke tint, stroke weight, rotation, and the item's shearing angle. Regarding text boxes, also text size, typeface, text justification, vertical text alignment, letter spacing, and line height are considered. In further versions of the evolutionary engine, other properties must be used, both for page items and the pages themselves, e.g. using page margins and grid rulers to position items in a more organised manner.

Variation. As variation is concerned, both crossover and mutation operations are applied. In the current version of the system, crossover operations consist in passing whole page items to the offspring, not shifting individual item properties. Crossover executes as follows: the items contained in the first parent (P1) are iterated in random order. Each item (I1) has a 50% chance to be inherited by the offspring (with the same position, geometry and style). If I1 is not passed, the system tries to pick, from the second parent (P2), a random item (I2) that has not been inherited yet. If no such item exists (e.g. all have been inherited already), I1 is inherited instead. In that sense, the limit number of items contained by the offspring is respectively limited by the minimum and the maximum number of items per page, in the initial population.

As mandatory items must not disappear during the crossover, an exception has been created so such items can only shift with ones of their kind. That is, the user must label similar items with equal labels, e.g. a mandatory text box containing a title must have the label "title" across all pages, so titles can shift with each other. Thus, if a mandatory item I1 is not passed to the offspring, rather than a random I2, the system will look for an item I2 which label is equal to the label of I1. If no such item I2 exists, I1 is passed instead.

Mutation-wise, each position, geometry, and style properties referred to in Sect. 3.1 have a 1% chance to be changed randomly. Values of type *number* or ones in arrays of *numbers* (Fig. 1) can be randomly incremented or replaced by a new random value. Constants are randomly picked from fixed sets of values. So far, also colours are picked from a fixed list of seven values - black, white, magenta (pink), yellow, red, green, and cyan (blue). In future developments, any colour might be allowed.

Fitness Assignment. In the past publication [10], the evolutionary engine was tested by evolving posters that approximate the aesthetics of existing ones, using the negated MSE (i.e. the similarity) between PNG versions of the generated and target posters.

The approach in the present paper focuses on directing the generation towards dissimilar visual solutions compared to existing posters. To achieve that, as briefly mentioned in previous sections, an auto-encoder was trained to recall common visual features of images present in a data set of 4620 GD posters from various visual styles, designed by multiple designers worldwide. In theory, the poorer the ability of the model to reconstruct a given input poster (and any other image or even noise), the more visually distinct it might be from the posters in the data set, especially, from the most ordinary ones.

The posters were retrieved from typographicposters.com, an online platform for graphic designers to upload and share their poster designs. Besides its diversity in visual styles, this platform was selected to the detriment of others as the poster designs are posted raw, i.e. without background images or overlay effects that could skew the models' evaluation of the designs. Furthermore, the platform includes the work of multiple renowned graphic designers.

To assess fitness, the generated posters are first exported in the PNG format and then passed as input to the auto-encoder. The output image is then compared with the input one using MSE, returning a difference value between them. The bigger the difference, the better the fitness, i.e. if the model cannot reconstruct the input image, it might not be similar to any image in the data set, therefore, it might be more innovative compared to the training set.

In this paper, we are only concerned about testing the dissimilarity value of the generated posters compared to the posters in the data set, i.e. no other metrics are being applied to control the quality of the designs or to assess whether the message is being communicated. Even so, we believe that controlling legibility, page balance and other features must be beneficial to drive the system towards more production-ready posters.

4 Experimental Setup and Results

For testing the aforementioned hypothesis, experiments were set up, first, regarding the training of the auto-encoder and, secondly, on evolving posters using the trained auto-encoder for fitness assignment.

4.1 Memorising Posters

Table 1. Architectural summary of the auto-encoder developed.

Layer (type)	Output shape	Param #
Conv2D 1	[null, 200, 200, 32]	896
MaxPooling2D 1	[null, 100, 100, 32]	0
Conv2D 2	[null, 100, 100, 16]	4624
MaxPooling2D 2	[null, 50, 50, 16]	0
Conv2D 3	[null, 50, 50, 8]	1160
MaxPooling2D 3	[null, 25, 25, 8]	0
Conv2D 4	[null, 25, 25, 4]	292
MaxPooling2D 4	[null, 12, 12, 4]	0
Conv2D 5	[null, 12, 12, 2]	74
MaxPooling2D 5	[null, 6, 6, 2]	0
Conv2D 6	[null, 6, 6, 4]	76
UpSampling2D 1	[null, 12, 12, 4]	0
Conv2D 7	[null, 12, 12, 8]	296
UpSampling2D 2	[null, 24, 24, 8]	0
Flatten 1	[null, 4608]	0
Dense 1	[null, 5000]	23045000
Reshape 1	[null, 25, 25, 8]	0
Conv2D 8	[null, 25, 25, 16]	1168
UpSampling2D 3	[null, 50, 50, 16]	0
Conv2D 9	[null, 50, 50, 32]	4640
UpSampling2D 4	[null, 100, 100, 32]	0
Conv2D 10	[null, 100, 100, 3]	867
UpSampling2D 5	[null, 200, 200, 3]	0
Total params:	23059093	
Trainable params:	23059093	
Non-trainable params:	0	0

Fig. 2. Examples of images from the training set - posters gathered from typographicposters.com centred into frames of 200×200 pixels with a black background.

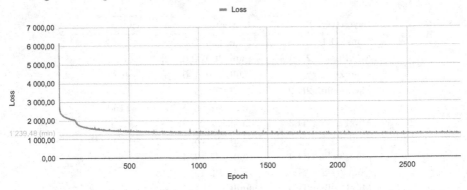

Fig. 3. Training loss of the developed auto-encoder along 2886 epochs; data size: 4620; test-data size: 30; batch size: 100; a minimum loss of 1239.48 was achieved.

The experiments to develop the auto-encoder have started with the implementation of a conventional architecture [2], originally trained on the MNIST data set [3]. However, as the latter model was originally created for smaller resolution images and with an intent different to ours, it reviled not to be adequate for our purposes. For instance, contrary to more conventional auto-encoders, which are often meant to generalise from the training data, our model should slightly overfit the data to be able to fairly reconstruct (recall) only the training images or images that are similar to the training ones, i.e. the more divergent the input images are compared to the training examples, the poorer the reconstruction must be. For that reason, often by trial and error, several alterations were made to the base architecture until a satisfactory result was achieved.

The final auto-encoder, which architecture is summarised in Table 1, was trained using *TensorFlow.js*, the *Adam* optimiser [5], MSE for loss calculation and

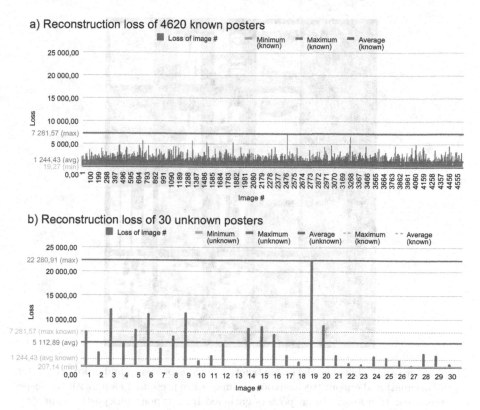

Fig. 4. Reconstruction loss of a) 4620 known and b) 30 unknown posters.

a batch size of 100. As mentioned before, the training data consisted of 4620 PNG posters from multiple graphical styles, retrieved from typographicposters.com. 30 additional posters from the same source were used as test data.

For passing the 4620 posters into the model, these were scaled down to a maximum size of 180 pixels (width or height, whichever is bigger). Also, for considering their different formats and orientations, these were centred into a frame of 200 × 200 pixels, with black background (Fig. 2).

As presented in Fig. 3, the training occurred along 2889 epochs (did not converge), resulting in a final training loss of 1239.48. This relatively high value might justify the fact that the model could not reconstruct in perfection the training posters (refer to the "KNOWN" posters of Fig. 5). Despite this, to distinguish training and non-training images, the current results can still be considered satisfactory, as the losses of posters used in training (referred to as known posters) revealed to be typically smaller than for posters that were not used in training (referred to as unknown posters). For instance, known and unknown posters regard 1244.43 and 5112.89 average losses, respectively (see Fig. 4 a and b, respectively). This suggests that known posters were considered more similar

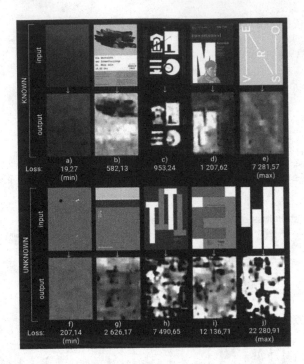

Fig. 5. Comparison between the reconstructed versions of known and unknown posters. Input (original) and output (reconstructed) images are presented along with the respective reconstruction losses. Input posters gathered from typographicposters.com

Fig. 6. Images from the training set, created out of posters gathered from typographicposters.com: a) posters containing big colour surfaces (red); b) posters containing full-width contrasting headers; c) posters containing yellow top corners. (Color figure online)

to the ones in the data set (which is true as these are the same), so the model might have successfully memorised a considerable part of their visual features.

Furthermore, by visually analysing the reconstructed posters in Fig. 5, the assumption that the model can better reconstruct known posters in detriment to unknown ones can be inferred. One can also observe that the model sometimes reasonably reconstructed parts of unknown posters. A possible yet speculative reason for that is that such visual features can be commonly found in some of the posters of the data set, e.g. Figure 5.f was reasonably well reconstructed probably because it is common to find posters composed of wide colour (red) stains (see Fig. 6.a). Similarly, the yellow header of Fig. 5.g was reasonably well reconstructed probably because contrasting full-width headers and yellow top corners are common (see Fig. 6.b and 6.c, respectively). Based on the latter assumption, the model might not only be able to distinguish posters that are different from the memorised ones but also able to reward posters that differ the most from memorised trends (i.e. some common visual features).

4.2 Evolving Dissimilar Posters

Table 2. Initialisation settings of the evolutionary engine.

Parameter	Value
Population size	50
Tournament size	2
Elite size	1
Prob. inheriting an item	50%
Prob. running a mutation method	1%
Mandatory items	Title, Date, Text-body
Maximum Generations (MG)	500
Termination criteria	Achieving MG

For setting up the experiments on evolving dissimilar posters, the system was configured as presented in Table 2. First, simulating the use process of a real user, three manually-created and little-stylised posters (unknown) were manually selected to be evolved (Fig. 7). Figure 8 showcases an initial population of 10 individuals, automatically generated out of the 3 pages of Fig. 7. During the experiments, populations of 50 individuals were generated (see Table 2).

In the system we based on [10], to approximate existing layouts using MSE as a fitness metric, 100 generations were enough until no significant fitness gains were achieved. Since the problem might be more challenging in the present experiments, 500 generations were set up to allow more time for optimisation. For statistic representativeness, 30 runs were performed. Figure 9 presents the fitness evolution of the best individual of each generation along with the 500 generations for each run.

Fig. 7. Unknown (manually created) pages selected to be evolved.

Fig. 8. Example of an initial population of 10 individuals, automatically generated out of the 3 selected pages of Fig. 7.

As the initial population posters were unknown to the auto-encoder, it could be predicted that the evolutionary process would start from relatively well-fitted individuals (4843.44 minimum fitness) compared to the minimum training loss (1239.48) (see Fig. 3), i.e. currently, the posters from the initial population were already considered different from the ones in the training set. However, considering fitness alone, the results suggest the system was successfully able to improve the posters even better regarding their dissimilarity to the training set.

Thereafter, phenotypes were analysed. By looking at Fig. 10.a, in which a selection of non-edited last-generation posters is showcased, one can consider the system can benefit from further fixes. Mainly, so the mandatory elements (i.e. text content) will not be out of the page, hiding behind other items, or having both text and background in the same colour (often, black). Such behaviours could be expected as there were no restrictions otherwise. Nevertheless, the legibility issue is not the focus of this paper and must be addressed in further developments as described before, e.g. using Optical Character Recognition (OCR) models to assess which percentage of the input text can be detected in the generated posters. Also, the disposition of page items was not driven by any organisational constraints, so more structured compositions would be almost

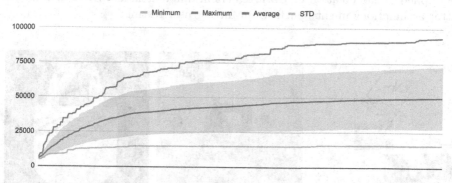

Fig. 9. Average, minimum, maximum and standard deviation (STD) of the fitness of the best individual of each generation, for 30 runs, 500 generations each.

impossible to achieve. Such organisation methods must be addressed in future developments as well.

Therefore, the generated posters were manually post-edited by a professional graphic designer from our research group (Fig. 10.b). Often, post-edition changes consisted of recolouring text boxes and geometrically transforming some of the items, sometimes deleting items, and always trying to minimise stylistic changes. No items were added. Also, final arts were created based on the post-edited posters (Fig. 11). Therefore, by comparing Fig. 10.a and 10.b, and considering that legibility, visual pleasantness, or conceptual inputs were not concern by the system, it might be reasonable to argue that such an approach as the potential to aid the exploration of visual ideas for aiding the creation of GD posters.

Fig. 10. Generated individuals (selected from last generations): a) non-edited, b) manually post-edited.

Moreover, considering Fig. 11, one can find that the generated ideas can already be applied in the creation of GD artefacts of different kinds. Even though, the latter assumptions might benefit from further user testing.

Fig. 11. GD artefacts based on the graphism of the posters of Fig. 10b: a) and f) posters; b) and d) book covers; c) and e) tote bags.

5   Conclusion

Enhancing aesthetics is of the utmost importance for making GD artefacts stand out over competing ones. However, due to the democratisation of GD and the increasing demand, often, designers tend to adopt trendy visual solutions, lacking in innovation and eye-catchy features.

This paper presented an evolutionary system that aims to explore innovative GD aesthetics, focusing on the creation of two-dimensional artefacts, e.g. posters or book covers. The system implements a state-of-the-art evolutionary engine [10] based on a conventional GA, previously tested for approximating the page balance of given target posters, using the negated MSE between the generated and target posters as a fitness metric.

Intending to drive the system towards innovative solutions, this paper presented a method for exploring designs that are dissimilar from ones in a given data set. An auto-encoder was trained using 4620 PNG posters, posted at typographicposters.com by designers worldwide. In theory, the bigger the loss returned by the auto-encoder in the reconstruction of a given poster, the bigger its dissimilarity to the posters in the data set. The returned loss was used for assigning fitness in the aforementioned evolutionary engine and experiments were carried out. The results suggested the viability of the approach for evolving GD posters that visually deviate from a given set of images, i.e. that can be deemed as new compared to the latter.

Nevertheless, we believe the presented approach can be further developed to create more robust systems, for example, by improving the data set and the architecture of the developed auto-encoder, and by including additional aesthetic metrics to the calculation of fitness, such as for assessing legibility, page balance, visual style and others.

In that sense, future developments must comprise: (i) including posters created by non-designers in the training set of the auto-encoder (e.g. less visually pleasing examples); (ii) further optimising the architecture of the auto-encoder; (iii) finding a method for continually retraining the auto-encoder (e.g. learning the generated posters); (iv) including a method for evaluating legibility; (v) implementing a grid system and a snap-to-grid method; (vi) including a method for evaluation of visual balance; (vii) including an evaluation method for exploring solutions of a given graphical style, e.g. another auto-encoder might be considered, by clustering the data set into different styles and approximating the aesthetics of a chosen one; (viii) creating a method for mapping keywords to the mutation operators that must more likely perform, limiting the search space according to given concepts; (ix) creating methods for considering the hierarchy of the page items; (x) including further technical and user testing.

References

1. Campbell, N.D.F., Kautz, J.: Learning a manifold of fonts. ACM Transactions on Graphics **33**(4), 1–11 (Jul 2014). https://doi.org/10.1145/2601097.2601212, https://dl.acm.org/citation.cfm?doid=2601097.2601212
2. Chollet, F.: Building autoencoders in keras (2016). https://blog.keras.io/building-autoencoders-in-keras.html. Accessed 1 Feb 2023
3. Cireşan, D.C., Meier, U., Masci, J., Gambardella, L.M., Schmidhuber, J.: High-performance neural networks for visual object classification (2011). https://doi.org/10.48550/ARXIV.1102.0183, https://arxiv.org/abs/1102.0183. Accessed 1 Feb 2023
4. Geigel, J., Loui, A.: Using genetic algorithms for album page layouts. IEEE MultiMedia **10**(4), 16–27 (2003)
5. Kingma, D.P., Ba, J.: ADAM: a method for stochastic optimization (2014). https://doi.org/10.48550/ARXIV.1412.6980, https://arxiv.org/abs/1412.6980. Accessed 1 Feb 2023
6. Kitamura, S., Kanoh, H.: Developing support system for making posters with interactive evolutionary computation. In: 2011 Fourth International Symposium on Computational Intelligence and Design, vol. 1, pp. 48–51. IEEE, Piscataway, USA (Oct 2011). https://doi.org/10.1109/ISCID.2011.21

7. Klein, D.: Evolving layout: next generation layout tool (2016). https://evolvinglayout.com. Accessed 17 Dec 2018
8. Loh, B., White, T.: SpaceSheets: interactive latent space exploration through a spreadsheet interface. In: Workshop on Machine Learning for Creativity and Design. 32nd Conference on Neural Information Processing Systems (NIPS 2018), Montréal, Canada (2018)
9. Lok, S., Feiner, S., Ngai, G.: Evaluation of visual balance for automated layout. In: Proceedings of the 9th International Conference on Intelligent User Interfaces, pp. 101–108. IUI 2004, Association for Computing Machinery, New York, NY, USA (2004). https://doi.org/10.1145/964442.964462
10. Lopes, D., Correia, J., Machado, P.: *EvoDesigner*: towards aiding creativity in graphic design. In: Martins, T., Rodríguez-Fernández, N., Rebelo, S.M. (eds.) EvoMUSART 2022. LNCS, vol. 13221, pp. 162–178. Springer, Cham (2022). https://doi.org/10.1007/978-3-031-03789-4_11
11. Lopes, D., Correia, J., Machado, P.: Adea - evolving glyphs for aiding creativity in typeface design. In: Proceedings of the 2020 Genetic and Evolutionary Computation Conference Companion, pp. 97–98. GECCO 2020, ACM, New York, USA (2020). https://doi.org/10.1145/3377929.3389964
12. Lopes, R.G., Ha, D., Eck, D., Shlens, J.: A learned representation for scalable vector graphics. In: Proceedings of the IEEE/CVF International Conference on Computer Vision (ICCV). IEEE, Piscataway, USA (Oct 2019)
13. Machado, P., Cardoso, A.: All the truth about NEvAr. Appl. Intell. Special Issue Creative Syst. **16**(2), 101–119 (2002)
14. Machado, P., Cardoso, A.: Computing aesthetics. In: de Oliveira, F.M. (ed.) SBIA 1998. LNCS (LNAI), vol. 1515, pp. 219–228. Springer, Heidelberg (1998). https://doi.org/10.1007/10692710_23
15. Oeldorf, C., Spanakis, G.: LoGANv2: conditional style-based logo generation with generative adversarial networks. In: 2019 18th IEEE International Conference On Machine Learning And Applications (ICMLA), pp. 462–468 (2019)
16. Önduygu, D.C.: Graphagos: evolutionary algorithm as a model for the creative process and as a tool to create graphic design products. Ph.D. thesis, Sabanci University (2010). https://research.sabanciuniv.edu/24145/, https://www.graphagos.com/portfolio/
17. Rebelo, S., Martins, P., Bicker, J., Machado, P.: Using computer vision techniques for moving poster design. In: 6.° Conferência Internacional Ergotrip Design. UA Editora, Aveiro, Portugal (2017)
18. Rebelo, S., Fonseca, C.M., Bicker, J., Machado, P.: Experiments in the development of typographical posters. In: 6th Conference on Computation, Communication, Aesthetics and X. Universidade do Porto, Porto, Portugal (2018)
19. Toivonen, H., Gross, O.: Data mining and machine learning in computational creativity. Wiley Int. Rev. Data Min. Knowl. Disc. 5(6), 265–275 (Nov 2015). https://doi.org/10.1002/widm.1170
20. Wang, W., Zhao, M., Wang, L., Huang, J., Cai, C., Xu, X.: A multi-scene deep learning model for image aesthetic evaluation. Signal Process.: Image Commun. **47**, 511–518 (2016). https://doi.org/10.1016/j.image.2016.05.009

Improving Automatic Music Genre Classification Systems by Using Descriptive Statistical Features of Audio Signals

Ravindu Perera(✉) ⓘ, Manjusri Wickramasinghe ⓘ, and Lakshman Jayaratne

University of Colombo School of Computing, 35 Reid Avenue, Colombo 07, Sri Lanka
`ravinduramesh@gmail.com`, `{mie,klj}@ucsc.cmb.ac.lk`

Abstract. Automatic music genre classification systems are vital nowadays because the traditional music genre classification process is mostly implemented without following a universal taxonomy and the traditional process for audio indexing is prone to error. Various techniques to implement an automatic music genre classification system can be found in the literature but the accuracy and efficiency of those systems are insufficient to make them useful for practical scenarios such as identifying songs by the music genre in radio broadcast monitoring systems. The main contribution of this research is to increase the accuracy and efficiency of current automatic music genre classification systems with a comprehensive analysis of correlations between the descriptive statistical features of audio signals and the music genres of songs. A greedy approach for music genre identification is also introduced to improve the accuracy and efficiency of music genre classification systems and to identify the music genre of complex songs that contain multiple music genres. The approach, proposed in this paper, reported 87.3% average accuracy for music genre classification on the GTZAN dataset over 10 music genres.

Keywords: Music Information Retrieval · Music Genre Identification · Music Genre Classification · Descriptive Statistical Features · Digital Signal Processing

1 Introduction

Music can be divided into many categories mainly based on rhythm, style, and cultural background. These categories are what we call music genres. Music genres are categorical labels created by human musical experts and are used for categorizing, describing, and even comparing songs, albums, or authors in the vast universe of music. There are a lot of music genre frameworks that segment songs into different genres by considering different facts such as the beat, energy, key music instrument, vocal tone, region, emotion, decade, and application of the music.

As the traditional music genre labeling process is done by human music experts, the number of manually generated rules increases and it may produce unexpected interactions and side effects. The traditional classification process is mostly implemented without following a universal taxonomy and this labeling process for audio indexing

© The Author(s), under exclusive license to Springer Nature Switzerland AG 2023
C. Johnson et al. (Eds.): EvoMUSART 2023, LNCS 13988, pp. 399–412, 2023.
https://doi.org/10.1007/978-3-031-29956-8_26

is prone to error. Human perception of music is dependent on a variety of personal, cultural, and emotional aspects. Therefore, its genre classification results may avoid clear definitions and the boundaries among genres are fuzzy [2, 3]. However, there is no complete agreement that exists in music genre definitions given by music experts and there are no strict distinguishing boundaries among music genres. Automatic Music Genre Classification (AMGC) is the classification of music into genres by a machine and as a research topic, it mostly consists of the selection of the best features and the development of algorithms to perform this classification [1–4].

Lots of facts make automatic music genre classification systems vital nowadays. Indexing and managing music databases, identifying and playing similar songs based on a genre preference, browsing and searching songs by music genre on the web, generating smart playlists, and choosing specific tunes among terabytes of songs are important tasks that are facilitated by automatic music genre classification systems. Another significance of automatic music genre classification systems is to make audio fingerprinting systems faster and more efficient by creating a database index with music genres. As described, music genre classification is an ambiguous and subjective task. Also, it's an area of research that is being contested, either for low classification accuracy, low efficiency, or the lack of clear definitions for music genres [2–4]. Therefore, the automatic music genre identification and classification process has emerged as a major research area and this research focuses on the automatic music genre identification and classification domain.

In this paper, a novel approach to develop an AMGC system is introduced by using a comprehensive selection of descriptive statistical features extracted from music audio signals and Machine Learning (ML) techniques. The best accuracy is given by the extra tree classifier with 200 estimators trained with the best 40 features selected using the information gain technique. A greedy approach is also proposed in this research to improve the accuracy and efficiency of music genre classification systems and to identify the music genre of complex songs that contain multiple music genres and to improve the efficiency of music genre classification systems. The final AMGC system developed in this research achieved 87.3% average accuracy on the GTZAN dataset over 10 music genres.

The paper is structured as follows. A brief overview of related work is provided in Sect. 2. Feature extraction methods, feature selection strategies, and the development process of the novel AMGC model are described in Sect. 3. Section 4 discusses the results given by the experiments and evaluates those results. Finally, Sect. 5 and Sect. 6 provide the conclusions of this research and an outlook on future work respectively.

2 Related Work

Music classification problems can be handled based on five different diverse perspectives in the Music Information Retrieval (MIR) domain according to Correa and Rodrigues [5]. Those five different diverse perspectives are the audio content-based approach, the lyrics-based approach, the symbolic content-based approach, the community meta-data-based approach, and the hybrid approach.

Audio content-based approaches can be considered as Digital Signal Processing (DSP) based approaches because they use features in audio signals to analyze and understand music. A review of audio content-based music reveals five major areas in the MIR domain. Those areas are music genre classification, mood classification, artist identification, instrument recognition, and music annotation [6].

Symbolic content-based approaches extract features from different symbolic data forms that describe music in a high-level abstraction. Lyrics-based approaches can be considered as Natural Language Processing (NLP) approaches. Community meta-data-based approaches aim to get the information embedded in an audio file such as song title, artist name, track length, instruments used, etc. Hybrid systems utilize multiple features arising from previous approaches.

Since the seminal study was done by Tzanetakis and Cook [7] in 2002, Music Genre Classification (MGC) has been a popular topic in the MIR community. Mckay and Fujinaga [8] elaborated a paper on why should researchers continue efforts to enhance the area of AMGC. The issues they point out are related to ambiguity and subjectivity in the classifications and the dynamism of music styles. It takes a lot of expertise and time to manually classify recordings, and also there is limited agreement among human annotators when classifying songs by music genre. Some other issues they point out are very few genres have clear definitions and there is often significant overlap among them and new genres are introduced regularly, and the understanding of existing genres changes with time.

Many works of literature have proved that DSP and ML techniques can be used to automate the music genre identification and classification task. The research study, done by Tzanetakis and Cook [7] in 2002, included the creation of a dataset comprised of one-thousand song clips evenly distributed across ten genres. This collection, referred to as the GTZAN dataset, has become the most popular dataset for AMGC research, although not the only one. They proposed mainly three different feature sets to represent timbral texture, rhythmic and pitch content. Short-time Fourier Transform (STFT), Mel-Frequency Cepstral Coefficients (MFCCs), Wavelet Transform (WT), and descriptive statistical features of the audio signal were used to obtain feature vectors. With these vectors, they trained statistical pattern recognition classifiers such as simple Gaussian, Gaussian Mixture Model, and k-Nearest Neighbor by using real world audio collections. They achieved correct classifications of 61% for ten musical genres.

Serra, Gomez, and Herrera [9] have identified that the basic musical facets of a song are timbre, tempo, timing, structure, key, harmonization, and lyrics. Timbre, also known as tone color is the music facet that makes a difference in different sound productions even when they have the same pitch and loudness. Simply it is what makes a difference between a piano and a violin playing the same note at the same volume. Tempo is the speed or pace of the music which can be easily changed by playing the music at different speeds. Key, harmonization, and lyrics are the tonality, chords, and words of the music. All these features imply the genre of a specific song as described by Serra, et al. [9]. Amelie, Emmanouil, and Simon [10] have experimented to improve music genre classification systems using automatically induced harmony rules in 2010. This study focuses on low-level harmony features retrieved from high-level harmony of songs. Harmony is a high-level descriptor of music, focusing on the structure, progression, and relation of chords. They have reached 91.13% classification accuracy for three selected genres in the GTZAN dataset by using a Support Vector Machine model with harmony rules. Results reported an improvement in accuracy when using harmony-based features.

The highest classification accuracy of 78% for all ten music genres in the GTZAN dataset without using any deep learning approaches is reported in 2012 by Chathuranga and Jayaratne [11]. They have extracted both long-term features and short-term features from audio signals and used Classifier Subset Evaluation along with the Best First Search method for short-term feature selection and the Info Gain Attribute Evaluation filtering method for long-term feature selection. They have used two SVM models for short-term and long-term features and an ensemble model to get the final single result by applying late fusion for combining classifier outcomes. Chathuranga and Jayaratne [12] have applied the same approach for 6 music genres in the ISMIR-2004-Genre dataset and have achieved 81% accuracy. Both evaluations are done by using the ten-fold cross-validation technique because both datasets contain less than two-hundred data points and less number of data points is not enough to perform a train, validation, and test split.

Classifying the generated spectrogram of the music signal is also a good option for genre identification [13–15]. Hareesh [14] has created a dataset of seven music genres by extracting ten-second sound clips from a total of 2.1 million YouTube videos in the "Music Genre Classification using Machine Learning Techniques" research work. He has trained an ensemble model combining VGG-16 Convolutional Neural Network (CNN) [16] and Extreme Gradient Boosting (XGB) and achieved 65% accuracy. Another highlighted finding in this literature is that MFCCs are identified as an important feature for music genre classification.

Nirmal and Mohan [13] extracted spectrograms to classify songs into music genres by using a CNN. They have selected only three music genres: Blues, Classical, and Rock from GTZAN dataset and trained the MobileNets [17] and have achieved 67% accuracy. The need for a specific Deep Neural Network (DNN) for AMGC has been shown in this research. Deepanway and Maheshkumar [15] have introduced a novel DNN architecture to solve music genre identification and classification problems in 2018. They have used two datasets that are the GTZAN dataset and the ballroom dataset and could achieve 94.2% as the best accuracy for the GTZAN dataset up to now. They have used their proposed DNN model with a Multi-Layer Perceptron (MLP) model to create an ensemble model to achieve the best accuracy.

As an overview of the works of literature in the AMGC domain, the GTZAN dataset is the most popular dataset that researchers used to train, test, and verify their robust DSP and Machine Learning (ML) approaches. Unlike other music genre related datasets, GTZAN is a well-balanced dataset with 10 music genres. The best accuracy achieved by previous researchers is 94.2% accuracy in ten-fold cross-validation for the GTZAN dataset by using an ensemble model, created by combining a CNN Max Pooling LSTM model and an MLP model [15]. But this state-of-the-art model takes high computational resources and also takes a long time to produce an output. The model proposed by Chathuranga and Jayaratne [11], which has both balanced accuracy and efficiency, has 78% mean accuracy in ten-fold cross-validation for the GTZAN dataset. Therefore, it's important to address these research gaps to improve the accuracy and efficiency of AMGC systems.

3   Methodology

The step-wise procedure in developing a novel AMGC system to improve the accuracy and efficiency of audio fingerprinting systems has been broken down into four major phases namely,

1. Data collection and preparation
2. Feature selection
3. Model Training
4. Model Evaluation

3.1   Data Collection and Preparation

The GTZAN music genre dataset is selected for this work of research because it is a well-balanced dataset with thousand songs and it contains 10 music genres: Blues, Classical, Country, Disco, Hip-Hop, Jazz, Metal, Pop, Reggae, and Rock. All audio files were converted to the MP3 format because it's the most popular audio format and considering only the signal features in the human hearing range is optimum to build a solution for the music genre recognition problem because music genres are defined based on what humans hear. Then the drum track, the bass track, the vocal track, the music without vocals, and the music without main instruments (drum, bass, and articulators) were separated from the original music audio by using Spleeter [18], a source separation Python library created by the Deezer RnD team.

The audio signal features were extracted from each separated audio track and from the raw audio file also. When deriving the STFT from the temporal signal, the window length is taken as 2048 samples (~93 ms at 22 kHz SR) because increasing the window length will decrease the accuracy, and decreasing the window length will gradually improve the accuracy while abruptly decreasing the efficiency. The same window length is used to derive other time-series audio signal features also. Then the descriptive statistical features were derived over the entire song considering all ~ 93 ms long excerpts. In this research, both the PyAudioAnalysis Python library [19] and the Librosa Python library [20] were used to extract time series features from the audio signal and compared those two Python libraries to identify which Python library gives better features to imply the music genre of an audio signal.

The final labeled dataset was generated with descriptive statistical values of extracted time-series music audio features merging with other features extracted from music audio signals. The final labeled dataset contains the following set of features of each five separated music tracks.

- Percentages of each music key: G#, G, F#, F, E, D#, D, C#, C, B, A#, A.
- The average, minimum, maximum, interquartile range, and standard deviation of the zero crossing rates of the windowed signal.
- The average, minimum, maximum, interquartile range, and standard deviation of the energy of the signal.
- The average, minimum, maximum, interquartile range, and standard deviation of the spectral centroid of the windowed signal.

- The average, minimum, maximum, interquartile range, and standard deviation of the spectral bandwidth of the signal.
- The average, minimum, maximum, interquartile range, and standard deviation of the spectral contrast of the signal.
- The average, minimum, maximum, interquartile range, and standard deviation of the spectral flatness of the signal.
- The average, minimum, maximum, interquartile range, and standard deviation of the spectral roll-off of the signal.
- The average, minimum, maximum, mode, interquartile range, and standard deviation of the tempo of the signal.
- Beat Predominant Local Pulse (PLP) count of the signal
- The average, minimum, maximum, interquartile range, and standard deviation of each MFCC feature of the signal.

At the end of the feature extraction process, there are 564 descriptive statistical features in total. Two datasets were created at the end of this phase. One dataset is created by extracting the descriptive statistical features of the entire song. And the other dataset is created by extracting the descriptive statistical features considering only the first fifteen-second-long part of the song to analyze features for a greedy AMGC solution.

If a song contains different music genres in different parts and if we consider descriptive statistical features of the whole song to identify the key music genre of the song, it will output a different unrelated music genre for the result. As an example, if there's a 30 s long song that contains:

- A 10 s long Pop part with 130 average BPM.
- A 10 s long Hip-hop part with 90 average BPM.
- Another 10 s long Pop part with 125 average BPM.

Then the average BPM of the entire song will be equal to 115 BPM but that's the average BPM of a Rock song. Therefore, these complex songs can mislead current AMGC systems. This research proposes a greedy approach that considers a part of the song to identify the main music genre of a song. The greedy approach simply proposes to consider only the first fifteen-second part of the song to identify the music genre of the entire song. The aim of this proposed greedy approach is to improve the accuracy and efficiency of music genre identification systems and to identify the primary music genre of complex songs that contain multiple music genres.

The optimum time frame for this greedy approach is not experimented or tested in this research. Only an initiation for a greedy approach is proposed by considering only the first fifteen seconds of the song. The reasons to select the time frame as the first fifteen seconds are listed below. But the optimum time frame can be different and that research question is open for future works.

a. The selected dataset contains songs after trimming the dramatic intro or silent part that occurs at the beginning of some songs.

b. The technology used in radio broadcast monitoring systems can be used to detect the song by ignoring the dramatic intro or silence part at the beginning of the song [21, 22].

c. After manually analyzing a set of labeled complex songs that contain multiple music genres, the observation shows that the first fifteen seconds of a song contain the main music genre (label) of the song. Only a few hip-hop songs didn't align with this observation.

The standardized versions of these descriptive statistical datasets were also generated for several experimentations because some ML algorithms give better results for standardized datasets and some ML algorithms give better results for non-standardized datasets. Experimentations are done by using the following standardization methods because the data has the Gaussian distribution.

a. Removing the mean and scaling to unit variance.
b. Scaling each feature to the range between zero and one.

3.2 Feature Selection

The feature selection phase is an important step in this research because there are a large number of features in the final dataset. Reducing the number of features to the most relevant features increases the accuracy as well as the efficiency of the research outcomes. Feature selection techniques such as ANOVA, Gini impurity, correlation matrix, and information gain are used to find the most relevant features that imply the music genre of a song. This research uses existing knowledge in works of literature on music genre identification and classification domain also to identify the most relevant signal features that imply the music genre of a song.

A combined dataset is created by including features extracted by using both pyAudioAnalysis and Librosa Python libraries to find which Python library gives the most accurate features of audio signals. And this combined dataset has features extracted by considering the entire song and the first fifteen seconds of the song to find whether the features extracted for the greedy approach will give better correlations to the music genre of the song. Therefore, one of the following suffixes can be found in each column name of the combined dataset to identify whether that feature is extracted by using Librosa or pyAudioAnalysis and whether that feature is extracted from the entire song or from the first fifteen seconds of the song.

- -librosa-first 15 s
- -librosa-whole song
- -pyAudioAnalysis-first 15 s
- -pyAudioAnalysis-whole song

This combined dataset contains descriptive statistical features from the raw audio signals as well as descriptive statistical features from the music track-separated audio signals. Therefore, the feature selection phase will help to identify which music track gives the key features to identify the music genre of a song. The feature analysis and selection are performed before standardizing the combined dataset and also after standardizing the combined dataset to identify the key features of the standardized dataset and the non-standardized dataset also.

This feature selection with the combined dataset phase will help to answer the following questions.

1. What is the best signal feature extraction Python library between Librosa and pyAudioAnalysis for music genre classification systems?
2. Will it give more correlated descriptive statistical data for the music genre of a song if we extract data by considering only the first fifteen seconds part of the song?
3. What is the best music track to identify the music genre of a song?

3.3 Model Training

Multiple ML algorithms such as Support Vector Classifier (SVC), K-Nearest Neighbor (KNN), Decision Tree, Random Forest, Ada Boost, Extra Tree, and Naive Bayes classifier are applied to the labeled descriptive statistical features extracted from the GTZAN dataset. The variable parameters of each model were adjusted so as to obtain the highest and most consistent accuracy for the test dataset across multiple runs (each configuration was run five times, and the median accuracy was considered).

Four SVC models were implemented with four different kernels: linear, poly, RBF, and sigmoid. One KNN classifier was implemented with uniform weights and the Minkowski distance matric. Two decision tree models and two random forest classifiers and two extra tree classifiers were implemented with Gini impurity criteria and information gain criteria. All models have 200 estimators in the estimator tree. An Ada boost model is also implemented with 200 estimators.

Only the best fifty features identified in the comprehensive feature analysis phase were selected to train all the implemented ML models. All models were duplicated to evaluate the greedy approach for AMGC. For the greedy approach, the second fifteen-seconds-parts of each song were also considered to increase the number of data points. The target of this novel general popular music genre classification system is to improve the accuracy while improving the efficiency of current music genre classification systems to solve the speed-accuracy tradeoff, identified in state-of-the-art AMGC systems.

3.4 Model Evaluation

An accuracy and efficiency evaluation was conducted to prove that the proposed novel music genre classification system performs better than the state-of-the-art AMGC systems. The ten-fold cross-validation technique is used to evaluate the accuracy of the proposed novel music genre classification system because the GTZAN dataset contains only a thousand songs and it's not enough to perform a train, validation, and test split. As mentioned in Sect. 2, all previous researchers who used the GTZAN dataset have used the ten-fold cross-validation technique. Therefore, it will be meaningful and easy to compare the novel proposed music genre classifier with the state-of-the-art music genre classifiers.

4   Results and Evaluation

All the features in the combined dataset were ranked according to their importance. As mentioned in Sect. 3.2, the combined dataset contains features extracted by using both the Librosa Python library and the pyAudioAnalysis Python library. It contains features extracted by considering the entire song and the first fifteen seconds part of the song also. And it contains the features extracted from the original song and also features extracted from the separated music tracks (the drum track, the bass track, the vocal track, the music without vocals, and the music without main instruments).

The twenty most important descriptive statistical features given by the Gini impurity measurement are shown in Fig. 1. The same features were ranked by using the information gain measurement and the ANOVA measurement also. Figure 2 shows the twenty most important features given by the information gain measurement and Fig. 3 shows the twenty most important features given by the ANOVA measurement.

As mentioned in Sect. 3.3, Two SVC models were implemented to derive the co-efficiencies between the music signal features and the music genre of an audio file. The SVC model implemented with the descriptive statistical features derived using the Librosa Python library gave 34.5% accuracy. The SVC model implemented with the descriptive statistical features derived using the pyAudioAnalysis Python library gave 31.5% accuracy.

According to Fig. 1, Fig. 2, and Fig. 3, the features derived by using the Librosa Python library imply the music genre of a song better than features derived by using the

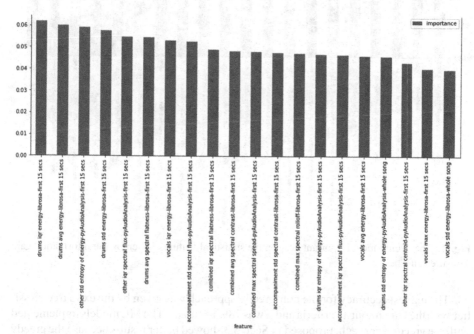

Fig. 1. The twenty most important descriptive statistical features, given by the Gini impurity measurement.

pyAudioAnalysis Python library. According to those charts, the drum track and vocal track separation technique can improve the AMGC systems and the drum track has the most correlation with the music genre of a song. The results given by the SVC models also imply that the features derived by using the Librosa Python library give better accuracy. Similar results were given for the standardized datasets and for the dataset created for testing the greedy approach also.

Although these two Python libraries allow extracting some different time-series signal features, only the Librosa Python library is used to derive features for the ML model implementation phase in this research because the Librosa Python library gives more correlated features and using two different methods for feature extraction takes more time and will reduce the efficiency of the overall AMGC system.

Accuracy evaluation of all implemented AMGC models was done by using the ten-fold cross-validation technique because it will be easy to compare the results with the state-of-the-art models and there are only one-thousand songs in the GTZAN dataset. The F1-score, precision, and recall matrices are equal to the accuracy of the ML model because the dataset is a well-balanced dataset with ten music genre labels.

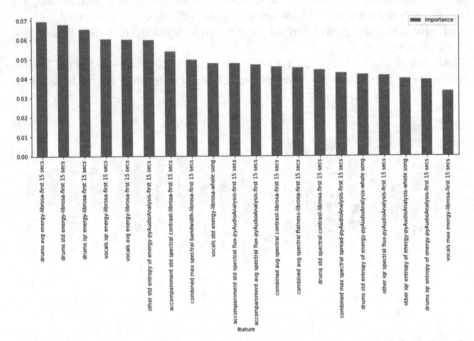

Fig. 2. The twenty most important descriptive statistical features, given by the information gain measurement.

The highest accuracy for the non-greedy approach was given by the extra tree classifier with the Gini impurity criteria and it was 78% accuracy. The ML models implemented with the greedy approach, proposed in Sect. 3.3 showed better results because the greedy approach can handle complex songs that contain multiple music genres and the greedy approach helps to increase the number of training data. The followings are summarized

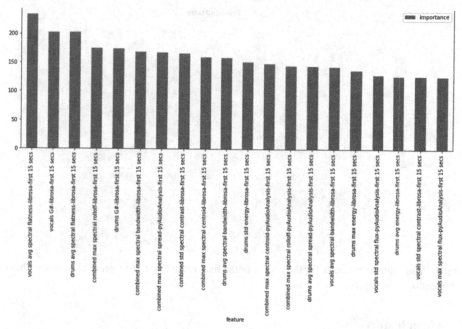

Fig. 3. The twenty most important descriptive statistical features, given by the information gain measurement

results of all ML models implemented to evaluate the greedy approach. All models were run five times and took the average accuracy.

- The SVC model showed 48% final accuracy
- The decision tree classifier with the Gini impurity criteria showed 63.75% final accuracy.
- The decision tree classifier with the information gain criteria showed 62.45% accuracy.
- The random forest classifier with the Gini impurity criteria showed 82% accuracy.
- The random forest classifier with the information gain criteria showed 81.9% accuracy
- The Ada Boost classifier showed 29.85% accuracy
- The extra tree classifier with the information gain criteria showed 87.3% accuracy.
- The extra tree classifier with the Gini impurity criteria showed 87% accuracy.

According to these results, the extra tree music genre classification model with information criteria is the best AMGC model implemented in this research. The results confusion matrix for the results of the extra tree music genre classifier is shown in Fig. 4.

The extra tree classifier with the information gain criteria implemented by using the descriptive statistical data extracted from the first fifteen seconds long part is a solution for the speed-accuracy trade-off in AMGC systems because it's more accurate than the state-of-the-art ML models and it can extract necessary features from the audio signal and predict the music genre of a song within 3 s in a computer with a 2.3 GHz CPU and a 12 GB RAM (a GPU is not required). Therefore, it's faster than the DNN models with CNN layers and state-of-the-art ensemble music genre classification models.

Predicted Label

	Blues	Classic	Country	Disco	Hip-hop	Jazz	Metal	Pop	Reggae	Rock
Blues	189	1	1	1	1	5	0	0	1	1
Classic	3	181	1	0	0	14	0	0	1	0
Country	5	1	177	3	0	7	0	1	2	4
Disco	0	1	2	178	6	0	3	6	3	1
Hip-hop	2	0	0	1	173	0	7	10	6	1
Jazz	3	8	1	0	0	188	0	0	0	0
Metal	6	0	0	2	2	0	185	0	0	5
Pop	0	0	12	6	3	0	0	166	6	7
Reggae	3	0	10	4	8	0	0	8	166	0
Rock	10	1	12	13	0	4	11	2	4	143

(Real Label)

Fig. 4. The confusion matrix for the results of the final extra tree music genre classifier with the information gain criteria.

5 Conclusion

This paper proposes a solution for the speed-accuracy tradeoff in AMGC systems found in the literature by identifying the descriptive statistical audio signal features that are most correlated with the music genre of a song and proposing a greedy approach for music genre identification. The proposed greedy approach is a solution for the music genre misclassification problem that occurs due to complex songs that contain multiple music genres within a song.

According to the results shown in Sect. 4, the most important music track of the song to identify the genre of the music is the drumbeat of the song. The energy of the drumbeat is the most important time-series feature that can be used to identify the music genre of a song. The most important descriptive statistical features of a music audio signal for the music genre identification task are the following.

- The standard deviation of the spectral contrast of the music of the song.
- The average, the standard deviation, and, the interquartile range of the energy of the drum music track.
- The average, the standard deviation, and the interquartile range of the energy of the base music track.
- The average spectral flatness of the drum music track.
- The average spectral bandwidth of the drum music track.
- The maximum energy and the average energy of the vocal track.
- The average and the standard deviation of the energy of the raw song.
- The average spectral flatness of the raw song.

The descriptive statistical features that imply the music genre of the first fifteen-seconds-part of the song are more than ninety-five percent similar to the descriptive statistical features that imply the music genre of the entire song. Therefore, a fifteen-second part of the song can imply the music genre of the entire song and it's a good solution to identify the music genre of complex songs that contain multiple music genres in several time frames. The highest accuracy for the non-greedy approach is 78% but the highest accuracy for the greedy approach is 87.3%. Therefore, the AMGC ML models that are implemented with the greedy approach showed better results also.

According to the results shown in Sect. 4, the AMGC ML model developed by using the extra tree classification algorithm with the information gain criteria and two-hundred estimators improves the accuracy and speed. The greedy approach used in the feature extraction phase is one of the main reasons for this improved accuracy and speed as discussed in Sect. 3.1, and Sect. 4. It solves the speed-accuracy tradeoff of state-of-the-art AMGC systems. The novel AMGC model that is proposed in this research outperforms state-of-the-art AMGC models with 87.3% accuracy with a good music genre identification speed. It can extract necessary descriptive statistical features from the audio signal and predict the music genre of a song within 3 s in a computer with a 2.3 GHz CPU and a 12 GB RAM.

6 Future Works

A possible improvement for AMGC systems can be done by implementing a time-series classification model to classify tempograms of songs and to classify energy spectrograms of songs. A self-supervised approach also can be tried to identify key parts of songs that imply the music genre of songs. It can improve the greedy approach also proposed in this paper.

As mentioned in Sect. 3.1, the optimum time frame for the proposed greedy approach is not experimented or tested in this research work. Only an initiation for a greedy approach is proposed by considering only the first fifteen seconds of the song. The proposed greedy approach can be improved by experimenting with different time frames of each song.

Another possible improvement for AMGC systems can be done by using the novel AMGC model proposed in this paper with a different number of the most correlated features. It will help to identify the optimum number of features and will increase the accuracy of the proposed music genre classification model. The same approach proposed in this paper can be tried with features extracted from the MARSYAS (Music Analysis, Retrieval, and Synthesis for Audio Signals) [23] which is an open source software framework for audio analysis.

References

1. Jeong, Y., Lee, K.: Learning temporal features using a deep neural network and its application to music genre classification. In: 17th International Society for Music Information Retrieval Conference (ISMIR), New York (2016)
2. Scaringella, N., Zoia, G., Mlynek, D.: Automatic genre classification of music content: a survey. In: IEEE Signal Processing Magazine, vol. 22 no. 2 (2006)

3. Anan, Y., Hatano, K., Bannai, H., Takeda, M.: Music genre classification using similarity functions. In: 12th International Society for Music Information Retrieval Conference (ISMIR 2011) (2011)
4. Arora, V., Kumar, R.: Probability distribution estimation of music signals in time and frequency domains. In: 19th International Conference on Digital Signal Processing, Hong Kong (2014)
5. Corrêa, D.C., Rodrigues, F.A.: A survey on symbolic data-based music genre classification. Expert Syst. Appl. Int. J. vol. 6, no. C (2016)
6. Fu, Z., Lu, G., Ting, K.M., Zhang, D.: A survey of audio-based music classification and annotation. IEEE Trans. Multimedia **13**(2), 303–319 (2011)
7. Tzanetakis, G., Cook, P.R.: Musical genre classification of audio signals. In: IEEE Transactions on Speech and Audio Processing (2002)
8. Mckay, C., Fujinaga, I.: Musical genre classification: is it worth pursuing and how can it be improved? In: 7th International Conference on Music Information Retrieval (ISMIR 2006) (2006)
9. Serrà, J., Gómez, E., Herrera, P.: Audio cover song identification and similarity: background, approaches, evaluation, and beyond. Stud. Comput. Intell. **273**, 307–332 (2010)
10. Anglade, A., Benetos, E., Dixon, S.: Improving music genre classification using automatically induced harmony rules. J. New Music Res. **39**, 349–361 (2010)
11. Chathuranga, D., Jayaratne, K.L.: Musical genre classification using ensemble of classifiers. In: Fourth International Conference on Computational Intelligence, Modelling and Simulation (2012)
12. Chathuranga, D., Jayaratne, L.: Automatic music genre classification of audio signals with machine learning approaches. GSTF J. Comput. (JoC) **3**(2), 1–12 (2013). https://doi.org/10.7603/s40601-013-0014-0
13. Nirmal, M.R., Shajee Mohan, B.S.: Music Genre Classification Using Spectrograms. In: International Conference on Power, Instrumentation, Control and Computing (PICC) (2020)
14. Bahuleyan, H.: Music genre classification using machine learning techniques. ArXiv abs/1804.01149 (2018)
15. Deepanway, G., Kolekar, M.H.: Music genre recognition using deep neural networks and transfer learning. In: Interspeech 2018 (2018)
16. Simonyan, K., Zisserman, A.: Very deep convolutional networks for large-scale image recognition. arXiv:1409.1556 (2014)
17. Howard, A., et al.: MobileNets: efficient convolutional neural networks for mobile vision applications arxiv:1704.04861 (2017)
18. Hennequin, R., Khlif, A., Voituret, F., Moussallam, M.: Spleeter: a fast and efficient music source separation tool with pre-trained models. J. Open Source Software **5**, 2154 (2020)
19. Giannakopoulos, T.: pyAudioAnalysis: an open-source python library for audio signal analysis. PLoS ONE **10** (2015)
20. McFee, B., Raffel, C., Liang, D., Ellis, D.: librosa: audio and music signal analysis in python. In: Python in Science Conference (2015)
21. Karunarathna, G.A.G.S., Jayaratne, K.L., Gunawardana, P.V.K.G.: Classification of voice content in the context of public radio broadcasting. Int. J. Adv. ICT Emerg. Reg. (ICTer) **12**(2) (2019)
22. Weerathunga, C.O.B., Jayaratne, K.L., Gunawardana, P.V.K.G.: Classification of public radio broadcast context for onset detection. In: European Journal of Computer Science and Information Technology (EJCSIT) by European Centre for Research Training and Development UK, 7(6):1–22 (2019)
23. Tzanetakis, G., Cook, P.: MARSYAS: a framework for audio analysis. Organ. Sound **4**(3), 169–175 (2000)

Musical Genre Recognition Based on Deep Descriptors of Harmony, Instrumentation, and Segments

Igor Vatolkin[1]([⊠]) [iD], Mark Gotham[2] [iD], Néstor Nápoles López[3],
and Fabian Ostermann[1] [iD]

[1] Department of Computer Science, TU Dortmund University, Dortmund, Germany
{igor.vatolkin,fabian.ostermann}@tu-dortmund.de
[2] Department of Arts and Sports Sciences, TU Dortmund University,
Dortmund, Germany
mark.gotham@tu-dortmund.de
[3] McGill University, CIRMMT, Montréal, QC, Canada
nestor.napoleslopez@mail.mcgill.ca

Abstract. Deep learning has recently established itself as a cluster of methods of choice for almost all classification tasks in music information retrieval. However, despite very good classification performance, it sometimes brings disadvantages including long training times and higher energy costs, lower interpretability of classification models, or an increased risk of overfitting when applied to small training sets due to a very large number of trainable parameters. In this paper, we investigate the combination of both deep and shallow algorithms for recognition of musical genres using a transfer learning approach. We train deep classification models once to predict harmonic, instrumental, and segment properties from datasets with respective annotations. Their predictions for another dataset with annotated genres are used as features for shallow classification methods. They can be trained over and again for different categories, and are particularly useful when the training sets are small, in a real world scenario when listeners define various musical categories selecting only a few prototype tracks. The experiments show the potential of the proposed approach for genre recognition. In particular, when combined with evolutionary feature selection which identifies the most relevant deep feature dimensions, the classification errors became significantly lower in almost all cases, compared to a baseline based on MFCCs or results reported in the previous work.

Keywords: Musical genre recognition · Deep neural networks · Transfer learning · Interpretable features · Evolutionary feature selection

1 Introduction

Many music classification tasks in the audio signal domain are nowadays solved with the help of deep neural networks. The price for achieving very high accuracy

C. Johnson et al. (Eds.): EvoMUSART 2023, LNCS 13988, pp. 413–427, 2023.
https://doi.org/10.1007/978-3-031-29956-8_27

of classification models is often a long training time, lower interpretability, and danger of overfitting due to huge number of trainable parameters. Traditional "shallow" methods built upon manually engineered features offer an alternative, although typically leading to a lower classification performance. However, in some scenarios, where a decision for a target class can be theoretically explained using some mid-level, semantic properties, there exists an opportunity to apply jointly deep and shallow classifiers trying to combine their advantages and reduce their individual drawbacks.

Recognition of musical genres or styles presents such a scenario. Genres, sub-genres, personal preferences, or mood-related tags are usually defined either by experts or listeners based on some more or less clear semantic properties, like the instrumentation, applied digital effects, details of harmonic structure, or characteristics of melodic lines. Then, it is possible to train deep neural networks only once for the prediction of these "mid-level" properties, and integrate a simple, fast, and potentially more interpretable classifier for the prediction of "high-level" target categories over and again. This procedure very well describes the situation where listeners define different personal categories selecting only a few tracks that either perfectly match or mismatch the target class.

In this work, we introduce a framework which integrates both deep and traditional classification models implementing a transfer learning approach. In the first step, deep models are trained with convolutional neural networks to predict harmonic, instrument, and segment statistics, using several annotated datasets. Then, these models are applied to extract predictions for another dataset of music pieces with annotated genres which serve as high-level musical categories. Based on these predictions, various statistics for time frames and complete tracks are saved as features. Finally, these features are used to train a shallow classification method (random forest or support vector machine) to predict genres. To estimate particularly useful dimensions for different genres, an evolutionary feature selection is additionally applied. The results show the high potential of the proposed framework. In combination with evolutionary feature selection, deep features achieve lower classification errors for all tested genres using both applied shallow classifiers, compared to the baseline using MFCCs, but also in almost all cases compared to results reported in the previous work.

The remainder of this paper is organized as follows. Section 2 presents some related work on deep learning and musical genre recognition. Section 3 summarizes deep semantic features used in our study. Section 4 describes the setup of our experiments. Section 5 discusses the results. In Sect. 6, the most relevant findings of the study are outlined and some ideas for future work are provided.

2   Related Work

Deep neural networks have been shown to be effective for many music information retrieval (MIR) classification tasks. Often, the architectures have been adopted from the image recognition domain [10,22]. In particular, convolutional neural networks (CNNs) [11] play an important role, with Mel frequency spectrograms as (image-like) 2D-input [5]. For example, the recognition of predominant instruments was addressed in [8] and segment recognition in [7].

Musical genre recognition is one of the most widely explored classification tasks in MIR [24]. However, it has some problematic issues, e.g., genre taxonomies may be very distinct [18], and frequently applied evaluation measures are not optimal [25]. Nevertheless, genres represent examples of high-level musical categories. Also for genre recognition, CNN-based approaches have been proposed and successfully applied [32].

Deep learning on small datasets still may be problematic because of too many trainable parameters, even when techniques like dropout layers or data augmentation may increase the robustness of classification models. Shallow classifiers usually are more suitable to this, if deep learners are not heavily customized [19]. The idea of combining deep and shallow classifiers for genre classification was presented in [28]. In the experimental study, however, only instrument statistics were integrated as deep features for genre recognition, as an alternative to low-level signal descriptors and an evolutionary based approximation of instrumental texture. In [16], deep features were also used for multi-modal genre recognition.

A more general concept to use mid-level predictions as features for the prediction of high-level categories can be found, e.g., in [1], where so called anchors were designed to measure similarities between music pieces, or in [30], where supervised models were trained to predict expert annotations like instrumentation, moods, or vocal characteristics.

Also, transfer learning has been applied to genre and tag prediction earlier [4, 26]. It always shows high potential whilst introducing additional difficulties, for instance, when deep features are too specialized for the source task. Nevertheless, transfer learning is a highly promising direction for future machine learning based classification tasks [33].

3   Deep Semantic Features

Figure 1 illustrates a general overview of our approach. CNN models which predict harmonic properties, instrument, and segments are trained using four datasets with respective annotations. Then, their predictions are used to calculate deep semantic properties for complete tracks or shorter time frames (171 harmonic, 328 instrument, and 30 segment properties, 529 dimensions in total) from the 1517-artists dataset [21] with genre annotations. These features are either all combined to create traditional classification models for genre recognition, or the most relevant dimensions are identified using evolutionary multi-objective feature selection, which simultaneously minimizes the classification error and the number of selected feature dimensions.

3.1   Harmonic Properties

Recently, the modeling of complex tonal relationships has received attention from the MIR community, for example, with a surge of models for automatic

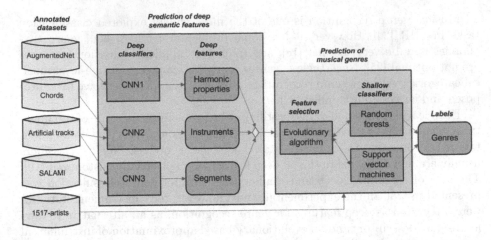

Fig. 1. Overview of data flow in the proposed classification framework.

Roman numeral analysis, which provide key-and-chord information simultaneously [13–15]. Sometimes, the tonal information is computed concurrently [14,15] and sometimes in a modular fashion [6,13].

Although most of these comprehensive automatic chord recognition models have been trained on symbolic music files, the input representations often resemble the information contained in an audio chromagram extractor [12]. Using this to our advantage, we adapted a recent approach, AugmentedNet [15], to operate with audio chromagram features instead of symbolic ones. The AugmentedNet model provides multitask outputs related to the harmonic rhythm, chord, and key information of the music. More specifically, in the latest version of the network[1] these include one output predicting the segmentation of the chords (harmonic rhythm), two outputs predicting changes of key (local key and tonicized key), and six outputs predicting diverse aspects of the chords (Roman numeral class, pitch class set, and a four-note arrangement of the chord as a bass, tenor, alto, and soprano notes). The four-note arrangement of the chord is a prediction of each individual note in the chord, arranged in ascending order from the bass, modelling each of those notes as a separate output in the multitask layout.

Using this model, we extract mid-level harmonic features to be used for genre classification. Figure 2 shows the architecture of the AugmentedNet.

Table 1 summarizes 171 statistics after AugmentedNet predictions, which are used as deep features for genre recognition. All of them are calculated for time frames of 4s with 2s step size, which are later used as inputs / classification instances in genre recognition.

---

[1] https://github.com/napulen/AugmentedNet, accessed on 31.01.2023.

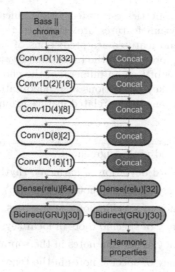

Fig. 2. Architecture of the AugmentedNet. The parameters (filter sizes for convolutional layers, activation functions, GRU layers) are provided in brackets. The numbers of neurons per layer are provided in squared brackets.

3.2 Instrument Predictions

For the prediction of instruments, two datasets were considered to train the neural networks. The first one consists of 5,000 samples and chords generated by mixing of individual instrument samples as described in [28]. 51 different instruments from within and beyond Western music contribute to these examples, many of them represented with several distinct instrument bodies. A CNN after [8] was trained with the Mel spectrograms to output a relative strength of each instrument in the mix (contribution to its overall energy). The second artificial audio multitracks (AAM) dataset [17] contains 3,000 artificially composed music tracks synthesized with real instrument samples using 31 instruments as a subset of instruments from [28]. Here, the Mel spectrograms were estimated only for 2s frames starting with annotated onsets. Figure 3 presents the architecture of the CNN for instrument recognition.

Table 2 summarizes 328 statistics after instrument predictions, calculated for time frames of 4s with 2s step size.

3.3 Segment Statistics

Classification models to predict musical segments are trained using the SALAMI dataset [23] and an artificial track dataset [29] using a CNN after [7]. While SALAMI boundary annotations do not contain additional information, the annotations of the artificial track dataset do list details of the changes in instrumental texture, tempo, and key between the musical segments, so that it was possible to

Table 1. Deep harmonic properties estimated for classification frames of 4s with 2s step size. The harmonic rhythm features are related to the chord segmentation; the bass, tenor, alto, and soprano features are predictions of each individual note in the chord, arranged in ascending order from the bass; similarly, the roman numeral feature is related to the specific class of Roman numeral of the chord; the local and tonicized key features are related with key predictions (e.g., modulations). Dim.: the number of all individual feature dimensions in the related feature group.

Features	Dim.
Predictions trained with AugmentedNet	
Mean and standard deviation of harmonic rhythm	1–2
Relative frequency of specific notes in the alto	3–24
Relative frequency of specific notes in the bass	25–47
Relative frequency of specific roots of local keys	48–71
Relative frequency of specific notes in the soprano	72–92
Relative frequency of specific notes in the tenor	93–112
Relative frequency of specific roots of tonicized keys	113–136
Relative frequency of specific roman numerals	137–160
Relative frequency of modes (major or minor)	161–162
Total number of different symbols	163–171

Table 2. Deep instrument features estimated for classification frames of 4s with 2s step size. Dim.: the number of all individual feature dimensions in the related feature group.

Features	Dim.
Predictions trained with chords	
Mean relative strength of 51 predicted instruments (acoustic and electric guitar, organ, piano and electric piano, viola, violin, etc.)	1–51
Standard deviation of the relative strength of 51 predicted instruments	52–102
Minimum relative strength of 51 predicted instruments	103–153
Maximum relative strength of 51 predicted instruments	154–204
Predictions trained with artificial tracks	
Mean relative strength of 31 predicted instruments (subset of 51 instruments)	205–235
Standard deviation of the relative strength of 31 predicted instruments	236–266
Minimum relative strength of 31 predicted instruments	267–297
Maximum relative strength of 31 predicted instruments	298–328

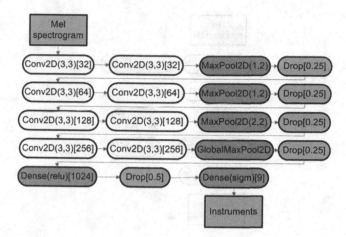

Fig. 3. Architecture of the CNN after [8]. The parameters (filter sizes for convolutional layers, activation functions) are provided in brackets. The numbers of neurons per layer are provided in squared brackets.

train four different CNN models using annotations of either individual boundary types or all segment boundaries. Figure 4 presents an overview of the CNN architecture used for segment boundary prediction.

Table 3. Deep segment statistics estimated for complete audio tracks. Dim.: the number of all individual feature dimensions in the related feature group.

Features	Dim.
Predictions trained with SALAMI	
Number of segments	1
Mean segment length	2
Standard deviation of the segment length	3
Maximal segment length	4
Minimal segment length	5
Mean deviation of segment length	6
Predictions trained with artificial tracks	
Segment statistics as for SALAMI, trained to detect all boundaries	7–12
Segment statistics as for SALAMI, trained to detect instrument boundaries	13–18
Segment statistics as for SALAMI, trained to detect key boundaries	19–24
Segment statistics as for SALAMI, trained to detect tempo boundaries	25–30

Table 3 lists 30 segment statistics derived from predicted boundaries. They are estimated for complete audio tracks, and the same values are assigned to all 4s classification frames.

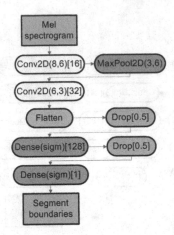

Fig. 4. Architecture of the CNN after [7]. The parameters (filter sizes for convolutional layers, activation functions) are provided in brackets. The numbers of neurons per layer are provided in squared brackets.

4   Setup of Experiments

For the recognition of musical genres, we use a publicly available dataset 1517-artists with 19 annotated genres [21].[2] Each binary genre classification task uses a *training set* of randomly selected 16 "positive" tracks from the selected genre and 18 "negative" tracks from all remaining genres (one track per genre). The number of tracks in the training set is explicitly selected to be rather low, as in the real-world situation, when a listener will try to avoid high efforts selecting many tracks to train an automatic music classification or recommendation system. For the evaluation of feature selection (see below), an *optimization set* of 228 tracks (12 per genre) is used. For the final independent evaluation of feature sets presented in the last iteration of evolutionary feature selection, a *test set* compiled from other 228 tracks (12 per genre) is taken into account. An artist filter was applied before the building of training, optimization, and test sets, so that all of them contain distinct tracks by different artists.

Classification models for genre prediction are trained with random forests [3] (with a default number of 100 trees) and linear support vector machines [31] using RapidMiner [9] integrated into the AMUSE framework [27]. While random forests train an ensemble of pruned decision trees, have only a few hyperparameters to setup, and are typically very robust to overfitting, support vector machines allow for linear separation between classes using transforms to more feature dimensions. Because of uneven distribution of positive and negative tracks in the test set in a binary classification scenario, the evaluation uses balanced relative error e_b which is defined as a mean error for positive and negative tracks:

---

[2] http://www.seyerlehner.info/joomla/index.php/datasets, accessed on 31.01.2023.

$$e_b = \frac{1}{2} \left(\frac{fn}{tp + fn} + \frac{fp}{tn + fp} \right), \tag{1}$$

where tp is the number of true positives (tracks correctly predicted as belonging to the genre), tn is the number of true negatives (tracks correctly predicted as not belonging to the genre), fp is the number of false positives (tracks wrongly predicted as belonging to the genre), and fn is the number of false negatives (tracks wrongly predicted as not belonging to the genre).

Because of the high overall number of 529 deep feature dimensions, some of them may be too irrelevant or noisy, depending on the particular genre to predict. Therefore, in an additional experiment, evolutionary multi-objective feature selection is applied to select the most relevant features after [30]. Evolutionary algorithms are a good choice here, as they explore a large number of different combinations of feature dimensions, applying a random mutation which selects or deselects individual dimensions to train the classifier. For the multi-objective optimization of two criteria, the minimization of e_b and the minimization of the number of selected feature dimensions, the S metric selection evolutionary multi-objective algorithm (SMS-EMOA) [2] was adopted. For each genre, 10 statistical repetitions are conducted, each based on 1,000 evolutionary generations, with a population size of 50 (number of feature sets under investigation). For further implementation details, we refer to [30].

5 Discussion of Results

Table 4 provides a summary of results with balanced relative errors for all 19 genres and different feature sets: individual deep feature groups, as well as their combination using all 519 dimensions, and results after evolutionary multi-objective feature selection. The top half of the table contains the test errors achieved by random forests, and the bottom half by support vector machines. As a baseline to compare with, we have also trained classifiers using a set of 13 Mel frequency cepstral coefficients (MFCCs) [20] which were originally developed for speech recognition, but have also been frequently used in music classification tasks. Further, we included in the table the best results from [28]; however, below we explain why it is hard to fairly compare them to the results of our study because of some differences in the experimental setup.

The results show that the performance of deep features varies very strongly and depends on the genre, and some errors are rather high (a random classifier would achieve an expected error of 0.5). However, please note that several challenges do exist in our application scenario. First, the training sets are quite small, as in real world situation where listeners wish to define a category without many attempts, and based on a limited number of selected prototype tracks. Second, the deep features are trained on other datasets, often with limitations. E.g., AugmentedNet is compiled only with classical music; artificial tracks used to train models for the prediction of instruments and segments are composed by an algorithm and thus do not perfectly represent commercial music. So, the

Table 4. Test e_b for 19 musical genre recognition tasks. [28]: the best results reported in that work (however, they are not completely comparable, see the text); MFCCs: Mel frequency cepstral coefficients; Harm: deep harmonic features listed in Table 1; Inst: deep instrument features listed in Table 2; Segm: deep segment features listed in Table 3; All: all deep features; All-FS: the best feature set after evolutionary feature selection. Bolded values are the best (smallest) for each genre in the current study. A bolded value using italic font marks a sole case where an error of [28] was lower than the lowest error in our study.

| Genre | Random forests | | | | | | |
	[28]	MFCCs	Harm	Inst	Segm	All	All-FS
Alternative and Punk	0.1928	0.2847	0.4861	0.3148	0.4375	0.3218	0.2431
Blues	0.3170	0.4028	0.3727	0.3495	0.4954	0.4259	**0.1921**
Childrens	0.3880	0.5069	0.5116	0.4329	0.3148	0.3102	0.2685
Classical	0.0929	0.1250	0.5231	0.0995	0.2106	0.2083	0.0833
Comedy and Spoken Word	0.2214	0.3333	0.3634	0.3125	0.2894	0.3125	0.2407
Country	0.2350	0.3472	0.4190	0.2199	0.3843	0.3403	**0.1273**
Easy Listening	0.2904	0.2894	0.4537	0.3542	0.3542	0.3866	**0.2245**
Electronic+	0.1487	0.3843	0.2731	0.0926	0.3472	0.2454	**0.0370**
Folk	0.2682	0.3935	0.4236	0.3449	0.5440	0.3264	0.1852
Hip-Hop	0.1240	0.3495	0.4954	0.1065	0.2477	0.2824	0.0880
Jazz	0.3123	0.3889	0.3681	0.3519	0.5231	0.4514	0.2523
Latin	0.3049	0.5069	0.5694	0.4028	0.5231	0.3704	0.2940
New Age	0.2349	0.3056	0.5139	0.2731	0.3773	0.3750	**0.1505**
R'n'B and Soul	0.2534	0.2731	0.4144	0.2500	0.4213	0.2616	**0.1898**
Reggae	0.1941	0.3194	0.5069	0.2454	0.4375	0.3912	0.1875
Religious	0.3759	0.4352	0.3634	0.3912	0.5093	0.3611	0.2523
Rock and Pop	0.2346	0.2870	0.5579	0.2894	0.6273	0.2963	**0.1343**
Soundtracks and More	0.2652	0.2708	0.5926	0.3079	0.4190	0.3750	0.2616
World	0.4059	0.3403	0.4144	0.5069	0.5046	0.4745	**0.2662**
	Support vector machines						
Alternative and Punk	*0.1656*	0.2593	0.4282	0.2546	0.5000	0.2639	**0.2060**
Blues	0.3030	0.4074	0.3449	0.2546	0.5000	0.2847	0.2153
Childrens	0.3366	0.5185	0.5162	0.4769	0.5000	0.5000	**0.1944**
Classical	0.0885	0.0903	0.4190	**0.0810**	0.5000	0.1574	0.0833
Comedy and Spoken Word	0.2360	0.3542	0.3519	0.3426	0.5000	0.2431	**0.1782**
Country	0.2247	0.3565	0.4352	0.2407	0.5000	0.2940	0.1319
Easy Listening	0.2980	0.2315	0.4514	0.4259	0.5000	0.5000	0.2477
Electronic+	0.1448	0.2245	0.3380	0.1412	0.5000	0.1806	0.0532
Folk	0.2621	0.3449	0.4190	0.3495	0.5000	0.4167	**0.1736**
Hip-Hop	0.1201	0.2431	0.5185	0.0671	0.5000	0.5000	**0.0810**
Jazz	0.2680	0.4190	0.2708	0.3588	0.5000	0.3356	**0.2338**
Latin	0.3168	0.4514	0.5509	0.4838	0.5000	0.5602	**0.2593**
New Age	0.2122	0.2685	0.4745	0.2894	0.5000	0.5000	0.1921
R'n'B and Soul	0.2594	0.3380	0.4514	0.2593	0.5000	0.4236	0.2014
Reggae	0.1872	0.2546	0.5301	0.2523	0.5000	0.3449	**0.1690**
Religious	0.3751	0.3981	0.4005	0.3935	0.5000	0.3912	**0.2269**
Rock and Pop	0.2390	0.2014	0.5648	0.2917	0.5000	0.4120	0.1389
Soundtracks and More	0.3108	0.2593	0.5231	0.3773	0.5000	0.3403	**0.2431**
World	0.3604	0.3472	0.4954	0.4306	0.5000	0.3495	0.2731

robustness of deep features is obviously sometimes limited, when they are calculated for other data. Another difficulty is that some genres like Childrens, Rock and Pop, or World are very ambiguously defined. Still, for 14 of 19 categories using random forests and 12 of 19 categories using support vector machines, deep features or their combination are better than models trained with MFCCs. Sometimes, integration of all deep features leads to larger errors compared to the errors of the individual groups, underlining the suggestion that simply using more features is not always the best solution (the curse of dimensionality).

When all features from individual feature groups are compared, instrument statistics seem to be the most relevant; the errors achieved with this group are smaller than from all other groups for 14 of 19 categories using random forests and 18 of 19 using support vector machines. With random forests, harmonic predictions are the best for genres Religious and World, and segment predictions for Childrens and Spoken Word. Segment descriptors did not work sufficiently with support vector machines which predict only one class in all cases. A potential explanation is that the number of dimensions was in that case too low for this classifier.

When all groups are combined (column "All"), the errors are lower than for individual feature groups only for 4 genres for both classifiers. This strengthens the suggestion that too many dimensions are irrelevant for a particular genre, and feature selection may help to identify the most relevant ones. This is confirmed by values in the column "All-FS", which report the smallest test errors achieved after evolutionary feature selection as described in Sect. 4. Only for Classical, the error achieved with deep instrument features is smaller than after feature selection. Even if this seems to be peculiar at the first glance, the explanation is that feature selection is strictly evaluated using an independent test set for a better measurement of its general performance, but the evaluation of feature sets during the optimization process is done on the optimization set. So, theoretically, if a very informative feature dimension for the test data would not belong to the very best dimensions for the optimization data, it will not contribute to the final output of the algorithm.

Finally, we may compare the results to the best reported errors of the previous work [28]. They show that in all but one case the smaller errors are achieved using our large deep feature set. However, it is important to mention that the comparison is not very fair, because in our study the application of feature selection helped to identify the most relevant feature dimensions, and theoretically some individual timbral or approximative descriptors from [28] could contribute to final feature sets. Further, despite of the same genre labels and similar setup of classifiers, the distribution of data was not the same: in [28], larger training sets of 72 tracks were used, and in our experiments we had to reserve enough artist-independent tracks for a separate optimization set for the evaluation of feature selection.

6   Conclusions

In this paper, we have applied a transfer learning approach, first training several deep convolutional neural networks to predict harmonic, instrumental, and segment characteristics, and then using two traditional shallow classifiers predicting genres based on those deep features. Such a combination of deep and traditional methods can save a lot of resources, as the extensive training of all deep models is done only once, but the prediction of semantic musical properties only one time per track, and the training and classification of genre recognition, which may be repeated over and again for different listeners and application scenarios, can be conducted significantly faster based on classifiers with only a few parameters.

The results showed that deep features are quite successful and contribute to relatively low errors in a challenging application scenario, where the sizes of training sets are very small, as in the typical real-world situation. However, there still exist a large number of noisy and irrelevant dimensions—partly because transfer learning may not always allow for the extraction of very robust characteristics—but also simply because different classification categories have very distinct properties. Thus, the application of feature selection which itself requires some costs, seems to be essential and complements the classification algorithm pipeline, leading to smaller classification errors for all 19 genres compared to MFCC baseline and all deep features, for 18 genres compared to individual deep semantic feature groups, and for 18 genres compared to errors reported in the previous work.

In future work, we plan to integrate more deep features, but also to improve their robustness. A first promising direction is to include more diverse genres into datasets involved to train deep models for prediction of semantic music properties. Additionally, more data augmentation methods can be applied to the annotated data for the training of deep features. Another relevant contribution to justify the application of deep learning would be a strict statistical comparison of "deep" and "shallow" features for the recognition of genres. However, such a comparison is not always straightforward: in preliminary experiments, we have already extracted shallow instrument and segment features using random forests and the same training data as for deep neural networks, but the performance was very poor, potentially, because the Mel spectrograms could not provide enough information for the training of this classification method. Another promising implementation will be to integrate more harmonic analysis features from AugmentedNet, which as a completely novel approach proved solid and successful for genre recognition.

Acknowledgement. The authors gratefully acknowledge the computing time provided on the Linux HPC cluster at Technical University Dortmund (LiDO3), partially funded in the course of the Large-Scale Equipment Initiative by the German Research Foundation (DFG) as project 271512359.

References

1. Berenzweig, A., Ellis, D.P.W., Lawrence, S.: Anchor space for classification and similarity measurement of music. In: Proceedings of the IEEE International Conference on Multimedia and Expo, ICME, pp. 29–32. IEEE Computer Society (2003)
2. Beume, N., Naujoks, B., Emmerich, M.T.M.: SMS-EMOA: multiobjective selection based on dominated hypervolume. Eur. J. Oper. Res. **181**(3), 1653–1669 (2007)
3. Breiman, L.: Random forests. Mach. Learn. **45**(1), 5–32 (2001)
4. Choi, K., Fazekas, G., Sandler, M., Cho, K.: Transfer learning for music classification and regression tasks. In: Proceedings of the 18th International Society for Music Information Retrieval Conference, ISMIR, pp. 141–149. International Society for Music Information Retrieval (2017)
5. Costa, Y.M., Oliveira, L.S., Silla, C.N.: An evaluation of convolutional neural networks for music classification using spectrograms. Appl. Soft Comput. **52**, 28–38 (2017)
6. Gotham, M., Kleinertz, R., Weiss, C., Müller, M., Klauk, S.: What if the 'when' implies the 'what'?: Human harmonic analysis datasets clarify the relative role of the separate steps in automatic tonal analysis. In: Proceedings of the 22nd International Society for Music Information Retrieval Conference, ISMIR, pp. 229–236 (2021)
7. Grill, T., Schlüter, J.: Music boundary detection using neural networks on combined features and two-level annotations. In: Proceedings of the 16th International Society for Music Information Retrieval Conference, ISMIR, pp. 531–537 (2015)
8. Han, Y., Kim, J., Lee, K.: Deep convolutional neural networks for predominant instrument recognition in polyphonic music. IEEE ACM Trans. Audio Speech Lang. Process. **25**(1), 208–221 (2017)
9. Hofmann, M., Klinkenberg, R.: RapidMiner: data mining use cases and business analytics applications. Chapman & Hall/CRC (2013)
10. Krizhevsky, A., Sutskever, I., Hinton, G.E.: ImageNet classification with deep convolutional neural networks. In: Proceedings of the 26th Annual Conference on Neural Information Processing Systems, NIPS, pp. 1106–1114 (2012)
11. LeCun, Y., Bengio, Y., Hinton, G.E.: Deep learning. Nature **521**(7553), 436–444 (2015)
12. Mauch, M., Dixon, S.: Approximate note transcription for the improved identification of difficult chords. In: Proceedings of the 11th International Society for Music Information Retrieval Conference, ISMIR, pp. 135–140 (2010)
13. McLeod, A., Rohrmeier, M.A.: A modular system for the harmonic analysis of musical scores using a large vocabulary. In: Proceedings of the 22nd International Society for Music Information Retrieval Conference, ISMIR, pp. 435–442 (2021)
14. Micchi, G., Kosta, K., Medeot, G., Chanquion, P.: A deep learning method for enforcing coherence in automatic chord recognition. In: Proceedings of the 22nd International Society for Music Information Retrieval Conference, ISMIR, pp. 443–451 (2021)
15. Nápoles López, N., Gotham, M., Fujinaga, I.: AugmentedNet: a roman numeral analysis network with synthetic training examples and additional tonal tasks. In: Proceedings of the 22nd International Society for Music Information Retrieval Conference, ISMIR, pp. 404–411 (2021)
16. Oramas, S., Nieto, O., Barbieri, F., Serra, X.: Multi-label music genre classification from audio, text and images using deep features. In: Proceedings of the 18th International Society for Music Information Retrieval Conference, ISMIR, pp. 23–30 (2017)

17. Ostermann, F., Vatolkin, I.: AAM: Artificial audio multitracks dataset (2022). https://doi.org/10.5281/zenodo.5794629
18. Pachet, F., Cazaly, D.: A taxonomy of musical genres. In: Proceedings of the 6th International Conference on Computer-Assisted Information Retrieval (Recherche d'Information et ses Applications), RIAO, pp. 1238–1245. CID (2000)
19. Pasupa, K., Sunhem, W.: A comparison between shallow and deep architecture classifiers on small dataset. In: Proceedings of the 8th International Conference on Information Technology and Electrical Engineering, ICITEE, pp. 1–6. IEEE (2016)
20. Rabiner, L., Juang, B.H.: Fundamentals of Speech Recognition. Prentice Hall, Upper Saddle River (1993)
21. Seyerlehner, K., Widmer, G., Knees, P.: Frame level audio similarity - a codebook approach. In: Proceedings of the 11th International Conference on Digital Audio Effects, DAFx (2008)
22. Simonyan, K., Zisserman, A.: Very deep convolutional networks for large-scale image recognition. In: Proceedings of the 3rd International Conference on Learning Representations, ICLR (2015)
23. Smith, J.B.L., Burgoyne, J.A., Fujinaga, I., De Roure, D., Downie, J.S.: Design and creation of a large-scale database of structural annotations. In: Proceedings of the 12th International Society for Music Information Retrieval Conference, ISMIR, pp. 555–560. University of Miami (2011)
24. Sturm, B.L.: A survey of evaluation in music genre recognition. In: Proceedings of the 10th International Workshop on Adaptive Multimedia Retrieval: Semantics, Context, and Adaptation, AMR, pp. 29–66 (2012)
25. Sturm, B.L.: Classification accuracy is not enough - on the evaluation of music genre recognition systems. J. Intell. Inf. Syst. **41**(3), 371–406 (2013)
26. van den Oord, A. and Dieleman, S. and Schrauwen, B.: Transfer learning by supervised pre-training for audio-based music classification. In: Proceedings of the 15th International Society for Music Information Retrieval Conference, ISMIR, pp. 29–34 (2014)
27. Vatolkin, I., Ginsel, P., Rudolph, G.: Advancements in the music information retrieval framework AMUSE over the last decade. In: Proceedings of the 44th International ACM SIGIR Conference on Research and Development in Information Retrieval, SIGIR, pp. 2383–2389. ACM (2021)
28. Vatolkin, I., Adrian, B., Kuzmic, J.: A fusion of deep and shallow learning to predict genres based on instrument and timbre features. In: Romero, J., Martins, T., Rodríguez-Fernández, N. (eds.) EvoMUSART 2021. LNCS, vol. 12693, pp. 313–326. Springer, Cham (2021). https://doi.org/10.1007/978-3-030-72914-1_21
29. Vatolkin, I., Ostermann, F., Müller, M.: An evolutionary multi-objective feature selection approach for detecting music segment boundaries of specific types. In: Proceedings of the 2021 Genetic and Evolutionary Computation Conference, GECCO, pp. 1061–1069 (2021)
30. Vatolkin, I., Rudolph, G., Weihs, C.: Evaluation of album effect for feature selection in music genre recognition. In: Proceedings of the 16th International Society for Music Information Retrieval Conference, ISMIR, pp. 169–175 (2015)
31. Yu, H., Kim, S.: SVM tutorial - classification, regression and ranking. In: Rozenberg, G., Bäck, T., Kok, J.N. (eds.) Handbook of Natural Computing, vol. 1, pp. 479–506. Springer, Berlin Heidelberg (2012). https://doi.org/10.1007/978-3-540-92910-9_15

32. Zhang, W., Lei, W., Xu, X., Xing, X.: Improved music genre classification with convolutional neural networks. In: Proceedings of the 17th Annual Conference of the International Speech Communication Association, Interspeech, pp. 3304–3308. ISCA (2016)
33. Zhuang, F., et al.: A comprehensive survey on transfer learning. Proc. IEEE **109**(1), 43–76 (2021)

Author Index

C. Johnson et al. (Eds.): EvoMUSART 2023, LNCS 13988, pp. 429–430, 2023.
https://doi.org/10.1007/978-3-031-29956-8

Printed in the United States
by Baker & Taylor Publisher Services

Printed in the United States
by Baker & Taylor Publisher Services